ANIMAL EARTH

Animal Earth online

You can watch videos and animations relating to this book at a dedicated YouTube channel: www.youtube.com/user/AnimalEarthBook

For Jane and Connie

Title pages: The shallow, marine habitats of the tropics support a glorious variety of animal life. Clearly visible in this image are craniates, sponges, cnidarians, annelids, echinoderms and molluscs. Beyond these, unseen in the cracks and crevices of the reef and in the sediment is a mind-bogglingly diverse range of other creatures – usually small, and representing many of the lesser-known animal lineages.

First published in the United Kingdom in 2013 by
Thames & Hudson Ltd, 181A High Holborn,
London WC1V 7QX

British Library Cataloguing-in-Publication Data
A catalogue record for this book is available from the British Library

ISBN 978-0-500-51696-6

Printed in China by Toppan Leefung Printing Limited

To find out about all our publications, please visit **www.thamesandhudson.com**. There you can subscribe to our e-newsletter, browse or download our current catalogue, and buy any titles that are in print.

ANIMAL EARTH

The Amazing Diversity of Living Creatures

Ross Piper

with 540 illustrations

The Animals

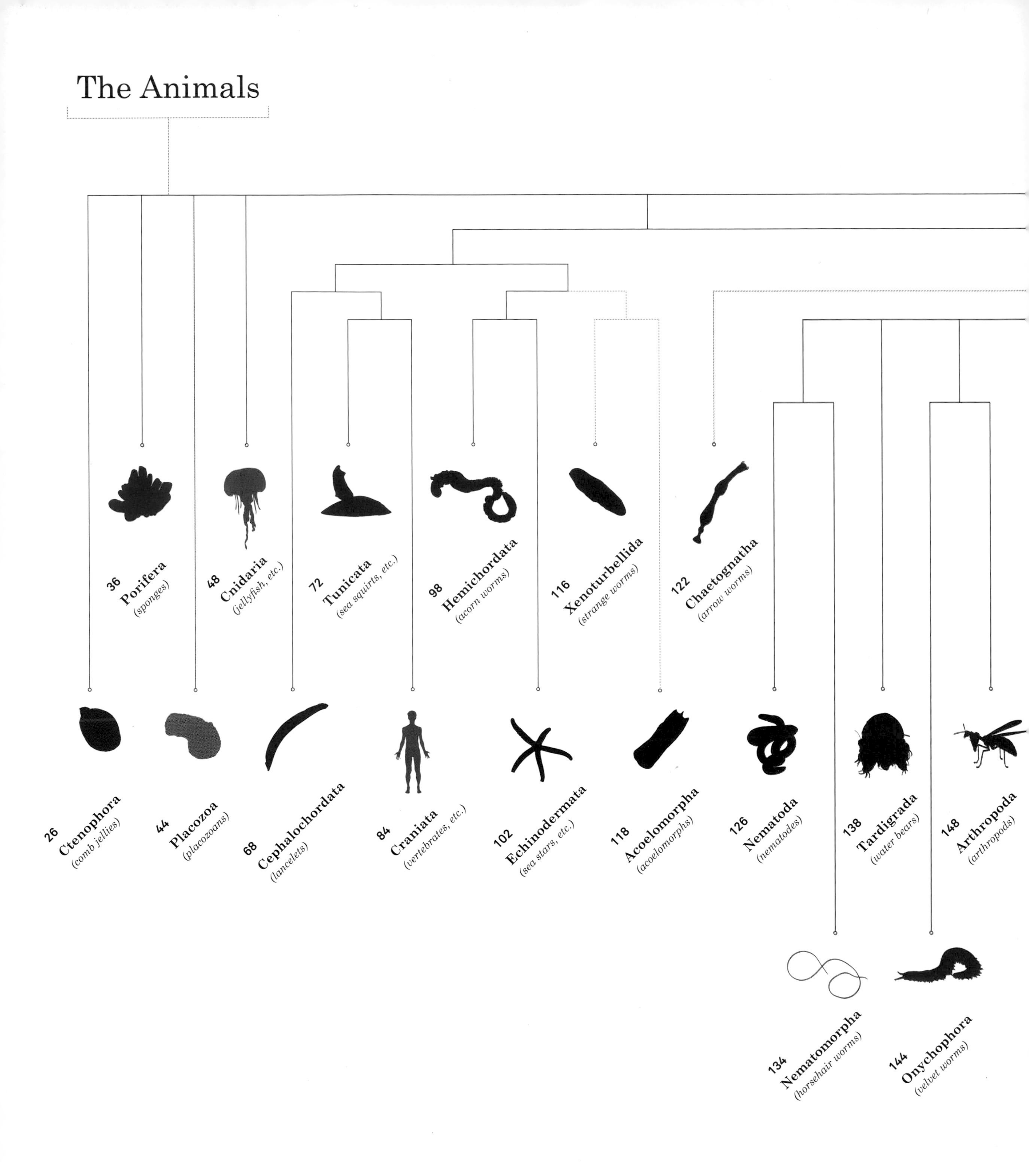

36 Porifera
(sponges)
48 Cnidaria
(jellyfish, etc.)
72 Tunicata
(sea squirts, etc.)
98 Hemichordata
(acorn worms)
116 Xenoturbellida
(strange worms)
122 Chaetognatha
(arrow worms)
26 Ctenophora
(comb jellies)
44 Placozoa
(placozoans)
68 Cephalochordata
(lancelets)
84 Craniata
(vertebrates, etc.)
102 Echinodermata
(sea stars, etc.)
118 Acoelomorpha
(acoelomorphs)
126 Nematoda
(nematodes)
138 Tardigrada
(water bears)
148 Arthropoda
(arthropods)
134 Nematomorpha
(horsehair worms)
144 Onychophora
(velvet worms)

Preface

'...from so simple a beginning endless forms most beautiful and most wonderful have been, and are being, evolved.'

Charles Darwin
***On the Origin of Species*, 1859**

A jumping spider, *Phidippus audax*.

To most people, the word 'animal' conjures up images of well-known and to our eyes impressive creatures like tigers, apes, crocodiles, eagles, elephants and sharks, or perhaps some commonly encountered smaller species such as snails or frogs. However, animals are amazingly diverse, much more so than natural history documentaries would suggest. The familiar animals – mammals, birds, crocodilians, lizards and their relatives, amphibians, cartilaginous fish and bony fish – account for only around 4 per cent of the roughly 1.5 million known animal species. If we pause to consider for a moment that there may actually be anywhere between 10 million and 200 million species of animal, then these familiar fauna, which are also relatively well studied, account for only between 0.03 and 0.6 per cent of the total.

Although there are a bewildering number of animal species, they are all offshoots from a relatively small number of lineages, the representatives of which all share a common body plan and evolutionary history. The purpose of this book is to provide a more-or-less equal summary of each of these lineages, providing the reader with a broad understanding of what animal diversity means, how we can make sense of this diversity, how the animals evolved, and the adaptations that have allowed them to become so successful.

The introduction sets the scene for this exploration of animal diversity by looking at where animals fit on the 'tree of life' and how the lineages are arranged on this tree (which also serves as a map for how the rest of the book is arranged). For each lineage, the common names (where applicable), species diversity and size range are given followed by a summary of defining characteristics, lifestyles, ecological roles and evolutionary relationships.

There are purposefully few mentions in the text of the negative impact of animals on humans – firstly because there are so many sources on this topic, and secondly because in focusing on the negative we reinforce the notion that humans are somehow above nature rather than part of it, and that the living world is simply there to be tamed and used as we see fit. Instead, this book concentrates on the mostly unseen positive impacts that maintain the balance of nature and make the living world what it is today.

As humans, we are in a very privileged position since we are the only species that can appreciate and marvel at the diversity of our fellow animals. However, we are becoming increasingly detached from the natural world, even though we utterly depend on it. With every passing year our activities drive other species to extinction; creatures unique in the universe with a rich and elaborate evolutionary history are annihilated. Every animal species, whether a gibbon swinging through the trees of a rainforest or a microscopic rotifer navigating the tiny channels between sediment grains on the seabed, has a right to exist. More pragmatically, every species is an integral component of earth's ecosystem and in eating and getting eaten they cycle nutrients and energy through the system to maintain the status quo. We can only piece together how an ecosystem works when we know all of its component parts and how they interact.

As intelligent beings it is our duty to protect and understand animal diversity not only for its own sake but also to maintain the natural systems that keep us alive, and because of what it can tell us about the incredible phenomenon of life.

Animal lineages

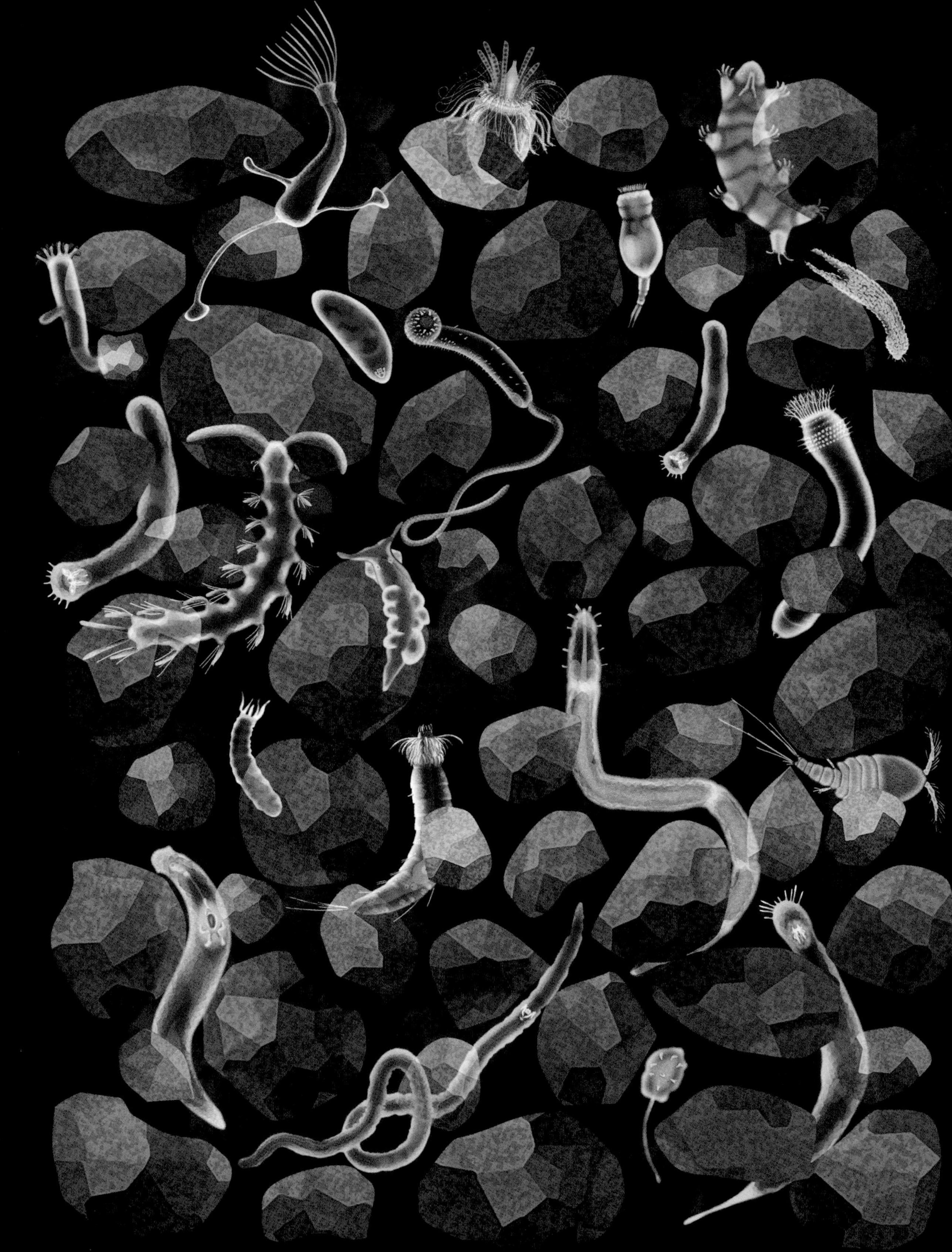

1 Most animals are small and rarely encountered (at least knowingly) by humans. Aquatic sediments, particularly those on the seabed, are alive with a glorious variety of minute creatures, collectively known as meiofauna. In this microcosm we can find representatives of at least 19 of the animal lineages – the most of any habitat.

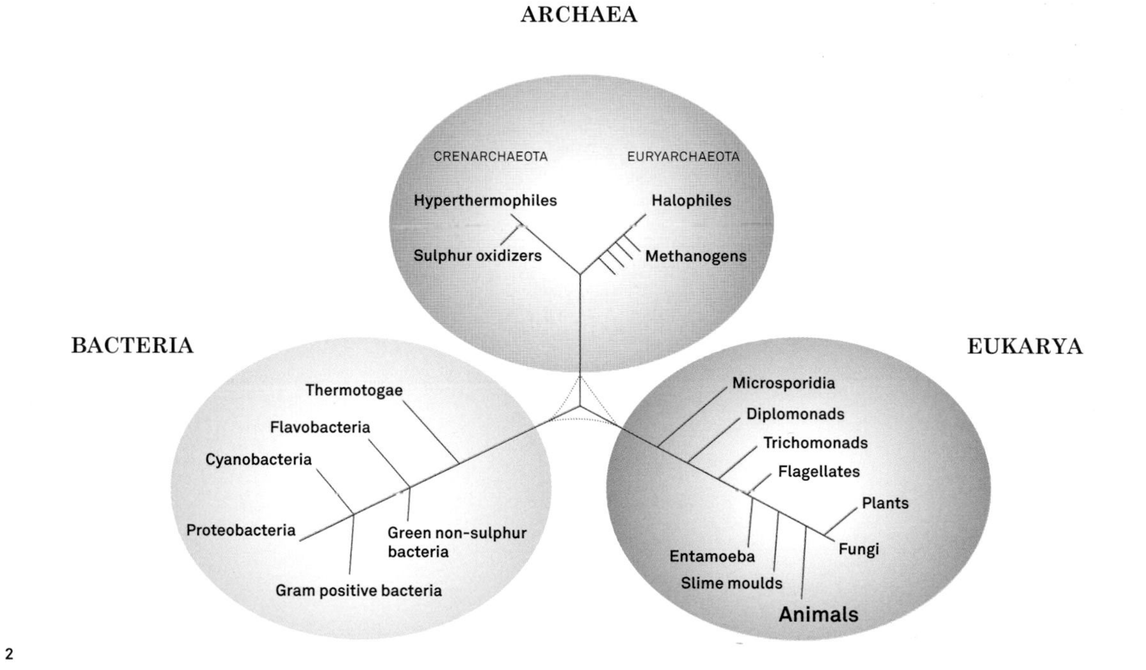

2

2 The tree of life. Cellular life forms can be grouped into three 'domains'. The dashed lines show that in the very early stages of the evolution of life genes were moving about between the three main branches.

Within the constraints of earth's physical environment, life has run riot, evolving into a staggering diversity of forms from a common ancestor that lived perhaps as many as 4 billion years ago. Life in all its diverse splendour is the single most interesting thing about our planet, but it is a phenomenon we take for granted, since we are just one expression of the evolutionary process surrounded by countless others. In two centuries of scientific endeavour humans have made great strides in understanding life – but really we have only scratched the surface.

Getting to grips with the diversity of life is key if we want to further our understanding. We need to know how many forms of life there are out there, how they are related and how they interact. Since all life on earth is ultimately descended from a single common ancestor, a 'tree', albeit a very bushy one, can be used to represent the diversity of life and the evolutionary relationships of extinct and living species. Evidence suggests the very earliest organisms that would later go on to give rise to all the life that we know today were actually sharing genes, a process known as horizontal gene transfer (which we can still see today when a bacterium acquires the genes for antibiotic resistance from another bacterium).

The two major branches on this 'tree' are cellular life and viral life. Cellular life is made up of three branches: the so-called domains of Archaea, Bacteria and Eukarya [2]. The first two are exclusively single-celled organisms, their seemingly simple structures belying an incredible degree of genetic diversity. The only branch where we find multicellular organisms is the Eukarya. Zooming in on this domain we see yet more branches, one of which is the animals, the organisms that are the focus of this book.

THE ANIMALS

What is an animal? Broadly speaking, it is a multicellular life form that feeds on other organisms, moves around for at least part of its life and responds to stimuli. The latter can be anything from a flatworm retreating from the light to a bull elephant responding to the pheromones produced by a female of his species.

A single-celled organism that lived perhaps more than 1 billion years ago was the common ancestor of all the animals that have ever lived, and in adapting to the world around them these multicellular organisms have diverged into a huge variety of species. Some 1.5 million species of animal have been formally identified, yet it is estimated the total number of species could be anywhere between 10 and 200 million. This is a huge range, but it underlines how little we know about the world around us and the branch of life that we ourselves belong to.

To make sense of all this animal life scientists have analyzed genes, development, morphology and lifestyles to define 35 major lineages [4], the representatives of which all share a defining body plan and evolutionary history [BOX 1]. In the Linnaean classification system these lineages correspond to phyla, but

BOX 1 ***Body plans***

At face value, animal diversity can appear to be beyond comprehension – there are just so many species. But variation on a few distinctive themes can account for all of this diversity, allowing us to group all the animals into one of 35 major lineages. One of these themes is the body plan. For example, with more than 1.2 million species the diversity of the arthropods is mindboggling, but all of them have a segmented body, an exoskeleton that is moulted in order to grow, and jointed appendages. These traits represent the arthropod body plan and all the diversity we see in this lineage is variation on this basic layout.

there are number of reasons why this is flawed, so its importance should be played down [BOX 2].

Regardless of its limitations, Linnaean classification is and will remain a familiar frame of reference for some time to come, but we have to be aware of its shortcomings and the superstitious dogma that influenced its design. For example, the main categories in the Linnaean classification represent the immutable levels of creation, and there are seven of them because the number '7' is supposedly a perfect number. Furthermore, Linnaean classification is influenced by anthropocentric notions such as the *scala naturae* (the great chain of being) and finalism. In the *scala naturae* all living beings were classified according to their degree of 'perfection', with humans at the summit [3]. Finalism is the idea that evolution is mysteriously driven towards the emergence of man. These ideas lead people to assume that organisms have an essence that precedes their existence. The evolution of life has not been a simple linear event, with 'primitive' forms always giving rise to more 'complex' ones, and the process has no 'goal'. This anthropocentrism has severely impeded our attempts to make sense of the diversity of life for a long time, and it is still ingrained today.

Evolving from a single common ancestor, all the major animal lineages appeared during a relatively small window of geological time, and even though the fossil record is far from complete we know that they were well established 500 million years ago. The first animal, the proto-animal, spawned five distinct branches of animal life: the Placozoa, Porifera, Ctenophora, Cnidaria and Bilateria [4]. The supremely adaptable bilaterian body plan permitted a riot of diversification and the evolution of all of the remaining animal lineages, which are divided into two main clades: the protostomes and deuterostomes (see the Summary Table on pp. 312–15 for the key characteristics of these groupings).

What is most incredible about our appreciation of animal diversity is that only a minority of the animal lineages are familiar even to biologists let alone everyone else [BOX 3]. The animals in the lesser-known lineages are nonetheless fascinating and ecologically important, and we cannot understand animal evolution without studying them.

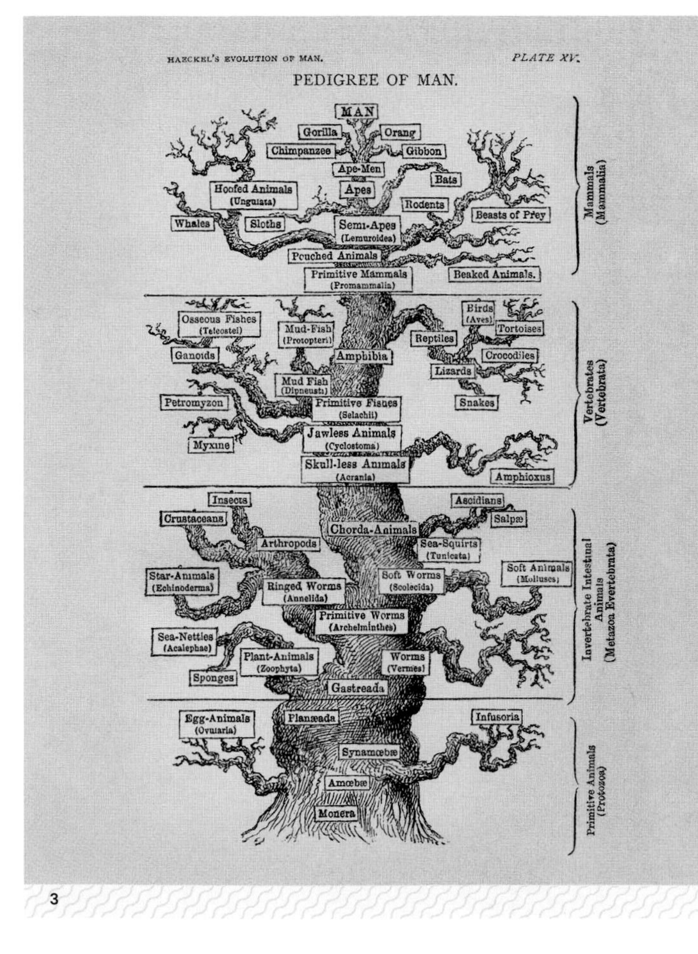

3

BOX 2 ***Making sense of diversity***

In looking at the incredible diversity of animal life, Linnaeus defined a hierarchy of organizational categories to classify these creatures – the familiar Kingdom, Phylum, Class, Order, Family, Genus and Species, to which a number of other categories have since been added. As seemingly neat as this hierarchy is, it is far from the most effective way to communicate diversity, since it was established a long time before Darwin and his contemporaries changed our fundamental understanding of the living world and our place in it.

The hierarchy works reasonably well when we look at a lineage in isolation, but it is meaningless when we compare lineages because there is no consistency in how the categories are defined from one lineage of animals to the next. For example, a Class, Order or Family of arthropods is not comparable to a Class, Order or Family of annelids or craniates.

The designation of a group of animals as one of these categories or the other is more a consequence of historical factors in the field rather than their biology. The Linnaean hierarchy also coerces us to 'pigeonhole' rather than to think about the evolutionary history of a group of animals, which is a fundamental aspect of biology.

3 *The Pedigree of Man* was Ernst Haeckel's 1874 attempt at making sense of animal diversity. In some ways it was quite forward thinking, but it retained the misguided view that evolution progresses towards a more perfect form, with humans at the pinnacle.

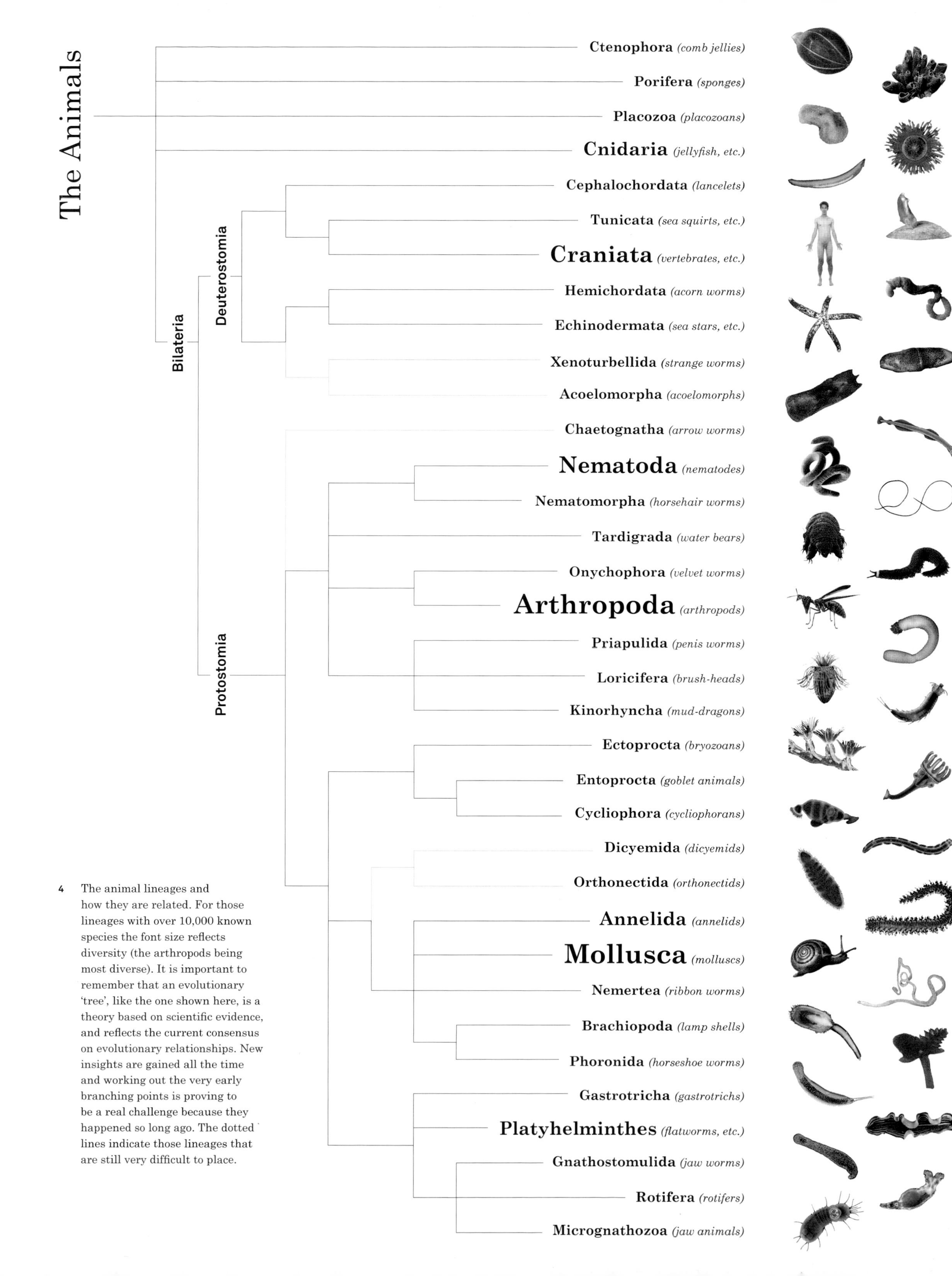

4 The animal lineages and how they are related. For those lineages with over 10,000 known species the font size reflects diversity (the arthropods being most diverse). It is important to remember that an evolutionary 'tree', like the one shown here, is a theory based on scientific evidence, and reflects the current consensus on evolutionary relationships. New insights are gained all the time and working out the very early branching points is proving to be a real challenge because they happened so long ago. The dotted lines indicate those lineages that are still very difficult to place.

5

BOX 3 ***Appreciating animal diversity in all its forms***

Humans have made great strides in understanding the natural world, but there are a number of reasons why we have only scratched the surface of animal diversity. Firstly, we are drawn to creatures we feel an emotional attachment to, either because of the way they behave or because of the way they look. Inevitably, these are closely related species, such as primates and other mammals. The fact is that many animals are faceless, so we have trouble forming any sort of emotional bond with them and much of what they do seems very alien indeed. It is much easier to identify with a lion and the challenges she faces nurturing her cubs than it is to form any sort of emotional link with a faceless crustacean that spends its adult life attached to the eye of a fish.

Size is also an important factor. We are big organisms, vastly bigger than most other animal species. To appreciate the smaller majority requires specialist equipment, such as microscopes. The photographs in this book show that microscopic animals can be very complex, but this incredible detail is effectively invisible to the naked eye.

5 This freshwater copepod crustacean is only around 1 mm (0.04 in.) long, tiny in human terms. However, it is a veritable goliath to the ciliates (single-celled organisms) that have established a colony on its abdomen, just between its two egg sacs.

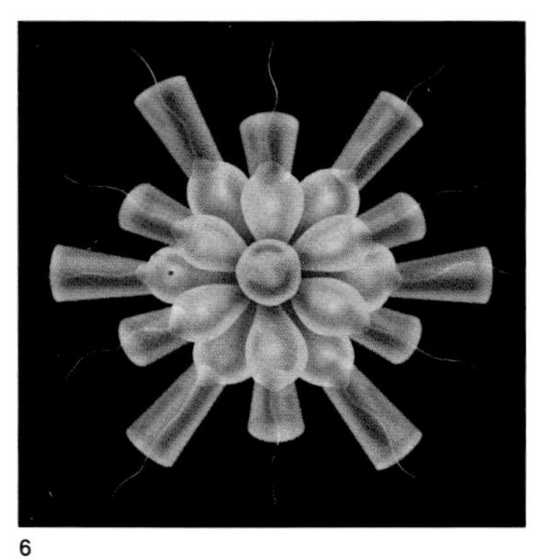

6

6 Choanoflagellates that remained together following cell division formed small, floating colonies. These tiny assemblages represent the beginnings of multicellularity and the common ancestry of all the animals that have ever lived.

The story of animal evolution is one of innovation. If we strip all the major animal body plans back to their bare essentials we see that there are a few key morphological innovations that each stimulated waves of rampant differentiation and specialization. It is these innovations that underpin the animal diversity we see around us today. Several key processes generated and tweaked the innovations, driving the diversification that ultimately allowed the exploitation of a huge range of habitats.

KEY INNOVATIONS

Multicellularity and organization

At least 1 billion years ago a marine single-celled organism was in the process of reproducing by dividing in two; however, instead of the two cells going their separate ways they remained joined. Further divisions ended with the same result, eventually forming a small floating colony of cells [6]. This seemingly insignificant event is actually one of the most important evolutionary leaps of all time, because every animal on the planet descends from this little floating assemblage of cells. How is this so? What were the benefits of colonial life for these organisms – these proto-animals?

Firstly, there was the advantage of size, since the predatory single-celled organisms of the time were simply too small to deal with colonial creatures. Secondly, living together in a tight little unit made it possible for individual cells to take on specialized functions. Some could devote their resources to propulsion, others could specialize in dealing with food and yet more could focus on producing cells that would give rise to more colonies. The die was cast and multicellularity became all the rage. Over huge stretches of time and generation upon generation the complexity of these colonies grew and cells evolved to fulfil ever more specialized functions until there came a time when the cells were all so interdependent that they could not live without each other [7,8]. This trend of specialization would lead to the organization of cells into discrete tissues, tissues into organs, organs into systems and individual animals into colonies.

The gut

Following on from the emergence of multicellularity was the evolution of a central space or tube where food could be dealt with – the gut [9]. Prior to this innovation the very first animals had to rely on cells taking in particles of edible matter, breaking the complex molecules into their constituent components and releasing these for absorption by the other cells. Like the living placozoans, the earliest animals of the seabed may have simply lain on top of their food, exuded digestive secretions and absorbed the resultant nutritive soup. By progressive in-pocketing of their body wall the proto-animals

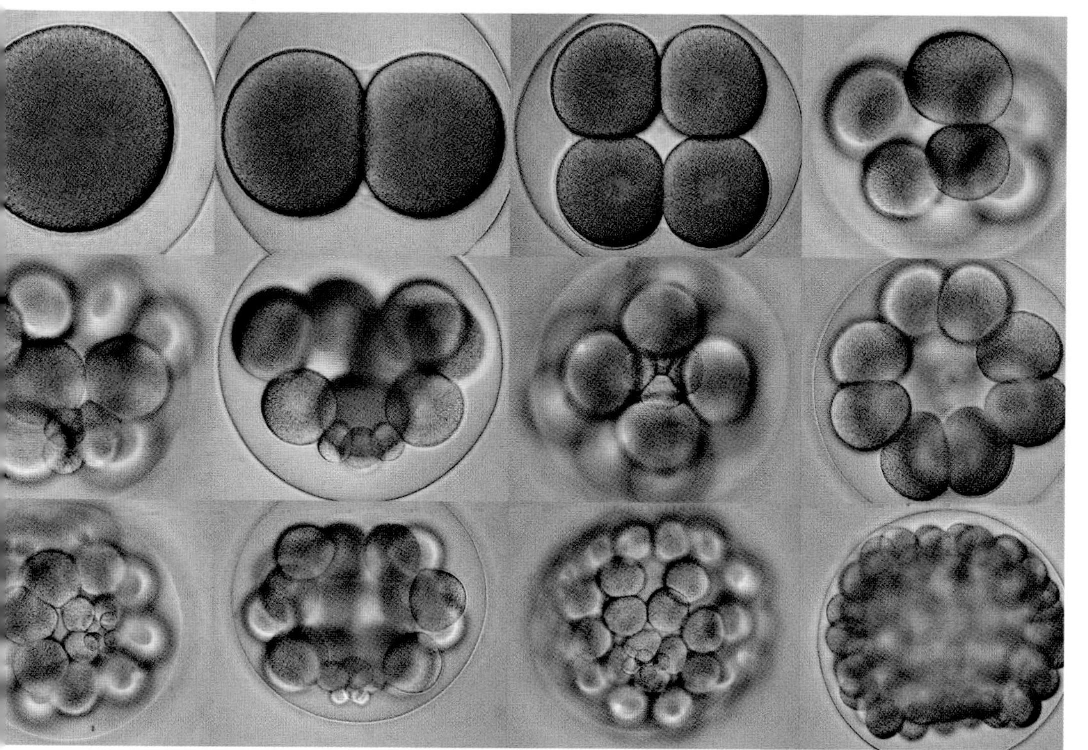

7

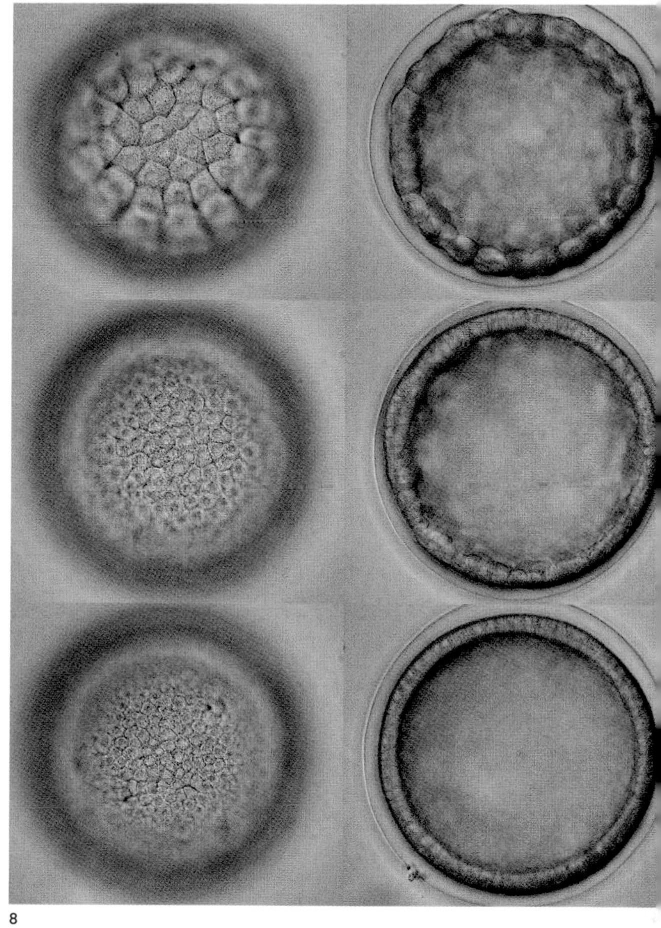

8

7 The embryological development of all animals is a reflection of their ancient ancestry. Cells divide but crucially remain joined, later differentiating to fulfil specialized functions. This image shows the early stages in the development of an echinoderm (*Clypeaster depressus*) from a unicellular entity to an embryo with 108 cells.

8 Leading on from the 108-cell stage, cell division continues apace.

9 The development of the echinoderm through gut. The first pore to form goes on to become the anus, while the second becomes the mouth.

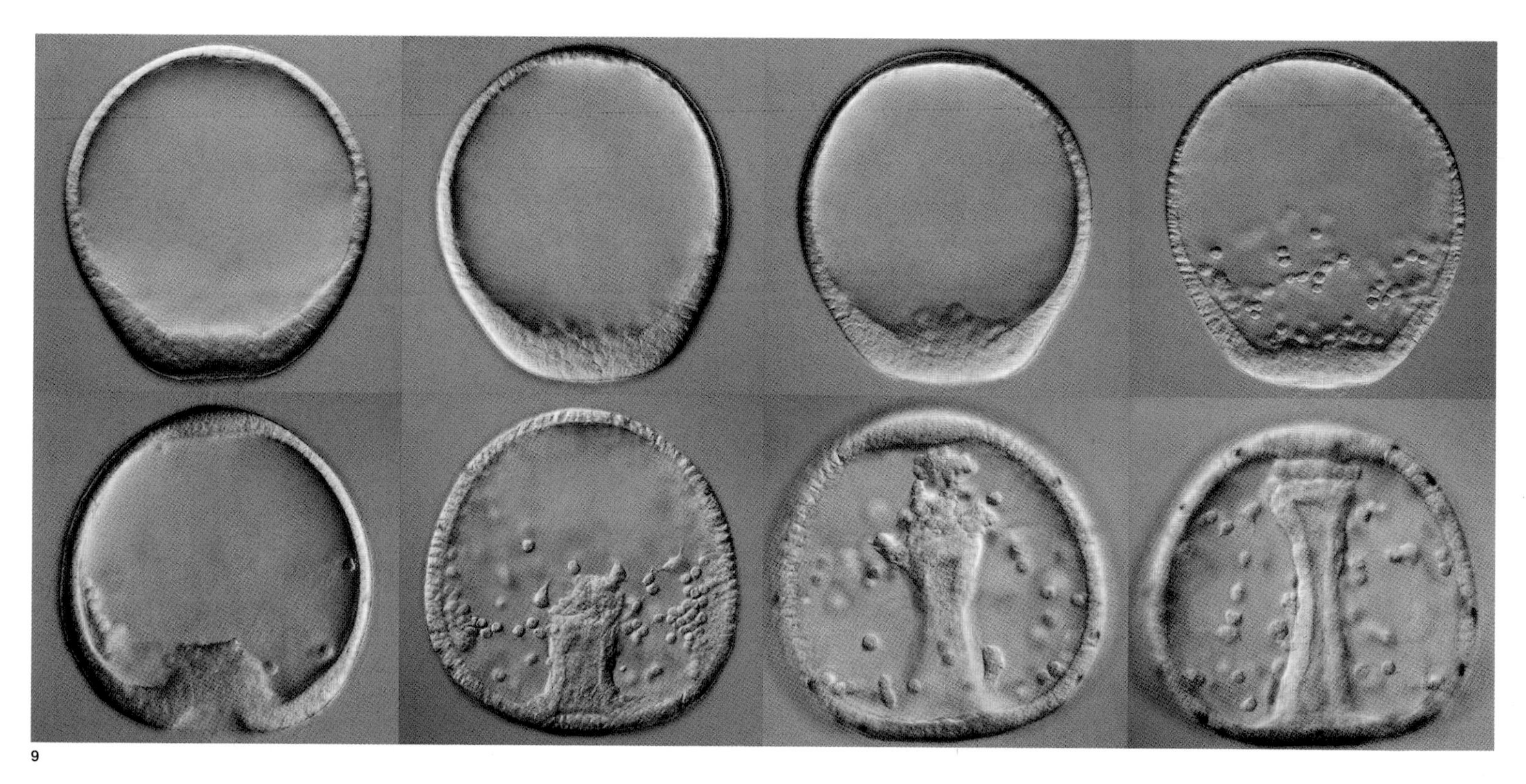

9

may have evolved a nascent gut; a small space for the more efficient digestion of food. The cells lining this space could collectively release digestive enzymes without them being overly diluted by seawater, competitors could not access the food once it was in the 'gut', and the products of digestion could be contained and ultimately absorbed. Improving the efficiency of digestion effectively increased the amount of energy the early animals could extract from a given quantity of food. By degrees, this allowed greater growth, activity and ultimately reproductive success, which in turn paved the way for ever more well-developed, differentiated guts and the evolution of other structures for food acquisition and digestion, and the disposal of waste.

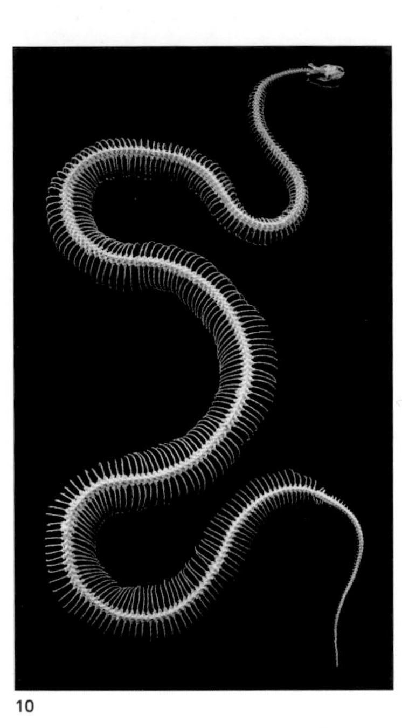

10

10 Segmentation is a defining characteristic of the craniate body plan. Although this is not always obvious, it is clear in the skeleton of a *Boa constrictor*.

11 Numerous repeating units make up the body of this annelid worm (*Eunice* sp.).

12 Even during the very early stages of embryological development, segmentation is evident in this centipede (*Strigamia maritima*), which is coiled around its reserve of yolk.

Cell layers and cavities

A few of the animal lineages (the Placozoa, Ctenophora and Cnidaria) have a body made up of two cell layers. In the ctenophores and the cnidarians an inner cell layer (the endoderm) lines a multifunctional space, while the second layer makes up their outer surface (ectoderm). In all the other animal lineages these two layers sandwich a third layer, the mesoderm, which is derived from cells migrating from the endoderm and ectoderm or entirely from out-pocketing of the endoderm during the early stages of embryological development.

This third cell layer is important because it represents a body of cells that have been released from being the outer or inner surface of the animal and the functions this entails. The cells of the mesoderm could specialize, providing the basis for other structures dedicated to particular functions, such as muscle, connective tissue and circulatory and excretory systems. They also made possible the evolution of a body cavity (the coelom), an internal partitioning that opens the door for yet more specialization. Prior to the evolution of a coelom, the muscles that evolved from the mesoderm could not act independently of the gut (and vice versa), due to these structures rubbing against each other, compromising both digestion and movement. A body cavity allowed further specialization of the gut as well as better, more efficient ways of moving around.

Segmentation

Segmentation is a common trait amongst the animals that appears to have evolved independently on a number of occasions. It is often clearly visible, especially in animals such as earthworms and centipedes; in other animals, including craniates, it may be less obvious [10–12]. But segmentation is one of the most important body plan principles in the animals, arguably on the same order of significance as partitioning of living material into cells, organelles into nucleate cells, and nucleate cells into multicellular organisms.

The significance of segmentation depends on the animal lineage in question, but in some (i.e., annelids) it may relate to partitioning of the body cavity and more efficient burrowing. This body cavity provided the basis for the evolution of a hydrostatic skeleton (see below), such that the pressure of fluid inside the body could keep the body wall rigid – extremely important for burrowing and moving around freely in marine sediments. Separating the ancient annelid body into a number of discrete segments allowed increased control and flexibility of the hydrostatic skeleton, making writhing and burrowing movements much more efficient.

In the annelids and the other segmented animals, partitioning of the body into a series of repeating units laid the foundations for further specialization and for an explosion of diversification. Distinct segments, appendages sprouting from individual segments, and segments uniting into groups all presented huge scope for developing the morphologies and lifestyles of animals. This in turn demanded more complex nervous systems to control and synchronize the movements of these specialized units, and promoted the development of large and complex brains by uniting the ganglia (nerve cell clusters) of multiple segments.

A further advantage of segmentation is that it confers a degree of redundancy in the body plan of an animal. If a segment of a body or appendage is damaged the animal can carry on more-or-less as normal.

Respiratory, circulatory and excretory systems

The cells of every animal require oxygen to release energy from their food, and in doing so they generate metabolic waste: carbon dioxide and water. This process of aerobic respiration provides the fuel for countless other cellular processes that in turn generate more by-products. Water also steadily diffuses into animal cells from the outside. Left to accumulate, this excess water and waste would disrupt the delicate balance of the animal's internal environment, so it has to be eliminated. How this problem is dealt with depends on the size of the animal.

11

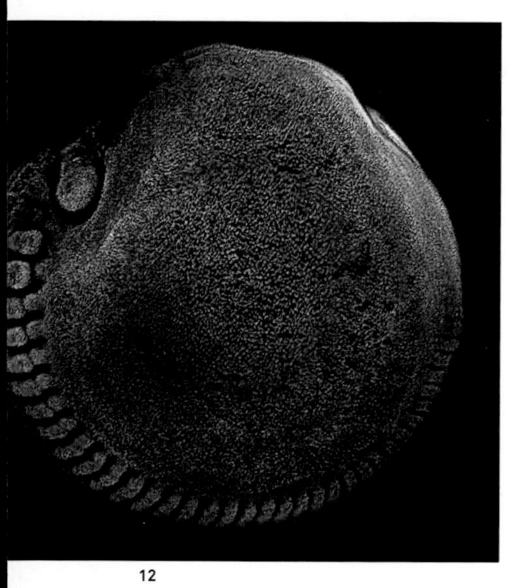

12

None of the cells of a small, aquatic animal is very far from the outside world; oxygen and waste can therefore freely diffuse in and out from the surrounding seawater without the need for dedicated respiratory, circulatory and excretory systems. But with increasing body size and greater complexity, a threshold is reached where simple diffusion is no longer sufficient in servicing the needs of all the cells. Some animals, such as the larger, free-living flatworms, get around this problem by being exceedingly flat, so that diffusion is still effective in all the cells of the body. But in others the implications of a larger body size for gas exchange and excretion have been met with the evolution of dedicated structures to ventilate every part of the body (e.g., gills, open/closed circulatory systems [13]) and to eliminate excess water and waste (e.g., kidneys and their associated structures). The emergence of these innovations in combination with others (e.g., skeletons) set the scene for the evolution of ever larger animals [BOX 4].

Skeletons

Skeletons come in a variety of forms and fulfil many functions, most notably giving the body support, providing a surface for the attachment of muscles and protecting the animal from damage. The simplest type of skeleton is the fluid contained by the body wall of an animal, which is actually very effective in giving the body some rigidity. These fluid or hydrostatic skeletons, as they are known, work in the same

BOX 4 ***Size and growth***

Size in the evolution of the animals is extremely important since it has implications for all sorts of basic processes, including gas exchange and excretion.

A larger body is more 'expensive' to produce and maintain, especially in terms of the various systems that are needed to get around the problems of ventilation, nutrient distribution and excretion; however, being big has a number of key advantages. Firstly, larger animals have a lower surface area to volume ratio, allowing them to regulate their internal environment more efficiently than smaller animals. Secondly, the larger an animal is, the fewer predators it will have. Finally, the greater efficiency of larger animals and their relative invulnerability means that they can dominate the habitats in which they live.

Individuals of soft-bodied species without any form of supporting scaffold simply can't grow big. But many animals, such as corals and siphonophores (see Cnidaria, pp. 48–67) and bryozoans have evolved an interesting way around this problem: they have become modular. Unlike solitary animals that grow by a general increase in body size, a modular animal grows by replicating itself over and over again, forming a colony of genetically identical modules. You only have to look at a reef of large coral colonies to see how modular animals can dominate habitats as effectively as large, solitary animals.

These modular animals, particularly some of the cnidarians, have mindbending and outlandish life histories that are so very different to our own they are beyond intuitive comprehension. Aging and death, for example, which come all too quickly in a solitary animal, are hugely postponed by the repair and replacement of modules, allowing some colonial creatures to reach ages of at least 4000 years. If a solitary animal sustains serious damage it is doomed, but colonial animals are extremely resilient.

Modular animals also force us to think more deeply about one of the basic levels of biological organization, since the colony, rather than the modular units, can be considered to be the 'individual'.

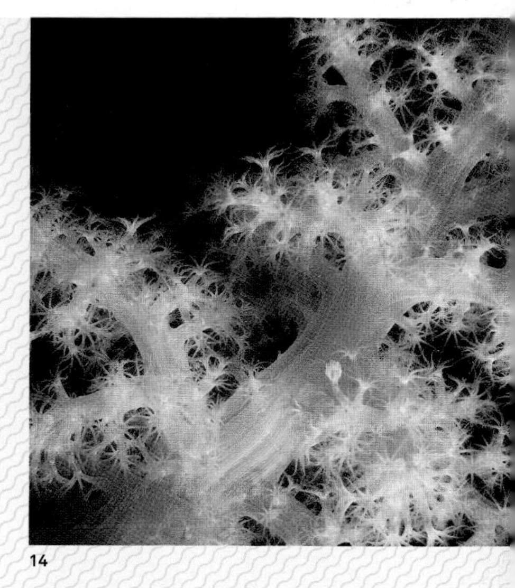

14

14 The tree-like growth form and multiple polyps of a soft coral, *Gersemia fruticosa*. Unlike the stony corals, which are protected by a calcareous skeleton, many of the soft corals use noxious chemicals to defend themselves.

‹ **13** Increasing body size in the animals drove the evolution of systems that could ventilate all the cells in the body. The gills of this larval great crested newt (*Triturus cristatus*) increase the surface area via which gases can be exchanged.

way that an inflated balloon holds it shape. A mesh of muscles both displace this fluid to allow localized swelling and therefore movement (think of an earthworm moving through the soil) and reinforce the body wall to stop these localized swellings from becoming blow-outs. This type of set-up is very effective when it comes to burrowing and many soft-bodied animals with a worm-like form have this type of skeleton [BOX 5].

The other major type of skeleton is solid, and can be external (e.g., the exoskeleton of an arthropod or the shell of a snail) or internal (e.g., the bony human endoskeleton). There are numerous driving forces behind the evolution of these structures, but the exoskeleton may have evolved primarily as a response to predation. As diversification gathered pace, predatory animals became ever more effective in locating and dispatching their prey, a selective pressure that prompted the emergence of shells and tough exoskeletons. Lots of animals have some form of shell, although this type of defensive structure is at its most diverse in the molluscs. The first shells may have been nothing more than a tough covering of protein, but an evolutionary arms race soon developed with the defences of the prey keeping pace with the weapons of the predators. Over time the shells of animals have become a marvel of nature, scaffolds of protein reinforced with crystals of calcium carbonate – but the arms race means the predator's weapons, such as the powerful pincers of a crab, the chitinous beak of a cephalopod or the acidic secretions and drilling radula of some predatory snails, are equally sophisticated. In some lineages, especially the arthropods, the exoskeleton became more than just a protective covering – it became part of a sophisticated musculoskeletal system, providing points of attachment for muscles, which in turn allowed the evolution of hinged, moveable extensions of the body, such as specialized mouthparts, sensory appendages, limbs, some types of gills and even wings.

A solid internal framework, such as your endoskeleton, may have started out as a solution to a metabolic problem, namely as a repository of calcium and phosphorous in active animals with cells hungry for these elements (see Craniata, pp. 84–97). Regardless of the endoskeleton's origins, internal deposits of inorganic materials and proteins were soon co-opted as an internal, supporting scaffold. Like the exoskeleton, the proto-endoskeleton became part of a musculoskeletal system ultimately underpinning the evolution of limbs.

Perhaps most crucially of all, the articulated exoskeleton (in the case of arthropods) and the

15

bony endoskeleton (craniates) made a terrestrial existence possible and heralded an explosion in the diversity of these two lineages as they adapted to this virgin domain (see below).

Muscles

Typically, animals have to pursue resources and mates, so they need to move. Life forms have evolved numerous ways of getting around, including simply growing, changes in internal body pressure, rotating flagella or beating cilia. Up to a certain body size these methods are adequate and are used by many smaller animals as their main form of propulsion (e.g., the beating cilia of a rotifer or a flatworm). But the evolution of contractile fibres into a specialized musculature made up of coordinated bunches of contractile cells underpinned a great wave of diversification [15]. This musculature could change the shape of the body, bend the body, twist the body, move appendages and provide propulsion. Furthermore, as the energy demands of these muscles grew, they themselves became part of more refined systems for gas exchange and the movement of nutrients and waste around the body, driving the evolution of ever more elaborate bodies.

In terms of propulsion, the ability of muscle cells to contract rapidly has been combined with specialized appendages to enable some animals to move extremely swiftly and with precision in water, on land and in the air. For example, some gastrotrichs 'row' through water using muscle-actuated bristles (see p. 276); the force generated by a cheetah's skeletal muscle allows it to run at speeds of around 30 m (100 ft) per second; and muscles drive the 300 wing-beats per second of a hoverfly that make these insects such masters of the air.

15 Specialized groups of contractile cells (muscles) allow rapid movement, and have been harnessed for propulsion, feeding, digestion, sound production, hearing and vision. The musculature of this aquatic crustacean is visible in red (*Daphnia* sp.).

Senses and nerve centres

Animals are usually motile organisms that depend on other organisms for food. In order to move around and find food, they must be able to sense their environment and respond accordingly. Sensory cells, tissues and organs pick up information from the environment and nerves convey this information to effectors – cells, tissues and organs that underpin the response. Very simply, animals use light, chemical, pressure and electrical field sensitive cells to detect food, mates, enemies or refuges and use effectors, such as muscles, to move closer to or away from the stimulus. As well as sensing and responding to the environment the nervous system also fulfils another crucial role, that of coordination. Complex animals with a variety of cells, tissues, organs and organ systems need them to work in concert. One way this is achieved is via the nervous system, in which long bundles of nervous tissue (nerves) relay information from one part of the animal to another extremely quickly allowing fine control and coordination.

Nervous systems must have appeared very early on in the story of animal life, because their basic components and structure are remarkably uniform, regardless of how any two animals are related. Receptors, neurones and effectors are connected in circuit, but there are differences in the degree to which certain parts of the nervous system have become specialized. The key driving force in this specialization may have been the transition from a life spent floating around in the water to a life creeping around the seabed. In contrast to an animal that floats freely, one that moves around on the seabed can only do so in a two-dimensional plane, mouth first. This part of the body is first to encounter things, and thus the senses became concentrated here; the nervous system to serve this nascent 'head' became more elaborate and specialized, eventually forming the central nervous system. In three animal lineages (craniates, molluscs and arthropods) this process of 'cephalization' has resulted in the evolution of staggeringly complex brains that are the seat of elaborate behaviours, problem solving and even the intangible phenomenon we know as consciousness.

BOX 5 ***The worm form***

If you have a close look at the tree of animal life you will see that many of the animal lineages are distinctly worm-like. And even in those lineages that typically are not worm-like, there are animals that have evolved this form (such as the caecilians and limbless lizards among the craniates).

Why is this shape so common? Among smaller animals, a worm-shaped body has a distinct advantage: it allows them to move freely around in very tight spaces, such as the exceedingly narrow channels between sediment particles at the bottom of aquatic habitats and the spaces between the cells of another organism. In short, a worm-like body makes all sorts of lifestyles possible.

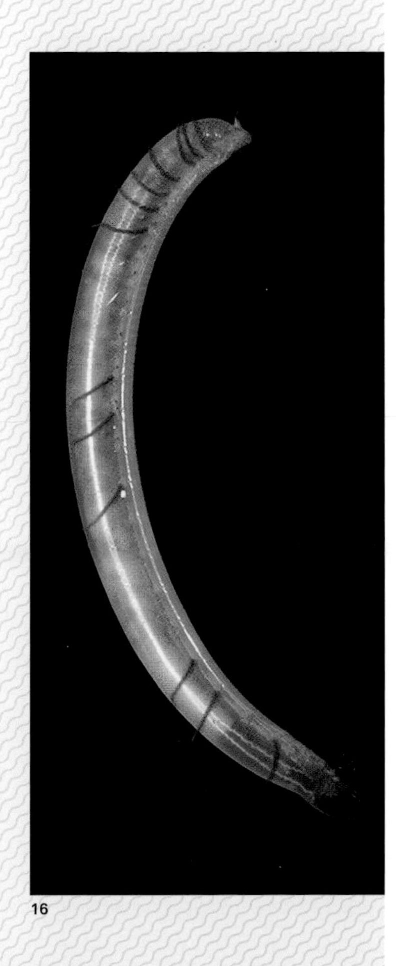

16

16 *Armandia* spp. and related annelids swim in loose sand and water.

KEY PROCESSES

Sexual and asexual reproduction

There is a tendency to assume that all animals reproduce via the combination of sperm and eggs, but there are many species that can also reproduce without any form of sex by splitting into two or more parts, or by budding off miniature versions of themselves. Others have reproductive organs and can produce gametes, but their egg cells can develop into embryos without being fertilized by sperm cells.

Animals that reproduce asexually can rapidly colonize suitable habitats because all of the individuals in a population can produce offspring and beneficial mutations that arise in the genome of an individual are passed on directly, allowing rapid evolutionary adaptation. In contrast, sexually reproducing species with separate sexes are at something of a disadvantage because only half the individuals (the females) in a given population can breed; there is also the risk of picking up pathogens during copulation. On the other hand, sexual reproduction allows the genes of two individuals to be shuffled and combined in all sorts of configurations, which also fuels evolutionary adaptation. An important driving force in the emergence of sexual reproduction is thought to be the perennial threat posed by parasites, since the mixing of two individual's genes may produce configurations that improve resistance to these enemies.

Development

The body of any animal grows and differentiates from a single cell into a multicellular collective via a sequence of developmental stages. Natural selection can and indeed does act on the variations in each of these stages, creating an effectively limitless potential for the evolution of new forms. One key aspect of this is the timing of events in the developmental sequence (heterochrony), a mechanism that underpins many evolutionary 'leaps'. Let us suppose the larva or juvenile of a particular species has a random genetic mutation (or mutations) that enables it to become sexually mature before it grows to the adult form. This unusual variant is smaller and morphologically simpler than its ancestors, but it is able to exploit a new niche; therefore natural selection favours it and a radically different offshoot of animals is established. This phenomenon is known as paedomorphosis because the new variant is 'child-like'; well-known examples include many animals of the meiofauna, dwarf males of a variety of marine animals, domesticated mammals (particularly dogs) and axolotls. Simplification, as can be seen in paedomorphosis, is an important part of animal diversification that crops up time and time again [BOX 6].

Similarly, another random mutation (or series of mutations) might result in an individual that matures past adulthood and develops traits that are unknown in its ancestors, thus propagating the formation of a new species. This is known as peramorphosis because the new variant is more 'adult-like', and again it is a common (but easily overlooked) phenomenon with well-known examples, such as extreme sexual dimorphism in certain insects (particularly beetles) and the narwhal tusk, which is a hugely elongated tooth.

HABITATS

The ease with which new iterations on a few themes can be generated has enabled animals to colonize just about every habitat there is and occupy the available niches in these habitats. There was a time when many places on earth were thought to be too extreme for animals to survive in, but as advances in technology have allowed us to explore the further reaches of the biosphere it turns out that there are few places where animals cannot survive and thrive. Animals can be found in abundance from the crushing depths of the ocean to the seemingly barren, ice-encrusted peaks of the highest mountains and everywhere in between.

Scale and accessibility make it difficult for us to appreciate the myriad habitats there are around us. A handful of sediment from the seabed is home to a bewildering variety of animals, but most are barely visible to the naked eye [1]; a bromeliad brimming with water in the canopy of an equatorial forest supports a whole community of animals, many of which live out their entire life high above the ground and out of sight; a swift flying in the sky plays host to a menagerie of superbly adapted parasites. It is microcosms like these that give the animal world its incredible, multilayered complexity.

Animals themselves, particularly larger species, are an ecosystem in their own right, offering all sorts of niches that are exploited by an incredible array of parasites [BOX 7]. These parasitic animals are often portrayed in a very negative way, but they are a vital part of a healthy ecosystem and in terms of biomass they can outweigh predators by a considerable margin.

BOX 6 ***Simplification***

It is important not to conflate evolution with increasing complexity, since in evolving to fill a particular niche an animal may lose many of its structural complexities. Structural simplification is a hallmark of animals that live sedentary or endoparasitic lifestyles, but it is the latter where this process has been taken to the extreme.

As adults, tapeworms live attached to the gut wall of their host (see Platyhelminthes, pp. 278–93). They evolved from free-living animals, creatures that were very probably active hunters with a complete gut, well-developed senses and a central nervous system. But in adapting to life inside other animals many of these complex structures were surplus to requirements and they were progressively lost. Surrounded by food, nutrients could simply be absorbed across the body wall, so the gut and its associated structures disappeared. They no longer needed to devote any energy to securing food, so their senses and central nervous system became ever-more simple.

Dicyemids, enigmatic animals that live inside the kidneys of molluscs, typically octopuses, have evolved beyond parasitism and actually benefit their host (see Dicyemida, pp. 204–05). It is thought that they descended from a more complex ancestor, but their complexities have dwindled away and now they have perhaps the simplest morphology of any animal. Their slender, worm-like form is made up of only 10 to 40 cells and there is no sign of a body cavity or differentiated organs.

Simplification has confounded attempts to understand the evolutionary relationships of the animals, especially so back when differences in morphology were all that scientists really had to go on. Furthermore, there was the assumption that evolution was a more-or-less linear event with simple 'primitive' forms at the bottom of the chain and complex, 'advanced' forms at the top. With this rationale, many groups of animals with pared down morphologies were wrongly assumed to be 'primitive' and near the base of the evolutionary ladder. For example, free-living flatworms (see Platyhelminthes, pp. 278–93), were once thought to have diverged little from the first animal that had three body layers and bilateral symmetry. However, new insights suggest the flatworms may have evolved from an ancestor with a more complex body, but in adapting to a specific niche they have secondarily lost some of their complex features.

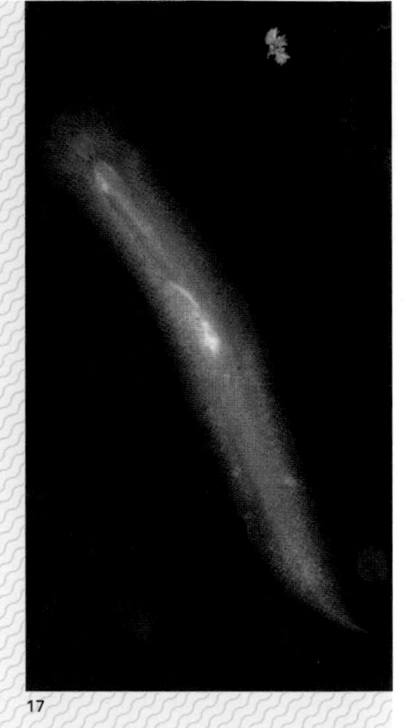

17

17 A marine free-living flatworm (unidentified species).

18

18 It was an animal very similar to the living coelacanth (*Latimeria chalumnae*) that gave rise to all land-living craniates. The fleshy fins evolved into weight-bearing limbs.

› **19** Animal diversity can be very high in localized patches of the deep sea. Here, 600 m (2000 ft) down in Bahamian waters, a community – including Venus flytrap anemones, a sea star, feather stars, basket stars and gooseneck barnacles – clings to a stand of dead coral.

On a large scale the most important habitats in terms of the variety of life they support are coral reefs, pristine equatorial forests and large lakes, the interstitial habitat (the labyrinth of channels between sediment grains in aquatic environments), and perhaps the deep sea [19]. The structural complexity of equatorial rainforests with their terrestrial, aquatic and aerial elements makes these habitats the most diverse in terms of species [20], but it is in the ocean where we find the greatest variety of the animal lineages. A mere handful of marine sediment can be home to more of these animal lineages than a whole equatorial rainforest, but the creatures of these interstitial habitats (meiofauna) have received very little attention because most of them are so small [1].

LEAVING THE OCEAN

Water is essential for life as we know it on this planet. This is because of its ability to act as a solvent – it is the ideal medium in which to dissolve a huge range of substances, from nutrients and messenger molecules to the waste products of metabolism. The suitability of water as a biological solvent relates to its physical properties, namely its ability to remain as a liquid over a very broad range of temperatures and the fact that it is less dense as a solid than as a liquid (which is why ice floats). Life on other planets where different conditions prevail could well be based on other solvents – such as ammonia, hydrogen cyanide, hydrogen fluoride or various hydrocarbons – but on earth, water is unbeatable in this role.

It is no surprise, therefore, that all life has its root in the ocean [18]. When it comes to the animals the vast majority of them are still chained to an aquatic way of life, indeed, it is only in marine habitats that we can find representatives of all the main animal lineages. There are certainly lots of animal species on land that live a truly terrestrial existence, but almost all of these belong to just two lineages: Craniata and Arthropoda.

Many representatives from other lineages live on land, but these have not truly mastered the terrestrial way of life. A huge number of nematode species, for example, could be said to live on land, but when we look at their lifestyles a little closer we can see that they have merely adapted to land-based aquatic habitats, such

21 Adult *Trichinella spiralis* in their host's intestine.

BOX 7 ***Animals as habitats***
Of all the places where animals live, the habitat that most vividly demonstrates the adaptability of these organisms is the body of another animal. This provides both shelter and food in abundance, so it is no surprise that a huge variety of species have become parasites, each on a spectrum of increasingly close associations.

Imagine the nest of a bird and all the matter that accumulates in it, including fluids from the hatched eggs and bits of skin, feather and faeces. The pungent aroma of this decaying material is a draw for all sorts of scavenging arthropods. These come to depend on the birds and their nests, but they do them no real harm (they might actually help by cleaning up the nest). However, once in such close association with another, larger animal, the transition to a more sinister way of life is fairly straightforward. Skin and bits of feather may start to be taken directly from the young birds, which can all too easily lead to blood feeding if the discrete nibblings are a little too deep. We can see the leading edge of this evolutionary trajectory in the huge variety of specialist bird ectoparasites, all of which evolved from nest-dwelling scavengers.

If an interaction is left to run its course over huge stretches of time, parasitism may eventually give way to an intricate mutualism, where both species are interdependent.

as the fluid-filled spaces between soil particles or the distinctly moist habitats inside other animals and plants. The same can be said of the molluscs, as although there are lots of slugs and snails on land they depend on habitats that are sufficiently moist, and they display behavioural adaptations to limit the loss of water through their permeable skin.

It is true that there are nematodes, molluscs and representatives of other lineages carving out a living in deserts, habitats very far from the watery beginnings of life, but they do so only by limiting their activity to when ephemeral rains bring moisture to these otherwise bone-dry landscapes. Some arthropods and craniates, on the other hand, can remain active in these environments all year round; it is these animals alone that have truly forsaken an aquatic habitat. That all but two lineages have failed to make a complete transition to life on land underlines what a massive step this was in the evolution of the animals. The features that the arthropods and craniates share that made this move possible are skeletons, limbs and water-proof cuticles – providing support for the body when the animal is no longer buoyed by water, and preventing the undue loss of precious fluids [18].

Limbs also provided the arthropods and some craniates with the adaptive plasticity that allowed them to take to the air. With gliding and powered flight these animals could evade their enemies more effectively, find food more easily and strike out in search of new habitats. The conquering of the land and the air are such huge milestones in the history of animal life that one might expect them to have occurred relatively recently, but these events actually took place hundreds of millions of years ago. Animals, specifically arthropods, first took to the land at least 500 million years ago and a mere 100 million years later, the blink of an eye in geological terms, insects were on the wing.

‹ 20 Primary, equatorial rainforest. The productivity and structural complexity of this habitat has helped drive a bewildering degree of speciation that is probably unparalleled anywhere else on earth.

Ctenophora

(comb jellies; sea gooseberries; sea walnuts; Venus's girdles)
(Greek *kteis* = comb;
pherein = to bear)

Diversity
c. 240 species

Size range
~5 mm to 1.5 m (with tentacles)
(~0.2 in. to 5 ft)

1 Almost all ctenophores are predators – and they are unfussy, eating virtually anything that they can fit in their mouths, from microscopic larvae and rotifers to the adults of small crustaceans and others of their kind. The anal pore is visible in this image (*Beroe abyssicola*).

2

2 The rhythmic pulses of the cilia usually run from mouth to anal pores (although reverse gear is possible). Hence comb jellies, unlike jellyfish, tend to swim forward mouth first (*Beroe* sp.).

Ctenophores are beguiling beasts. By day, they shimmer with all the colours of the rainbow and by night they emit intense flashes of bioluminescence. Ctenophores are known by a number of common names, including comb jellies, sea gooseberries and sea walnuts, mainly because of the size and appearance of the more frequently encountered species [1, 2, 6–12].

Exclusively marine, almost all comb jellies float freely in the open water. They propel themselves with bands of cilia (slender, hair-like protuberances), but are more or less at the mercy of the wind, tides and ocean currents. With bodies that look like little more than jelly-filled bags, they seem soft and delicate; in nature however, appearances often are deceptive; the ctenophores are actually voracious predators of small marine animals [1, 18].

FORM AND FUNCTION

Comb jellies are relatively simple creatures in that they have no organs and their body is composed of only two distinct cell layers – one, the epidermis, is the outer skin and the other, the gastrodermis, lines the central cavity. In between these layers is a thick, buoyant jelly known as the mesoglea, which contains cells of various types, some of which develop into smooth muscle cells as and when needed. .

Their central cavity is actually rather elaborate; it begins with a muscular mouth, followed by a pharynx and a stomach connected to a series of channels extending throughout the body. In many species the mouth is lined with large cilia, which act a bit like teeth, breaking up the prey a little before it reaches the muscular pharynx and the stomach beyond. However, instead of a true 'through gut', a tube with a mouth at one end and an anus at the other, this central cavity has anal pores at its far end [1]. Some of the indigestible matter from their prey is eliminated through these pores, but the bulk of it is simply ejected out of the mouth.

Most ctenophores propel themselves through the water in a smooth, stately fashion, albeit slowly, by the rhythmic beating of large cilia [BOX 1]. However, some species use muscular pumping to force a jet of water from their mouth. This enables them to move a bit faster than just their cilia allow, but the overall pace is still quite slow.

Stealth is important for an animal that catches its food by fishing. Many comb jellies have well-developed fishing apparatus consisting of a pair of branching tentacles that may be many times the length of the body [11]. These serve as excellent nets for catching the small marine animals that bob about in the plankton. The prey gets entangled in prehensile offshoots of the tentacles, which are studded with very sticky glue cells known as collocytes [13]. Glue cells of one sort or another occur in several animal lineages, but these collocytes are unique to the comb jellies. They stick to any creature they touch, and anything that struggles hard enough can draw out a spiral thread that will not let go, but unwinds and leaves the prey like a harpooned fish. Entangled and stuck fast,

BOX 1 ***Comb plates***

Ciliary locomotion is provided by eight thin bands of comb plates that run the length of comb jellies. Unique to these animals, each comb plate or ctene is composed of thousands of tightly packed cilia that, at up to 2 mm (0.08 in.), are probably the longest of any animal. It is the rhythmic wafting of these comb plates that scatters sunlight and gives these animals their iridescent shimmer as they move through the water.

Ctenophores are among the few animals that rely predominantly on cilia to get about rather than muscle. Although ciliary locomotion is far from swift, its one key advantage is that it is very smooth, so comb jellies can move around without disturbing the water and alerting potential prey.

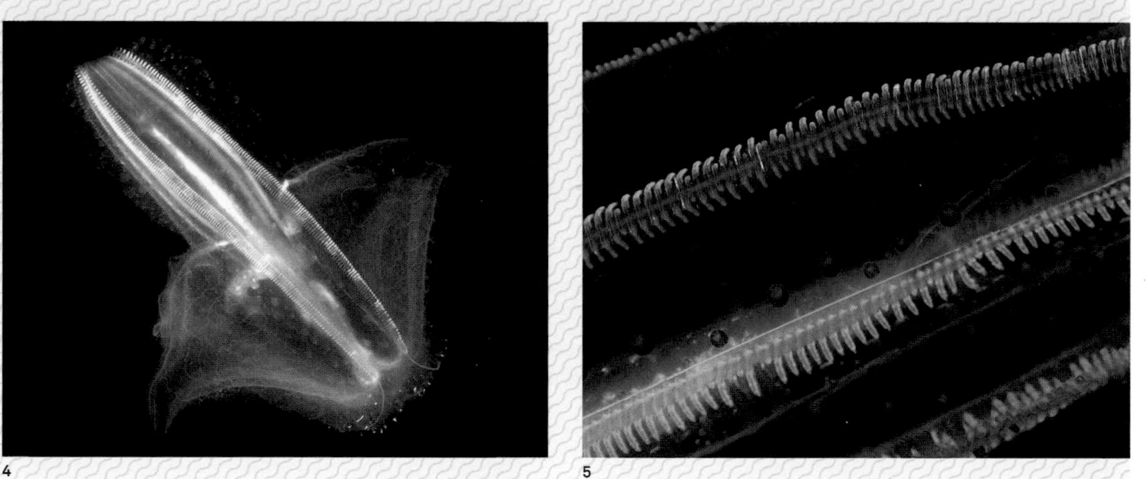

3 Close up of a comb jelly's comb plates – the characteristic arrays of large cilia that propel these animals through the water (*Bolinopsis infundibuliformis*).

4 Note the fin-like structures of this species, *Leucothea multicornis*.

5 With their continual beating, comb plates can scatter light to produce a rainbow effect (*Leucothea pulchra*).

the prey cannot escape and the hunter shortens the tentacle to bring its meal within range of the mouth. This passive capture of prey is undoubtedly effective, but many species are completely lacking in the tentacle department, so they simply swim at their prey and engulf it open-mouthed [1,2], or just let it swim inside [9].

With a life spent swimming slowly around in open water snaring or engulfing small animals, the comb jellies have had no need to evolve the hydrodynamic form that characterizes many fast-moving aquatic animals. Most of them are spherical or oval in form, but some have gone on a massive evolutionary tangent. The body of the so-called Venus's girdle is hugely elongated into what looks like a glittering, translucent cummerbund or ribbon [16]. These aberrant species use their cilia combined with muscular undulations of their bodies to propel themselves.

About one third of the known comb jellies – the platyctenids – have completely forsaken a pelagic existence in favour of creeping around on the seabed or clinging to the bodies of other benthic creatures [14,15,19,20]. These flattened species have lost the distinctive comb plates of their relatives and they bear something of a superficial resemblance to free-living marine flatworms (see Platyhelminthes, pp. 278–93); however, their well-developed tentacles give them away.

With their membranous forms, comb jellies have no need of gills or similar structures; gases simply diffuse freely in and out of their bodies.

They do possess a nervous system, albeit a relatively simple one with a net-like structure, and at the end of their body, opposite the mouth, they have what is known as an aboral 'organ', which is something of a very simple control centre [12]. The most obvious part of this 'organ' is a glassy or chalky structure known as the statocyst (see p. 120), which the animal uses to orient itself. This 'organ' may also be used to sense differences in light, pressure and chemicals in the water.

In addition to the shimmer from their beating cilia, most comb jellies can also faintly bioluminesce. This light emanates from the regions immediately underlying their comb plates, although some species can produce bioluminescent secretions. Bioluminescence is the rule rather than the exception in the comb jellies, but its significance is not well understood.

6

LIFESTYLE

Apart from the platyctenids that creep around on the seabed, ctenophores are animals of open water [14, 15, 19, 20]. They can be found in the shallow, warm seas of the tropics down to the bitterly cold depths of the ocean.

One species, *Lampea cancerina*, is known to be parasitic, but only during its immature stages when it takes advantage of salps (see Tunicata, pp 72–83). With the exception of this single species, all known ctenophores are active carnivores and just about any animal that is small enough to get entangled in their tentacles and which fits in the mouth is on the menu, although crustaceans, jellyfish, and tiny fishes make up the bulk of their diet [1, 18]. Indeed, they are important predators in marine ecosystems, but just how important has been difficult to gauge. Being very fragile, ctenophores are difficult to collect and handle, which for a long time was something of a stumbling block when it came to understanding their abundance and importance. Now, with constantly improving ways of investigating marine ecology, it is becoming clear these are very abundant animals, especially 400–700 m (1300–2300 ft) below the surface.

Their predatory voracity is clear to see when they end up in marine habitats where they are not naturally found. In the 1980s and 1990s, the western Atlantic species *Mnemiopsis leidyi* found its way to the Black Sea and Caspian Sea, probably in the ballast tanks of ships. It had no natural predators in these small seas, so its population boomed, reaching densities of 2200 individuals per sq. m (or 2600 per sq. yd) in some areas. It was not too long after *Mnemiopsis* became established that fish stocks collapsed, apparently a result of this predator eating fish eggs and larvae and competing with the fish for other prey.

With a body that is little more than a bag of gel, comb jellies are do not appear to be the most palatable morsels in the sea; however, lots of animals including fish and turtles will happily eat them. Some species even are specialist predators of others of their kind [8, 18].

Most ctenophores reproduce sexually and are hermaphroditic. Some species have testes and ovaries at the same time, while others alternate between being male or female. Eggs and sperm are normally shed into the water, and fertilized eggs develop into free-swimming larvae known as cydippids [17]. Some of the weird, benthic ctenophores can also reproduce asexually. Small pieces broken from the edges of their flattened bodies can give rise to fully formed individuals.

ORIGINS AND AFFINITIES

Not surprisingly, there are scant fossils of such soft-bodied, delicate animals. Until the 1980s we simply did not have a fossil record of the ctenophores, but then some creatures of the Devonian seas of around 400 million years ago were discovered in a German shale deposit. Ancient though these specimens may be, they bear the comb jelly hallmarks, which tells us that this body plan was already well established in this period and that if we want to find the origin of the ctenophores we must look much further back in time.

For a long time, most zoologists reasonably assumed that the comb jellies were evolutionary intermediates between the cnidarians and the platyhelminths, since the creeping, benthic members of the lineage are superficially similar to the flatworms [14, 15, 19, 20]. The ctenophores also used to be grouped with the cnidarians because of their apparent similarity to pelagic members of this group. But in light of DNA evidence we now know that neither of these theories is correct. The comb jellies are actually a distinct group of creatures that branched off on their own evolutionary trajectory a very long time ago. Their superficial resemblance to the cnidarians is a good example of convergent evolution – the phenomenon where two organisms that are not closely related come to resemble each other by adapting to similar niches.

6 *Beroe abyssicola*. Most, but not all, comb jellies are bioluminescent. The blue or green light is produced via a chemical reaction, but it can only be seen in the dark. The scattering of light by their beating cilia is often wrongly assumed to be bioluminescence.

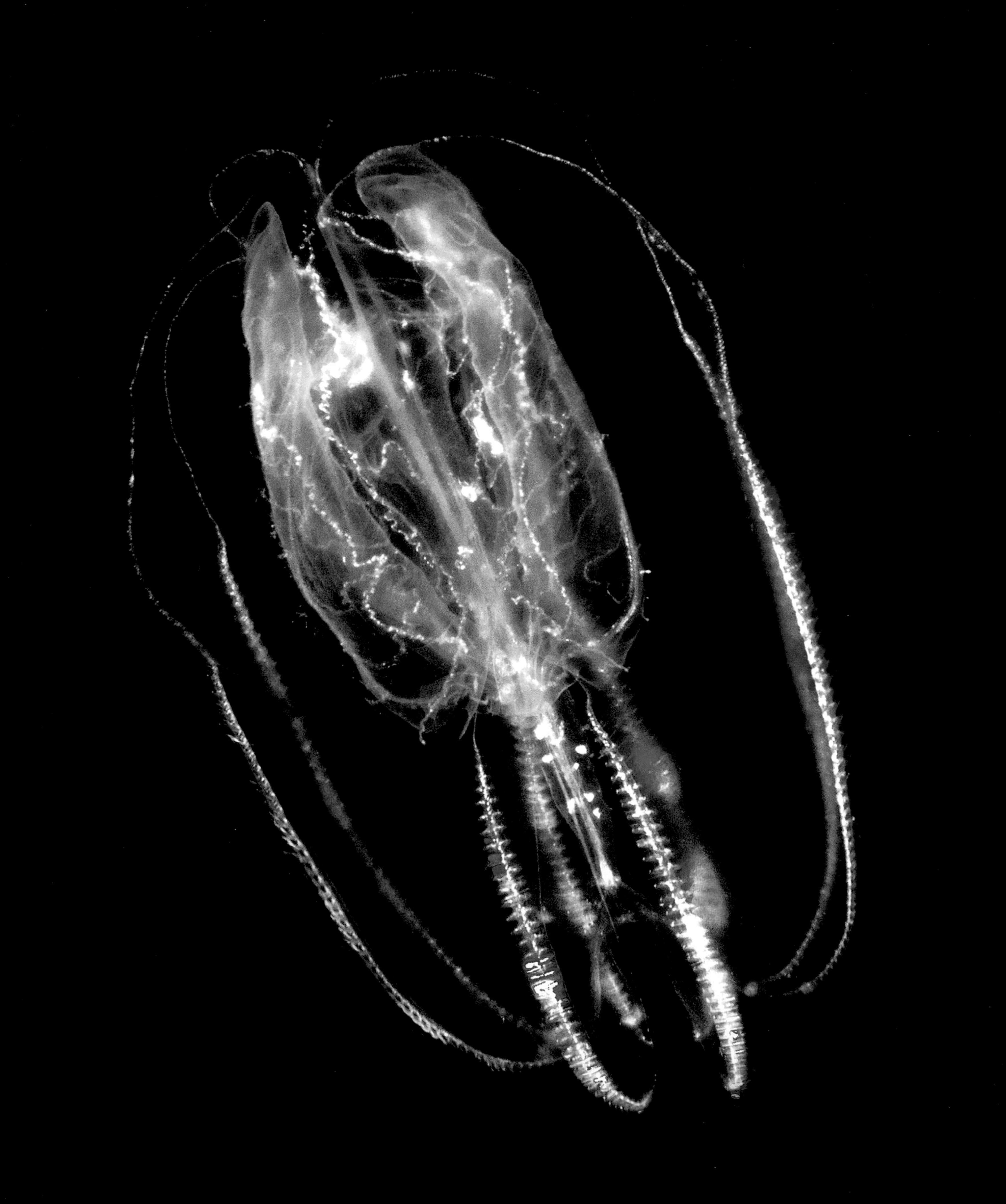

8

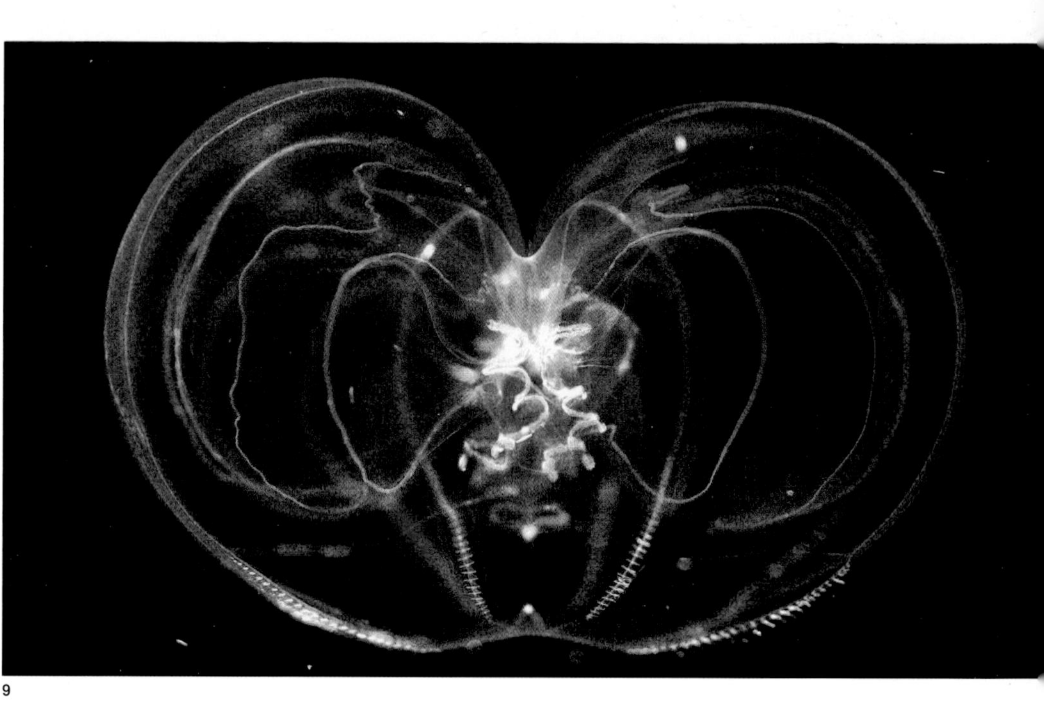

9

‹7 *Bolinopsis infundibuliformis*. This species of comb jelly, reaching up to 15 cm (6 in.) in length, has four long and four short comb plates. It is extremely fragile and individuals almost always tear and break up when handled.

8 Many comb jellies have a worldwide distribution. This species, *Beroe forskalii*, is found in open ocean and near shore, from surface waters down to depths of 500 m (1640 ft).

9 Unlike most other comb jellies that catch their prey by using muscles to suck in water, *Thalassocalyce* simply allows the prey (usually small crustaceans) to swim inside its bell where they stick to the mucus covered lining. The bell then snaps shut and the prey is digested.

10 This is one of the larger, shallow water comb jellies, reaching a size of around 25 cm (10 in.). Structures known as oral lobes help to trap prey and guide it into the mouth (*Leucothea pulchra*).

10

Overleaf

11 *Pleurobrachia pileus*. Note the two, retractable tentacles – these are one of the defining features of the ctenophore lineage, but they have been secondarily lost in some species.

12 The oral lobes of *Ocyropsis* spp. can be clapped together to give the animal a burst of speed if it needs to evade danger. The small dot at the 'top' of the animal is the sensory aboral 'organ'.

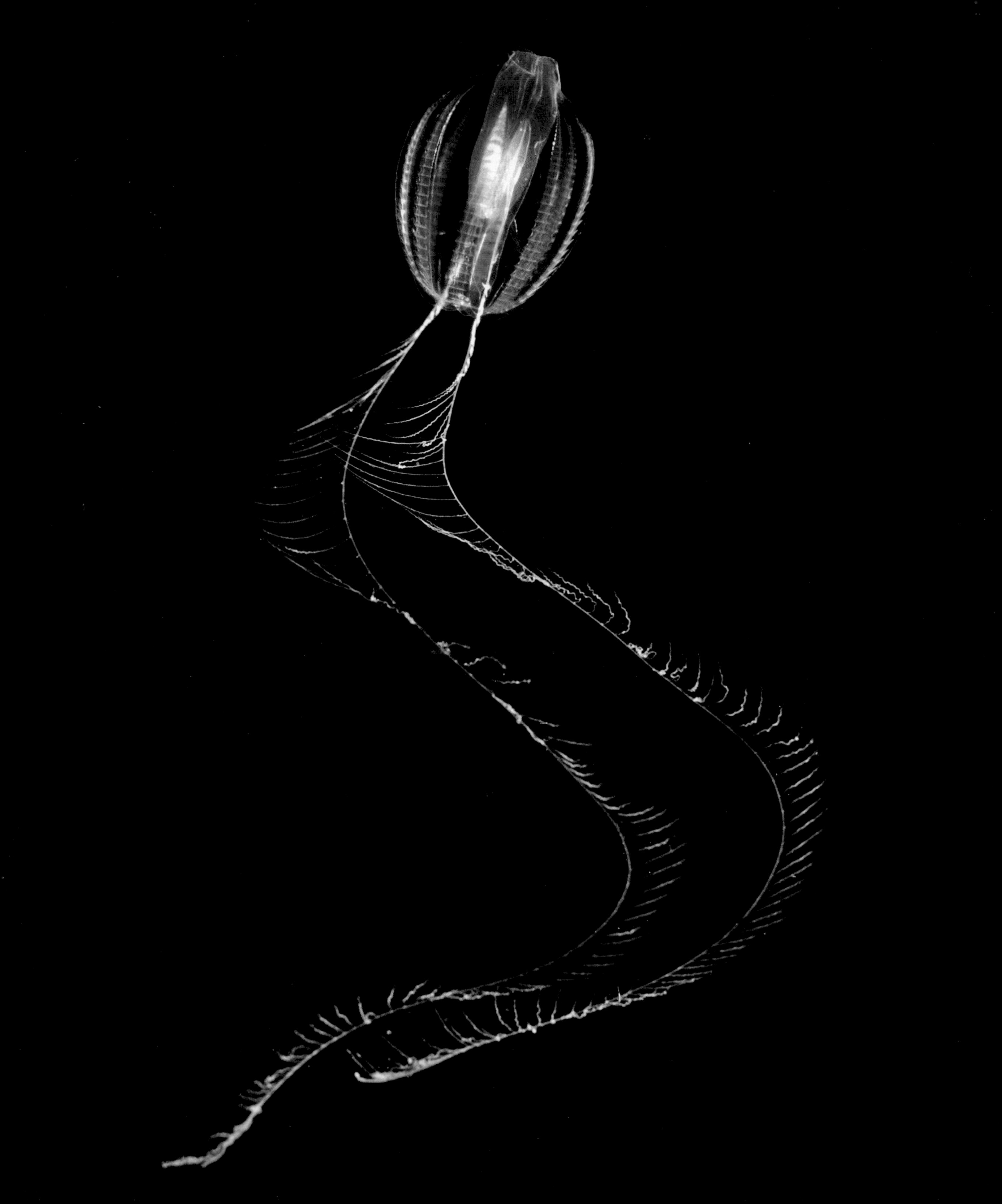

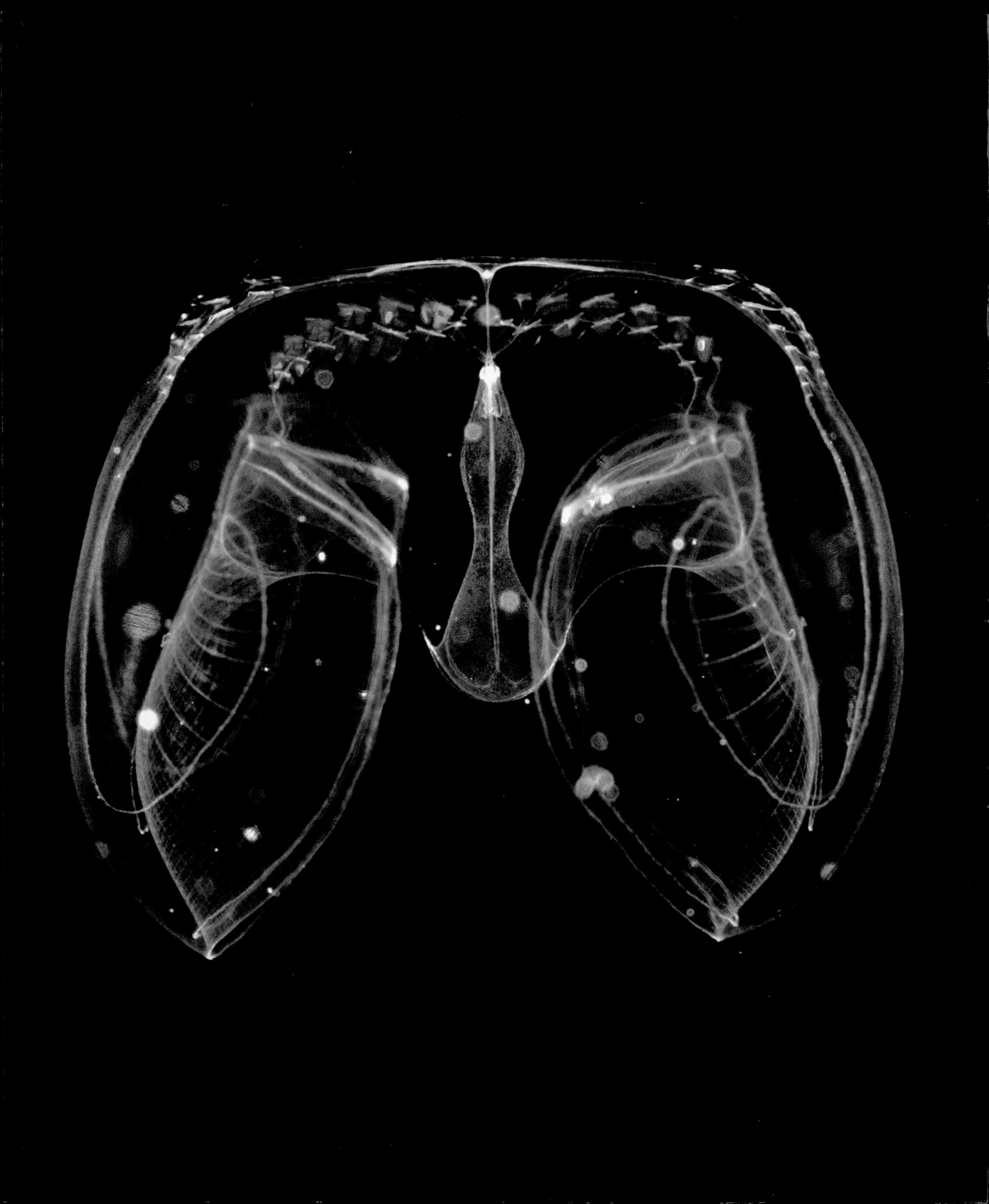

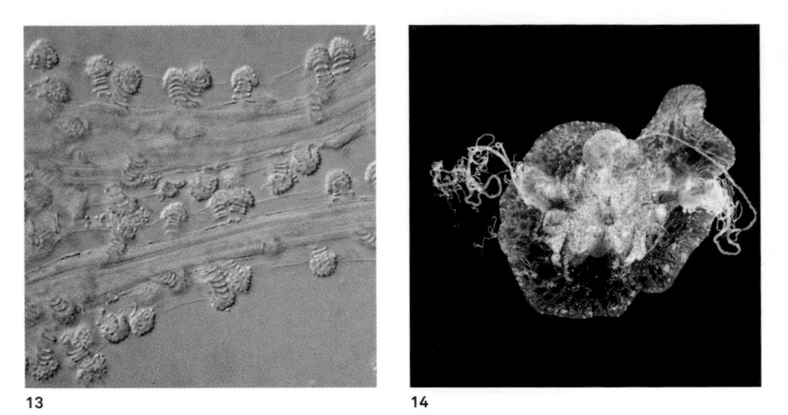

13

14

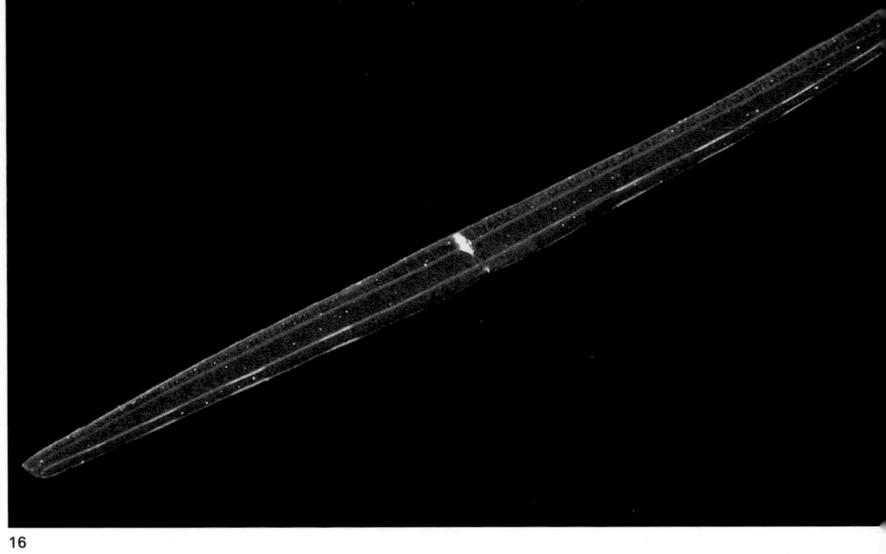

16

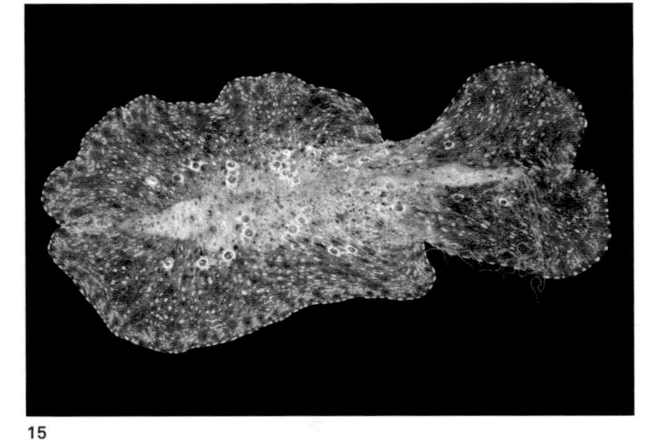

15

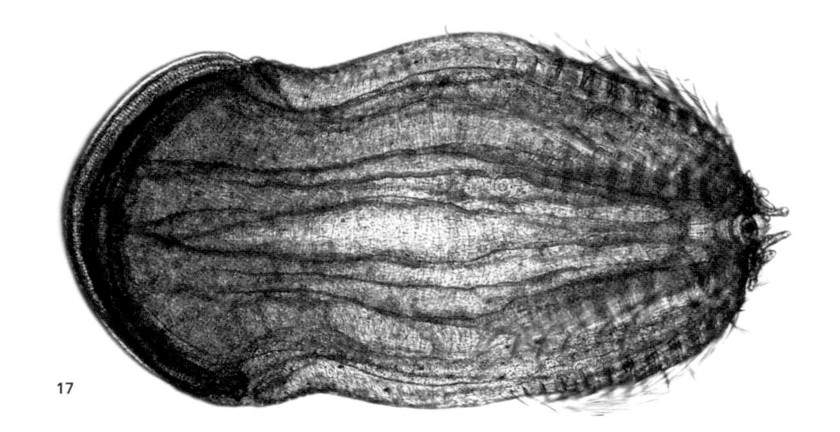

17

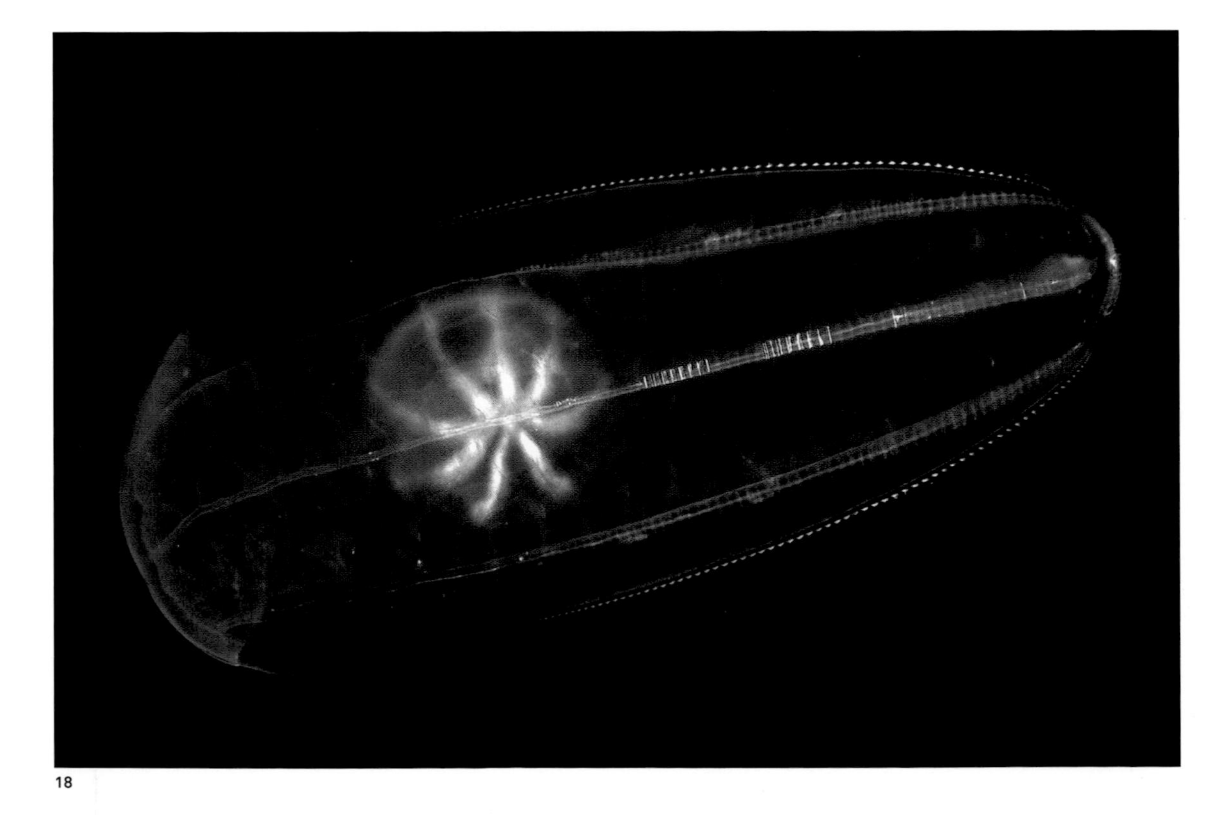

18

13 Glue cells studding the tentacle of a comb jelly. Each one is anchored by a spiral filament (*Vallicula multiformis*).
14 Platyctenids are comb jellies that have adapted to a life on the seabed. Note the partially extended tentacles (*Vallicula multiformis*).
15 With their tentacles almost fully retracted, the flattened platyctenids bear few superficial similarities to their pelagic relatives (*Vallicula* sp.).
16 Venus's girdle (*Cestum veneris*).
17 The planktonic cydippid larva of *Beroe ovata*.
18 Comb jellies are very effective predators. Here, a small comb jelly has fallen prey to a larger one, (*Beroe* sp.).

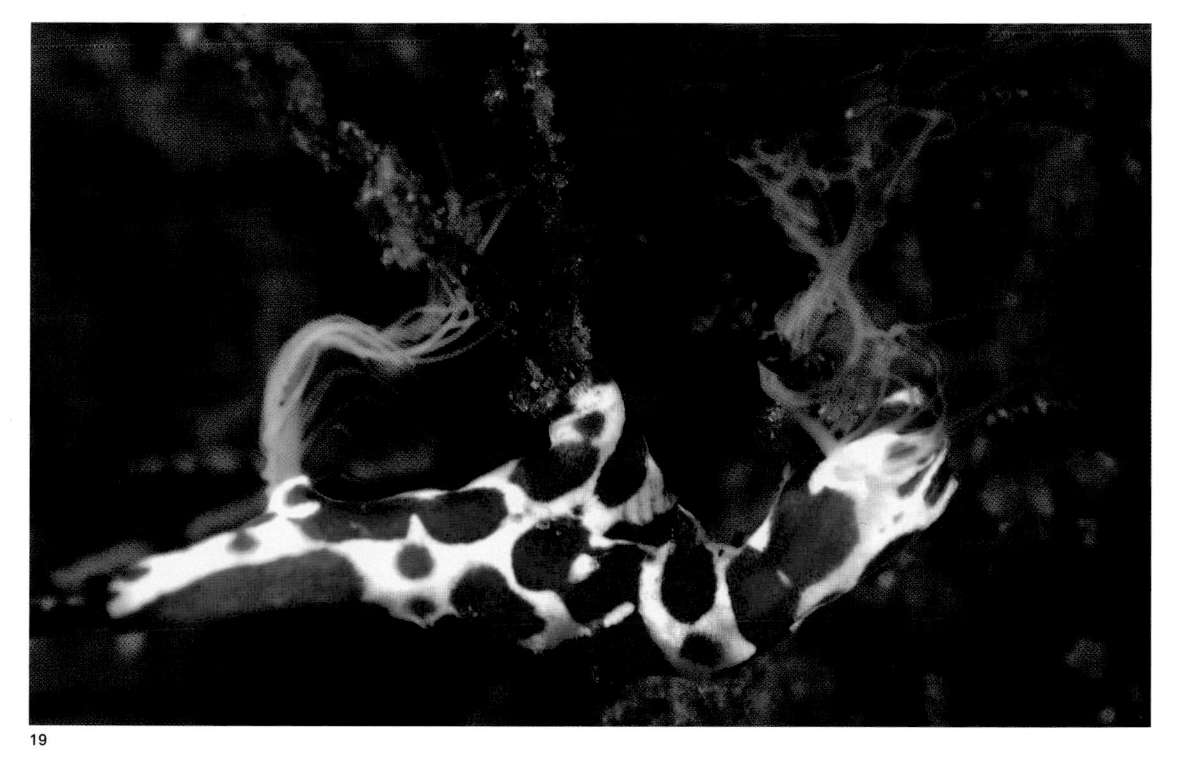

19

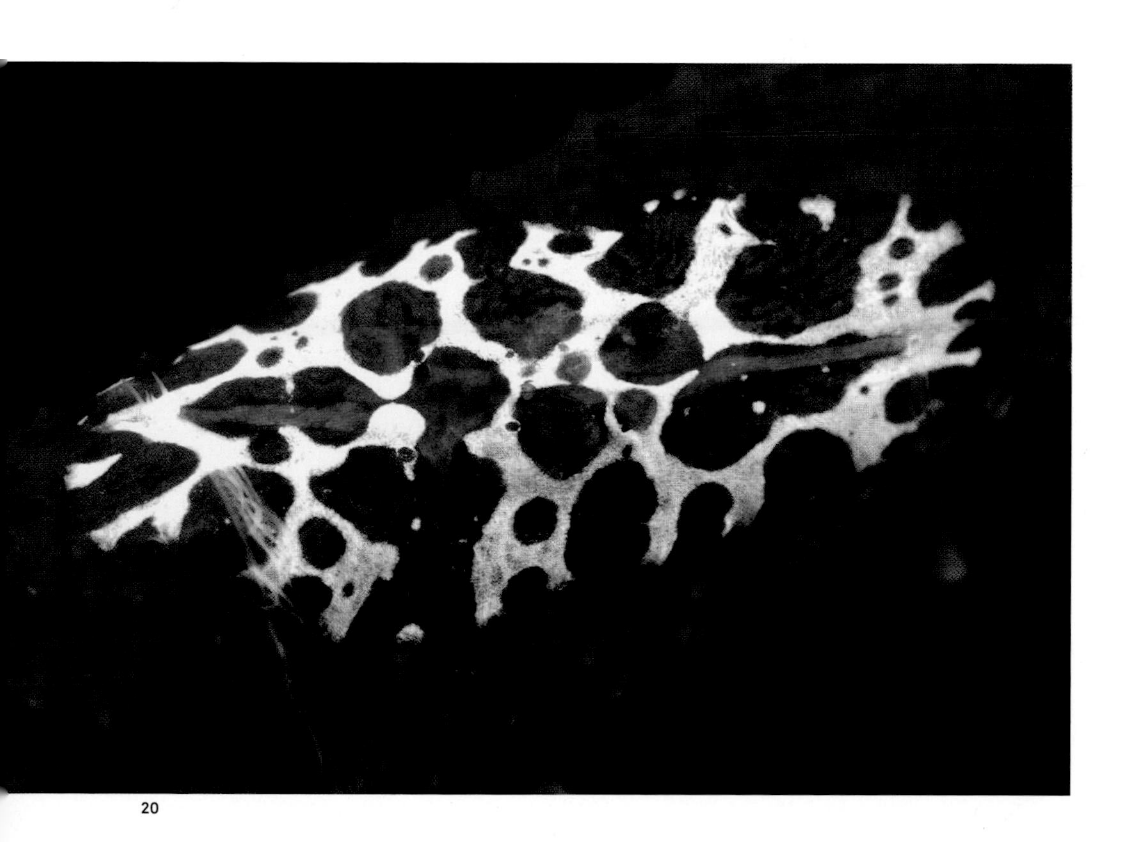

20

19 *Coeloplana astericola*, a colourful platyctenid. The tentacles that identify this creature as a ctenophore are clearly visible.
20 The benthic, flattened platyctenids mystified zoologists for a long time. It was wrongly assumed they represented a transitional form between comb jellies and free-living flatworms (*Coeloplana astericola*).

Porifera

(sponges)
Latin *porus* = pore;
ferre = to bear

Diversity
c. 8350 species

Size range
2 mm to 2 m
(0.08 in. to 6.5 ft)

1 The large, branching toxic finger sponge (*Negombata magnifica*). Completely sessile as adults, sponges are at the mercy of predators, a pressure that has driven the evolution of chemical defences. These, in combination with their tough and glassy skeletons render sponges rather unpalatable.

Animals in this lineage are what we commonly call sponges. They are plant-like in form, so much so that many people find it hard to imagine why zoologists say they are animals. Indeed, after puzzling over them Aristotle decided that they straddled the divide between animals and plants.

Sponges may look like simple animals, but in some ways they are startlingly sophisticated. They are widespread and successful, especially in the oceans, where they are found almost everywhere from shallow, tropical seas to the cold, dark depths of the Polar oceans. There are so many marine species, and they grow in such numbers, that they play an influential part in the oceanic ecosystem. There are even some 250 species that live in fresh water.

2

FORM AND FUNCTION

Sponges come in a variety of forms, but one of the most striking things about them is their lack of symmetry. In general, they are more or less shapeless. However, there are branching sponges, barrel-like sponges, tube sponges, fan sponges, flat, encrusting sponges, rock-boring sponges and sponges that look like lumps of rock [1, 2, 4, 5, 8, 9, 12]. Some of them have startlingly beautiful and interesting shapes and colours, but others are dull and amorphous.

All these shapes are made possible by an intricate internal scaffolding, a sort of skeleton that supports the soft tissues. No matter what shape the animal grows into, whether a vase, a crust, a lump or a branching fan, it can grow the skeleton to support it. This skeleton also permits the formation of a complex network of pores and internal channels – vital for the sponges' way of life.

Sponge skeletons are not all of the same type. Depending on the species, they are made of different substances, combinations of a protein known as spongin and inorganic reinforcement. Spongin is a form of collagen, the most important protein in animal connective tissue. The inorganic reinforcement is not mixed into the spongin just any old how – it is largely in the form of spicules, ornate crystals of calcium carbonate or silica extracted from the dissolved minerals in seawater [10, 11]. Some sponges take a short cut by incorporating sand grains into their spongin as crude, ready-made spicules.

The living tissues of a sponge are made up of anywhere between 10 and 20 cell types forming two distinct cell layers, each one cell thick. In between these layers there is a gel-like substance known as mesohyl. With little cellular differentiation to play with, sponges have no mouth, gut, nerves, muscles, respiratory or circulatory system. What they do have is a complex network of pores and canals through which water moves, the very pores that earned them the name Porifera. Lining these canals or special chambers are so-called collar cells (choanocytes). Each has a flagellum and they cooperate with their neighbouring choanocytes in wafting water through the channels with their flagella and in catching and engulfing food particles. Amoeboid cells that line these channels and move through the mesohyl are important in distributing this food to the rest of the cells in the sponge's tissues. These amoeboid cells have another, very special talent: they can develop into all the other cells that form the fully grown animal. We say that they are totipotent, much like stem cells. Depending on the requirements, they can give rise to eggs, sperm, nutrient storing cells, excretory cells and the cells that form the supporting scaffold of spongin and spicules.

LIFESTYLE

Most sponges are fixed to the spot, although there are a few species capable of creeping over the substrate at a dizzying speed of 1–4 mm (0.05–0.15 in.) per day. On the inside, however, sponges are constantly in motion. The frenetic activity of the collar cells' flagella is enough to pump their volume in water every five seconds or so. This pumping is aided by the structure of some species, which 'draws' water through the body of the animal, conserving energy. The amoeboid cells in the mesohyl are continually in motion too, creeping around the body of the animal.

Sponges need to be effective at pumping water because almost all of them are filter-feeders. The complex network of pores and channels increases the surface area available for trapping and engulfing the edible matter suspended in the continual flow of water. Individual cells digest this food and release the nutrients for absorption by cells throughout the body of the creature. There is more to it than just plain pumping though, as the sponge needs high speeds in some places and low speeds elsewhere. Water passing over the inner living membranes must move slowly so that their cells can capture incoming food. On the other hand, water pumped out of the

2 The creatively named elephant ear sponge (*Ianthella basta*). Sponges are a dominant component of shallow marine habitats. Even in the deep sea their diversity and abundance is impressive.

sponge carries waste and if the sponge does not live in flowing currents, it needs to propel this material a reasonable distance so that it is not simply drawn back in. All parts of a sponge's body are close to the water, so they do not need elaborate kidneys, respiratory, or circulatory systems like those of more complex animals; they have no problem when it comes to obtaining oxygen and getting rid of wastes such as carbon dioxide.

Some sponges have forsaken a filter-feeding existence for a carnivorous way of life because they live in places such as the deep sea, where very little food is suspended in the water [13]. These anomalous, carnivorous species lack the choanocytes and the complex system of pores and canals of their filter-feeding relatives. Some of these bear long threads studded with hook-like spicules. A hapless crustacean blundering into these threads gets inexorably tangled and is slowly drawn on to the body of the sessile carnivore as the threads shorten. The sponge slowly overgrows its prey and the ameboid cells gather to digest it.

To supplement the nutrients they obtain from filtering water, many species of sponges in shallow marine environments have struck up symbiotic relationships with photosynthetic organisms, including algae, protozoa and cyanobacteria. These live as partners or paying guests in the outer layers of the sponges' bodies. The sponges provide a fixed shelter, and in exchange they absorb the excess nutrients their symbionts produce, mainly carbohydrates, although some cyanobacteria produce nitrogen compounds as well.

The simplicity of sponges can be seen as a major disadvantage, but it allows them to do things that are impossible for more complex creatures. Their powers of regeneration, for example, are second to none. Even being chopped into tiny pieces is not much of a problem for them. The individual cells simply make a bee-line for one another to form clumps of cells and eventually tiny sponges. The ability of sponge cells to rearrange their relationships with their neighbours also allows a typical sponge to remodel its body extensively, fine-tuning its anatomy to filter edible matter from the water or to make the most of the light available for its photosythesizing symbionts.

The simplicity of the sponges is also an advantage when it comes to reproduction; they can clone themselves in a number of ways. Bodily fragments dislodged by wave action, strong water currents or a predator simply drift off and may settle elsewhere to give rise to a fully grown specimen. Budding also occurs in some species. These 'buds' break off, attach themselves to the seabed and grow into a fully formed individual. Intriguingly, some animals have exploited the regenerative abilities of sponges in a remarkable way [BOX 1].

Sponges also can produce exact copies of themselves by producing thousands of spore-like capsules called gemmules. The freshwater sponges are the ones that most commonly produce gemmules, since they present an effective means of dealing with the risk of freezing or desiccation. A gemmule consists of a tough shell of spongin fibres (sometimes studded with spicules) enclosing a cluster of amoeboid cells stuffed with nutrients. When the external conditions are right, a new sponge grows from this capsule; its cells may even mingle with those from other gemmules of the same species.

Sponges can also share and recombine their DNA through the means of sexual reproduction. With few exceptions, sponges are hermaphrodites – a single individual produces both sperm, derived from collar cells, and eggs, derived from amoeboid cells. At the right time of year sponges of the same species spawn by releasing sperm and sometimes eggs into the water [6, 7]. Fertilization can happen in the open water or within the channels and cavities of the sponge. Some species release the fertilized eggs, but in most they grow into flagellated larvae that leave the parent to drift among the plankton. After a few hours or days of drifting the larvae settle on the seabed and metamorphose into tiny, juvenile sponges.

After successfully finding a suitable spot in which to grow, some sponges can attain a great age. There are species that reach a size of around 1 m (3 ft), but they do so only at a rate of around 0.2 mm (0.008 in.) per year. If this growth rate is constant, such animals could be and incredible 5000 years old.

Being so numerous, marine sponges fulfil many important ecological roles in various environments. Their pores and canals provide smaller marine animals with abundant hiding places [15, 16]. Snapping shrimps (*Zuzalpheus* spp. and *Synalpheus* spp.) actually form social colonies in certain large species of sponges – the first known example of eusociality in a marine environment [14].

BOX 1 ***Sponge crabs***

The so-called sponge crabs carve out a piece of living sponge and hold it above their own back, using specially adapted hind legs. The sponge simply goes on growing and feeding, adapting itself to the crab's shape and size, meanwhile protecting its carrier and benefiting from the crab's messy eating habits and its movement through the water.

3 A sponge Crab (*Dromia dormia*) carrying a yellow sponge. Sponges have formed close relationships with a variety of other animals, particularly crustaceans. These seek refuge in sponges or afford protection in other ways. In exchange, the sponges get access to their partner's food and waste.

Some species of sponges are very good at eroding shell and coralline rock, and their activities shape seascapes over the ages. The huge quantity of oxygen produced by their photosynthesizing symbionts is also ecologically important. And in filtering suspended particles they help to keep the water clear, important for any organisms that depend on sunlight.

One reason why sponges are so abundant is that hardly any other creatures eat them. Though there are some predators that specialize in consuming them, they are few and far between (examples include the hawksbill turtle and a selection of fish, sea slugs and insects). Most other aquatic animals give them a wide berth. Those sponges with silica spicules are hardly more than glassy splinters embedded in gristle. Just handling such a species without gloves is a bad idea. Not only that, but many sponges produce noxious compounds that deter predators as well as preventing other organisms from settling on them or crowding them out [1]. Some even use these toxic compounds to kill corals to create new space to grow into. The huge variety of compounds produced by the sponges has attracted considerable interest from the pharmaceutical industry.

ORIGINS AND AFFINITIES

These bizarre beasts are a very ancient branch of the animal tree of life. Small fossils from Australia, 640–650 million years old, are believed to be sponges, and by around 580 million years ago they appear to have been flourishing. But neither the fossil evidence, nor comparative anatomy and molecular biology have yet settled the question of whether or not the sponges gave rise to any other lineages of modern animals. In fact some zoologists argue that the lineage we know as the Porifera actually represents four distinct lineages (demosponges, glass sponges, calcareous sponges and the homoscleromorph sponges), but there is growing evidence that this is not the case.

These sessile, filter-feeding animals have been a fixture of the marine environment for well over half a billion years, and it seems they evolved from colonial, single-celled organisms (*Choanozoa*) along a distinct path (see Introduction, p. 14). Indeed, sponge collar cells and their similarity to free-living, single-celled organisms shed some light on the murky, very early stages of animal evolution.

4

5

6

8

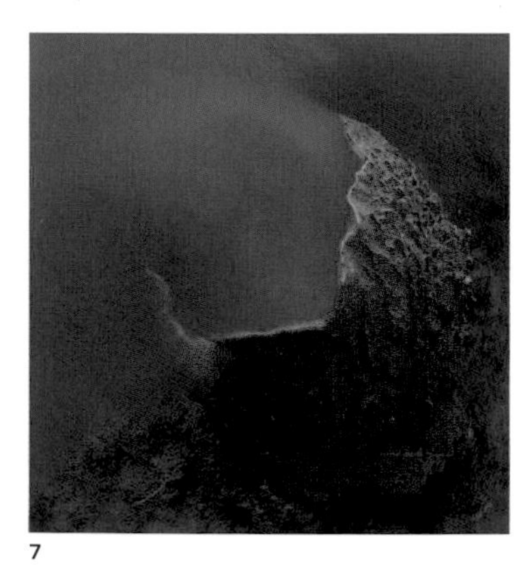

7

4 The sponge body plan, although very simple, allows them to form all sorts of shapes (*Agelas* sp.).

5 Many species of sponge have a low growing, encrusting form (several species, including *Haliclona* sp.).

6 Sponges are typically hermaphrodites, but a few species have separate sexes. Here, a female giant barrel sponge (*Xestospongia muta*) releases eggs into the water.

7 A male giant barrel sponge (*Xestospongia muta*) releases sperm.

8 Venus's flower basket (*Euplectella* sp.) is one of the so-called glass sponges, whose skeletons are reinforced with siliceous spicules.

9 One of the calcareous sponges (*Guancha arnesenae*), so-called because their skeletons are reinforced with calcium carbonate spicules.

9

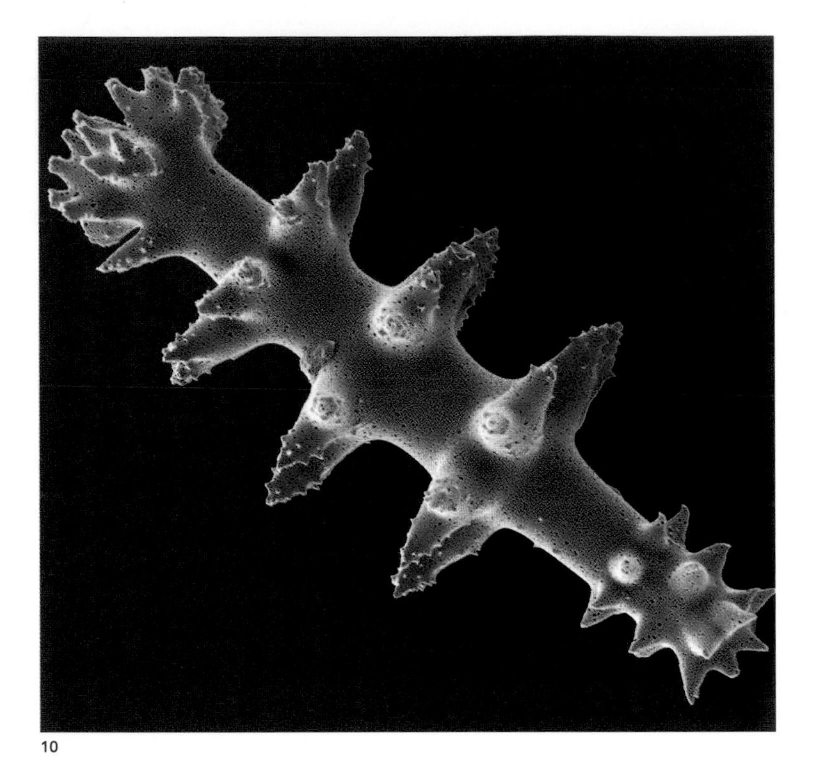

10

12

11

13

10 The skeleton of a sponge is reinforced with millions of ornate crystals known as spicules. Depending on the sponge in questions these spicules can be calcium carbonate or silica.

11 The siliceous spicules of *Tethya minuta*. Secreted by specialized cells, these structures confer rigidity and a degree of protection.

12 Deep-water glass sponge (*Caulophacus sp.*).

13 Some sponges have turned to a carnivorous way of life. This sponge (*Asbestopluma hypogea*) has managed to snare a crustacean.

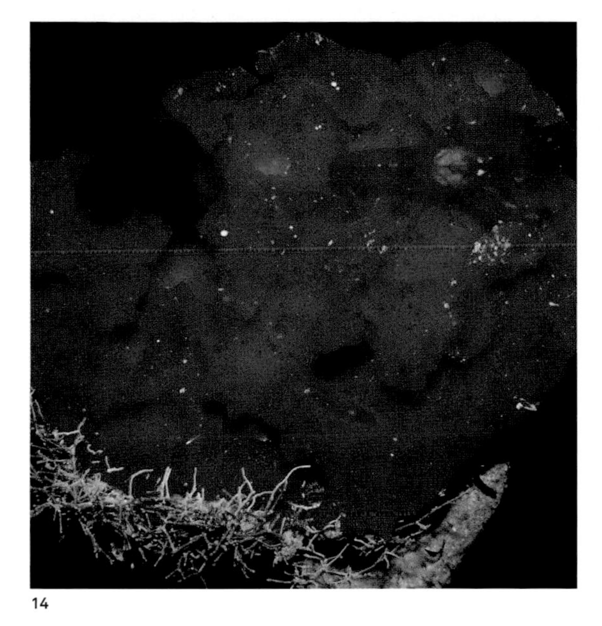

14

15

16

14 A sponge pistol shrimp (*Synalpheus dardeaui*) on its host sponge (*Lissodendoryx colombiensis*).

15 Sponges with their diverse growth forms add another layer of structural complexity to the seabed. This giant barrel sponge (*Xestospongia muta*) is a shelter, lair and focal point for all sorts of other marine creatures.

16 This outcrop of sponges is a handy refuge for a number of fish.

Placozoa

(placozoans)
(Greek *plakos* = flat;
zoion = animal)

Diversity
c. 8 species

Size range
1 to 3 mm (0.04 to 0.12 in.) across;
0.025 mm (0.001 in.) thick

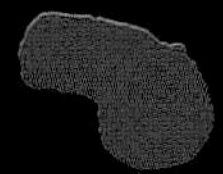

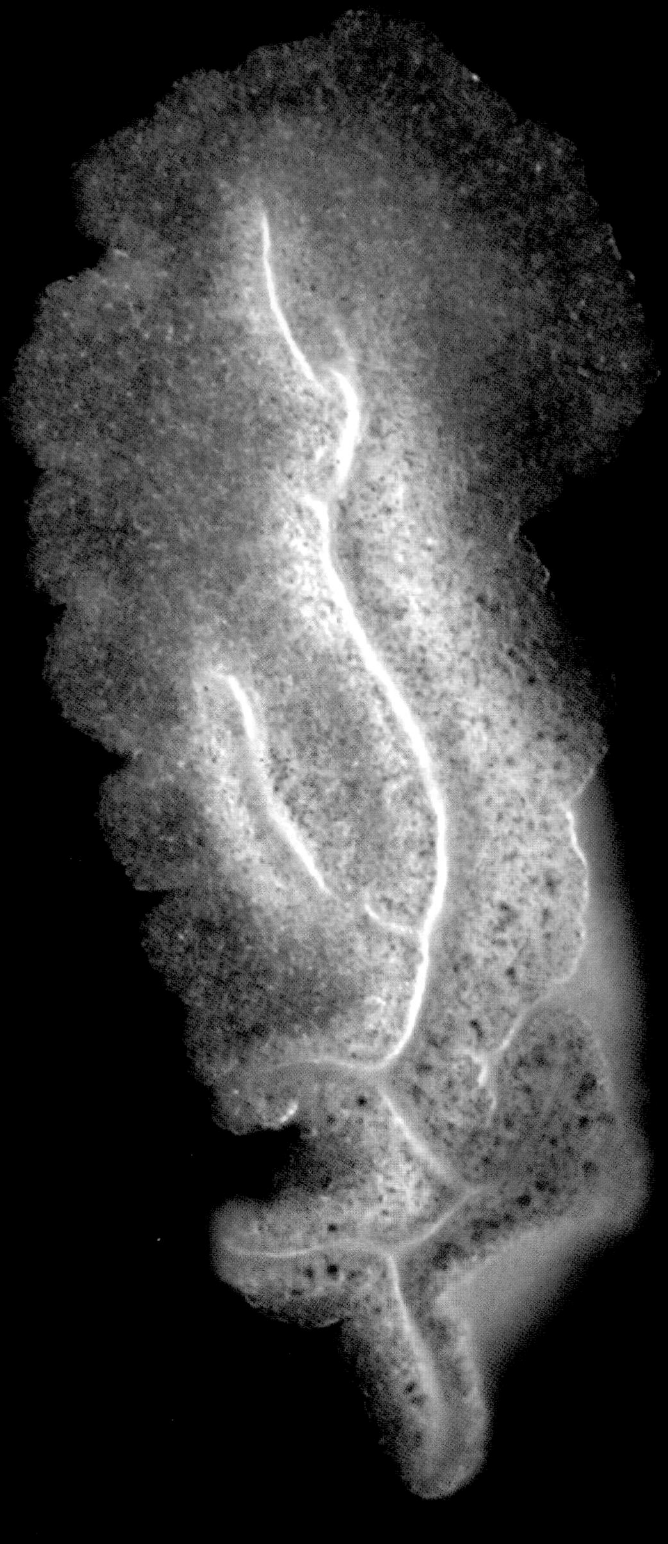

1 Placozoans might look like single-celled organisms, such as amoebae, but they are actually animals, albeit rather simple ones with an extremely thin body made up of around 1000 cells.

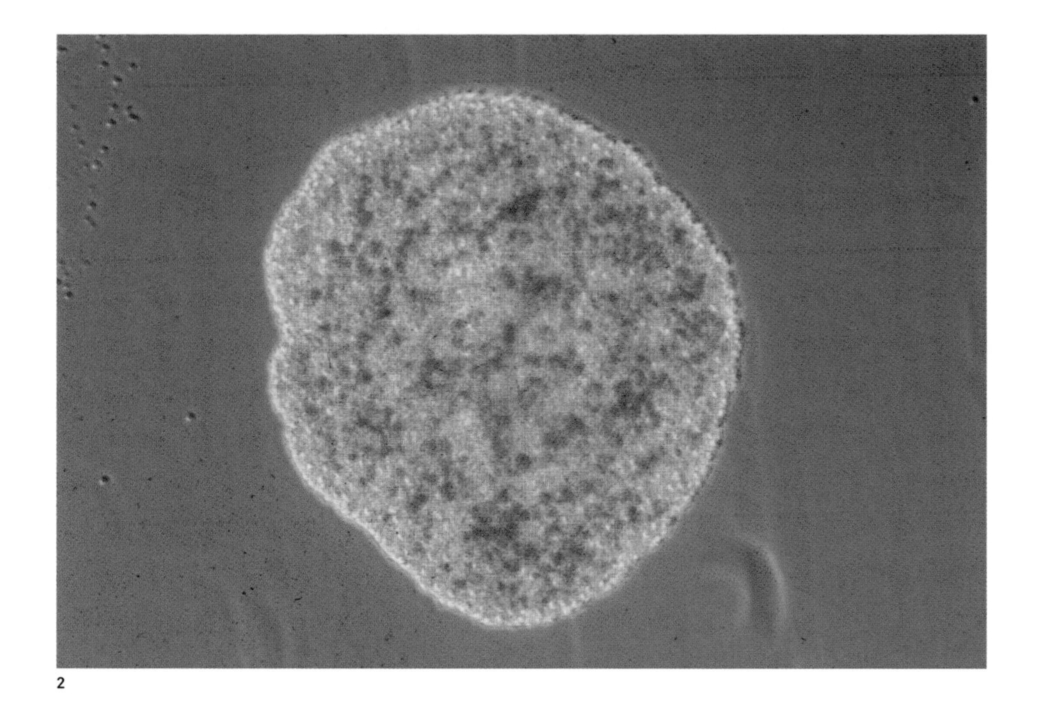

2

Placozoans are perplexing beasts to say the least. Denizens of marine environments, these tiny, extremely simple animals have no clear fossil record and nothing is known about how they interact with the other organisms of the marine realm since they have never been observed in their natural habitat. That sounds surprising, but they are so small and delicate that they are easy to miss. They have also bamboozled generations of zoologists who have tried to figure out how they are related to other animals.

For a long time it was thought that *Trichoplax adhaerens* was the sole placozoan, but several other species are now known and others undoubtedly await discovery.

FORM AND FUNCTION

Superficially, placozoans resemble the single-celled organisms known as amoeba, but they are undoubtedly animals, albeit very simple ones. Their body, if you can call if that, is little more than an asymmetrical splat [1, 2, 4] – an extremely thin gelatinous disc composed of around 1000 cells of four distinct types. This cellular differentiation is very meagre indeed when we consider that more complex animals, such as mammals, have 200 cell types. Even sponges have something like 10 or 20 types.

Placozoans have one type of epithelial cilia-bearing cell that forms the upper surface of their body and two types of epithelial cell in the lower surface – cilia-bearing cells and gland cells. Sandwiched between the upper and lower surface is the fourth kind: fibre cells that form a central, connective layer in these animals.

With so few cell types, differentiation in the placozoans is almost non-existent. They have no head, mouth or gut, and there is no sign of nervous, excretory, respiratory or reproductive systems. As if this were not enough, these flattened creatures also lack two key characteristics common to all other animals – the extracellular matrix (ECM) and basement membrane. The ECM supports living cells in a number of ways and when laid down in sheets it forms the basement membrane on which cells rest. In the absence of an ECM the cavity between the placozoan's upper and lower cell layers is filled with fluid similar in composition to seawater.

The cells on the underside of the placozoan bear cilia and it is these the animal glides around with. They can also move around in much the same way as an amoeba by contracting the network of connective tissue sandwiched between their upper and lower cell layers. There is no front or back end, so they can easily travel in any direction without turning; whichever way they go, it is forward. Occasionally they move in two directions at the same time, wrenching themselves apart in the process.

Being so thin, placozoans are very fragile and are easily torn apart even if handled gently, but this is compensated for by their remarkable powers of regeneration. Quite small fragments of one of these animals can each develop into a fully formed placozoan.

2 Nothing is known about how placozoans live in the wild as they are so small and difficult to find (*Trichoplax adhaerens*).

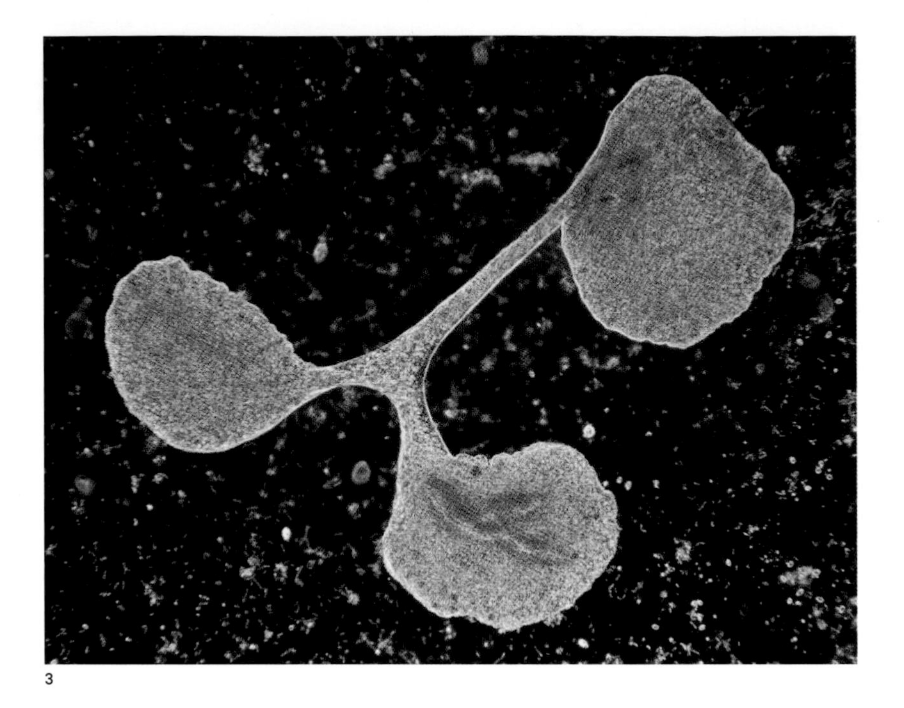

3

3 One way placozoans reproduce is to simply split into two or three daughter individuals.

LIFESTYLE

Being so small and translucent it would be extremely difficult to spot a placozoan in the wild, let alone follow it about to see what it does. It is no wonder that people who wish to study them simply collect sediment from the seabed and inspect it under the microscope, or just wait to see whether any of them begin to breed in an aquarium. That is more or less how they first were discovered in 1883: some were seen clinging to the glass of a marine aquarium in Austria, and subsequently more were discovered around the world.

Their structural simplicity places severe restrictions on how these animals can live. Placozoans are limited to creeping around on marine substrates in search of edible matter, namely algae, organic waste, and so on. Since they have no mouth or even a gut for that matter, they do not really eat – they just lie on top of their food, exude digestive enzymes from their gland cells and absorb the nutrients. Sometimes they may arch parts of their body to form a little pocket, an impromptu 'pseudo-stomach', for the more efficient digestion of their food, but that is about the extent of their feeding abilities.

Very little is known about how the placozoans reproduce. They seem to be capable of sexual reproduction (i.e., the production of sperm and egg cells), but there still is a great deal to find out about this aspect of their biology. Typically they reproduce asexually, either by splitting into two or three daughter individuals [3] or budding off small, spherical clones, each of which has a small complement of the four cell types found in the adult.

ORIGINS AND AFFINITIES

So where do the placozoans fit on the animal 'tree of life'? Are they close to the root of this 'tree' or did they evolve more recently via simplification of a more complex lineage of animals? In short, we do not know for sure yet, although scientists are edging closer to the answer by comparing the DNA of placozoans with that of other living creatures. It seems that they represent a very old and distinct branch of the animal family tree.

The placozoans may have evolved from free-floating colonies of cells that took to a benthic existence and in the process became increasingly flattened (see Introduction, p. 14). These seemingly insignificant creatures may even be a window through time to some of the very first animals that lived perhaps more than one billion years ago.

You might well ask why scientists are bothering to spend their time scrutinizing an animal that is so insubstantial. Apart from the obvious answer of pure scientific curiosity, one of the few points that practically all zoologists agree on is that some of our key understandings, both of the history of animal evolution in general, and the principles underlying the way in which cells combine and cooperate to form the bodies of multicellular organisms, are likely to emerge from the study of just such creatures.

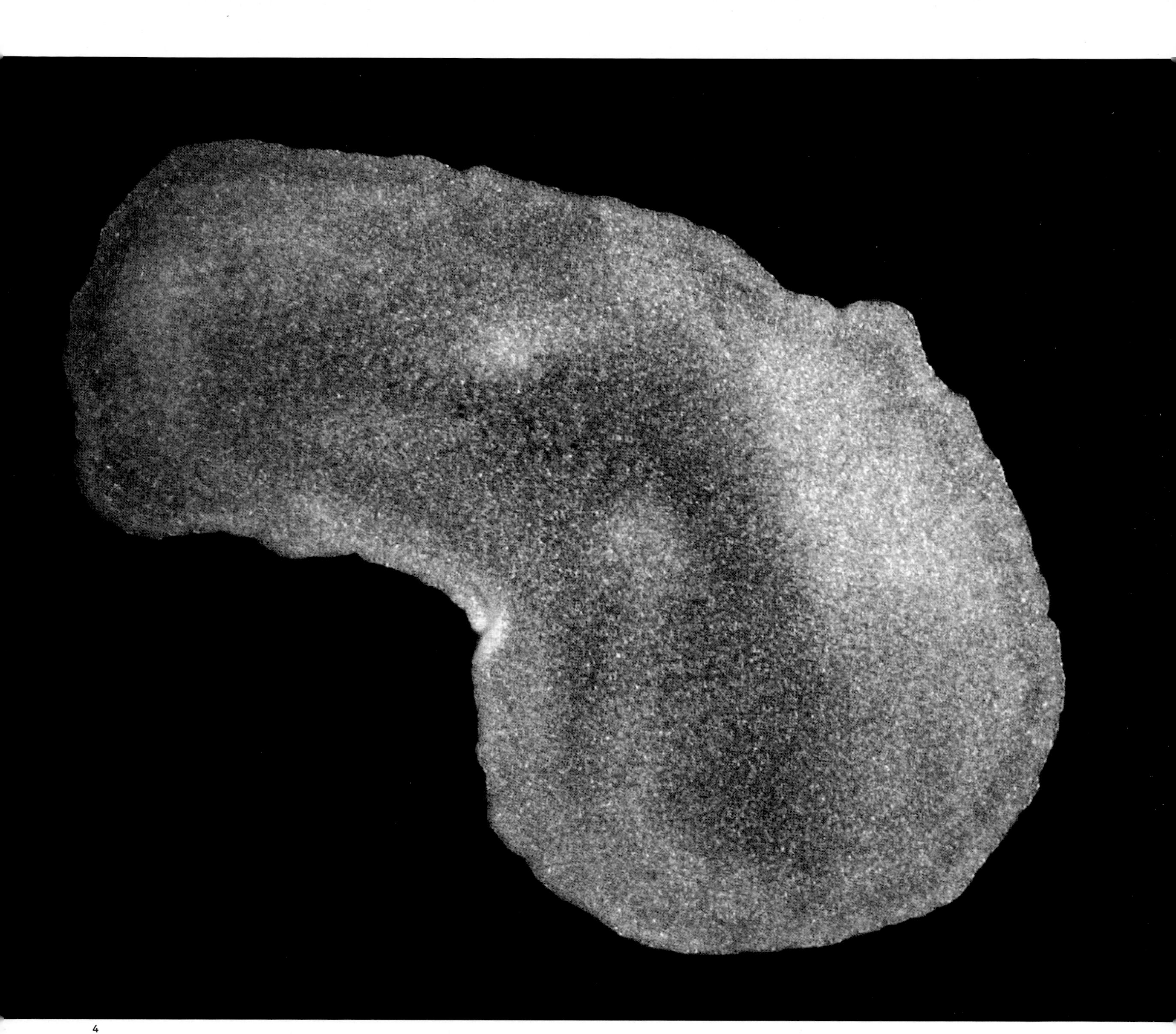

4

4 The general appearance and extremely simple form of placozoans has led to the assumption that they represent a link between single-celled organisms and animals, but this theory has now been largely discredited. Study of the placozoans may help to illuminate some of the very early events in the evolution of the animals, but we are still some way from understanding exactly where these bizarre creatures fit on the animal family tree.

Cnidaria

(jellyfish, anemones, stony corals, soft corals, sea pens, sea fans, sea daisies, hydra, myxozoa)
(Greek *knide* = nettle;
Latin *aria* = like or connected with)

Diversity
c. 12,500 species

Size range
0.01 mm (Myxozoa spores)
to ~40 m (colonial forms)
(0.0004 in. to ~130 ft)

1 A large sea anemone (*Cribrinopsis* sp.) displays the polyp growth form, which is typical of cnidarians. This polyp has a large, multifunctional, gut-like cavity enclosed by a solid body wall. The single opening to the cavity, surrounded by tentacles, functions as a mouth/anus.

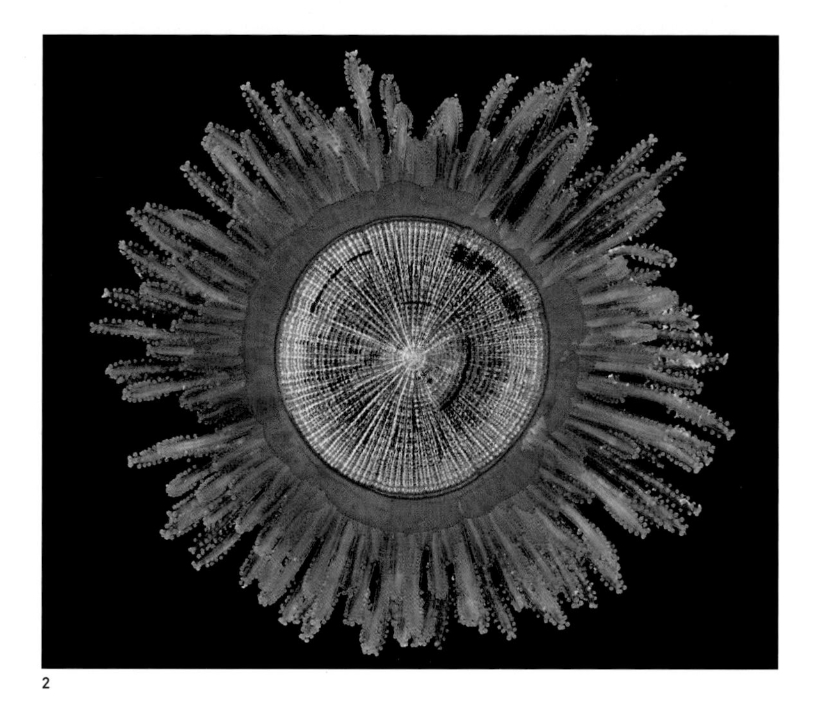

2

2 In the cnidarians, what looks like a single individual is often a colony of polyps with specialized functions. In this floating colony (*Porpita* sp.) there are polyps for providing buoyancy, feeding (tentacles), digestion and reproduction.

The cnidarians come in a beautiful and bewildering variety of forms with familiar (and some less familiar) names such as jellyfish, sea anemones, corals, sea pens, sea fans, sea daises, hydroids and hydras [1, 2, 19–22, 27, 34, 39]. These are exclusively aquatic animals and all but 20 or so species are sea-dwellers. They are found in all marine environments from the frigid waters of the poles to the warm ones of the tropics and from shallow, temperate seas to the deepest ocean trenches. They are at their most abundant and diverse in warm, shallow-water habitats.

Cnidarians are extremely important components of the marine ecosystem, preying on a huge variety of other animals and providing food for other creatures higher up the food chain. The activities of some species, notably the corals, have far-reaching impacts on the biosphere as a whole.

FORM AND FUNCTION

Cnidarians have three cell layers. The endoderm (gastrodermis) lines their inner cavity, while the ectoderm (epidermis) is their outer surface. Sandwiched between these two layers is a gelatinous, often translucent layer known as the mesoglea. This layer is the 'jelly' that gives some of these animals their common name. It contains collagen, providing a degree of rigidity, as well as wandering cells that fulfil various functions within the body. Both the endoderm and the ectoderm contain muscle cells, nerve cells and secretory cells. The body encloses a single, multipurpose space where life's functions can take place, such as gas exchange (in the manner of a lung), reproduction (like a gonad and brood pouch) and the processing of waste (like a kidney).

None of the cnidarian bodily structures have reached a degree of organization where they could be considered to be organs. Some of the cnidarian sensory structures are complex but they are composed of only sensory tissue, so, by definition, they are not organs. An organ, such as the eyes you are reading this with, is a complex arrangement of different tissues (e.g. sensory, blood, muscle).

With the exception of some extremely odd parasitic species [BOX 1], an individual cnidarian, which can take a polyp or a medusa form, is radially symmetrical with a number of tentacles arranged around a central mouth [23–26]. Polyps are typically sessile animals (such as corals), although some species, including the anemones, are able to move slowly around on the seabed. Others can perform acrobatic somersaults to get around; hydras are a well-known freshwater example [19]. Yet other kinds of polyps can use their tentacles as legs. In contrast, medusae, such as jellyfish, are typically pelagic animals that use muscular contractions of their dome-shaped body to swim in a mesmerizing and surprisingly powerful fashion [28–33, 38]. Many of the hydrozoans develop as polyps for part of their lives before budding off medusae that live a pelagic existence [21].

Not only are the cnidarians represented by these two different forms, but there are also

3

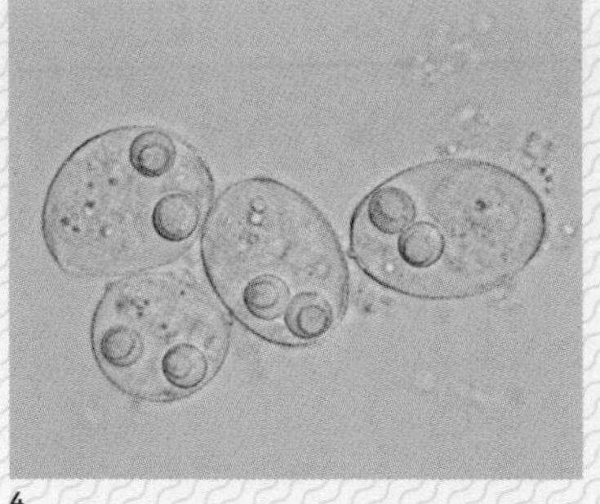

4

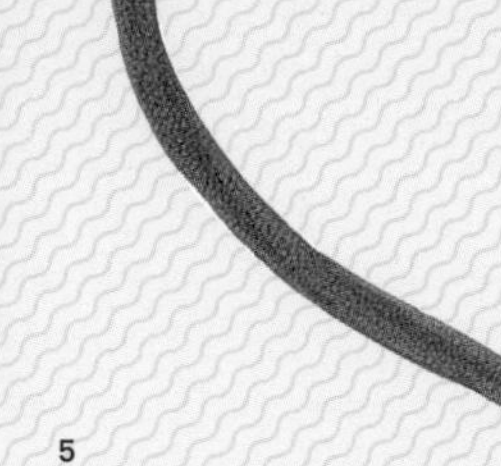

5

6

BOX 1 ***Parasitic cnidarians***

Organisms with the unattractive name 'myxozoans', meaning 'mucus animals', have confused scientists for well over a century. Around 2400 species of these tiny aquatic parasites have so far been identified, and for a long time the consensus was that they were not animals, but simply single-celled organisms.

It is now known these peculiar beings are cnidarians that have adapted to a parasitic existence, losing much of their structural complexity in the process. They form tiny multicellular spores equipped with polar filaments (modified cnidocysts – see **BOX 3**) **[3, 4]**. There are even a few species that have taken on a strange, worm-like form with no discernible top, bottom, front or back **[5]**.

Despite their extremely simple bodies, myxozoan life cycles are varied (and poorly known). The general theme is that spores swallowed by a suitable host discharge their polar filaments to attach to a host cell. Mobile cells creep along the filament and into the host cell where they will ultimately produce countless more spores **[6]**. Often two separate hosts (including worms, fish, bryozoans, turtles, birds and mammals) are involved, suggesting that the myxozoans took to a parasitic existence very far back in time.

Polypodium is another very odd parasitic cnidarian. It spends most of its life in the eggs of sturgeon and paddlefish. During its time in the host egg it develops from a single cell with a pair of nuclei into an inside-out larva and then into an elongated, inside-out stolon. All of these parasitic stages are surrounded by a single, protective cell that also functions in digestion. Just before the host releases its eggs, the *Polypodium* stolon everts its tissues so that it is no longer inside-out, revealing a number of tentacles. The central cavities of the parasite fill with the host's yolk and it eventually emerges as a free-living stolon **[7]** that fragments into smaller individuals **[8]** that multiply by splitting in two. Eventually, these free-living stages reproduce to continue this bizarre life cycle.

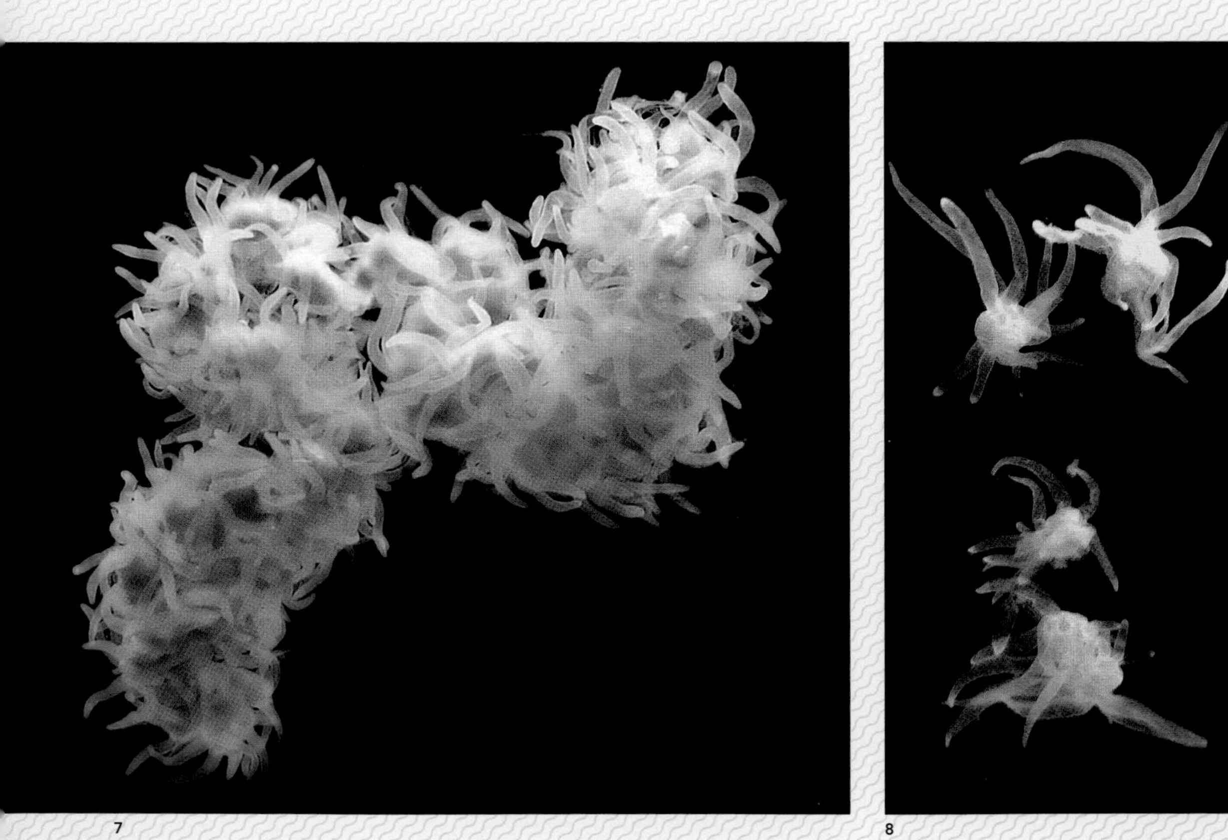

7

8

3 Myxozoan spores. The cnidocytes containing the coiled polar filaments are clearly visible (*Myxidium coryphaenoideum*).

4 Spores of another myxozoan (*Sinuolinea* sp.).

5 A worm-like myxozoan, (*Buddenbrockia* sp.). The muscles of this individual have been highlighted with a green stain.

6 Cysts in the muscle of the common carp (*Cyprinus carpio*) caused by the myxozoan *Myxobolus artus*. The exact life cycle of this parasite is unknown.

7 *Polypodium hydriforme* is a parasitic cnidarian that spends much of its life in the eggs of sturgeon and paddlefish. This is the stolon stage that has just emerged from a host egg.

8 *Polypodium hydriforme* stolon fragments give rise to these medusa-like, free living forms with 12 tentacles.

Overleaf

10 A siphonophore, the Indo-Pacific Portuguese man-of-war (*Physalia utriculus*).

11 The siphonophores are considered to be among the most abundant carnivores of the open ocean (*Rosacea* sp.).

solitary and colonial species. In the former, the juvenile polyp steadily grows, eventually becoming a solitary adult, such as an anemone or a jellyfish. In colonial forms, such as stony corals, the juvenile polyp grows only a little before replicating asexually to form another polyp (a zooid) that remains attached to the founding polyp [BOX 2]. Zooids give rise to yet more zooids and, depending on the species, a colony, perhaps hundreds of thousands strong, may be formed.

A pair of two-dimensional nerve nets, evolved from specialized epithelial cells, controls the muscular contractions that enable medusae to swim and polyps to constrict and bend their body. As well as controlling the muscles, the nerve nets process information from various sensory structures that detect stimuli, such as light, temperature, chemicals, vibration and gravity. Cnidarian sensory structures are generally individual cells or aggregations of cells; however, the deadly box jellyfish have sophisticated eyes, each of which is equipped with a lens and light sensitive pigments in a retina [31].

The surface of a cnidarian bristles with unique cells known as cnidocytes, the contents of which are used to great effect in capturing prey and defence [BOX 3].

LIFESTYLE

All cnidarians are carnivores. They use their cnidocysts to catch and subdue a range of prey, generally the small creatures that make up zooplankton. The discharged cnidocysts themselves are effective snagging weapons, but their ability to subdue prey is enhanced by toxins. Cnidocysts contain various toxins, typically neurotoxic in nature, that interfere with the nervous system of the victim, quickly paralyzing it. It is these toxins, combined with the physical penetration of the everted cnidocyst, that give these animals their sting. Those produced by some cnidarians are sufficiently potent to cause severe injury and death in much larger animals. Indeed, few animal venoms can match that produced by some of the box jellyfishes (such as *Chironex fleckeri*), which have killed many humans, especially in the Malay Archipelago [31].

Once the prey has been subdued, the tentacles bring it within range of the mouth and it is eaten. In addition to the toxins, there may also be enzymes that initiate the process of digestion before food is swallowed.

Some kinds of corals complicate the cnidarian diet picture, since an important part of their energy budget comes from photosynthetic algae that they maintain in their tissues;

BOX 2 *Colonial cnidarians*

Colonial growth is at its most elaborate in the cnidarians. The zooids in many cnidarian colonies are all morphologically identical, sessile individuals, but there are some species where colonial growth gives rise to free-living colonies made up of zooids with specialized functions. This way of life is best illustrated by the siphonophores [9–11], of which the Portuguese man-of-war is a well-known example. A man-of-war looks like a malformed jellyfish [10], but it is actually a single elongated polyp from which lateral zooids asexually bud to take on specific functions. The pedal end of the originator polyp (the part that anchors a normal polyp to the seabed) grows and folds in on itself to form a gas-filled float. At the opposite end there is a mouth and a single tentacle. The lateral sets of buds develop into zooids specialized for the collection of prey using cnidocyte-loaded tentacles, digestion and reproduction.

Some siphonophores also have specialized medusae zooids that propel the colony, or zooids that are little more than gelatinous flaps to streamline and protect the colony. Of all the colonial animals the siphonophores are unsurpassed both in terms of the degree of zooid specialization and the precision of organization.

9 What look like tentacles on this siphonophore are the feeding and reproductive zooids (*Physophora hydrostatica*). The bell-like medusae zooids at the top propel the colony.

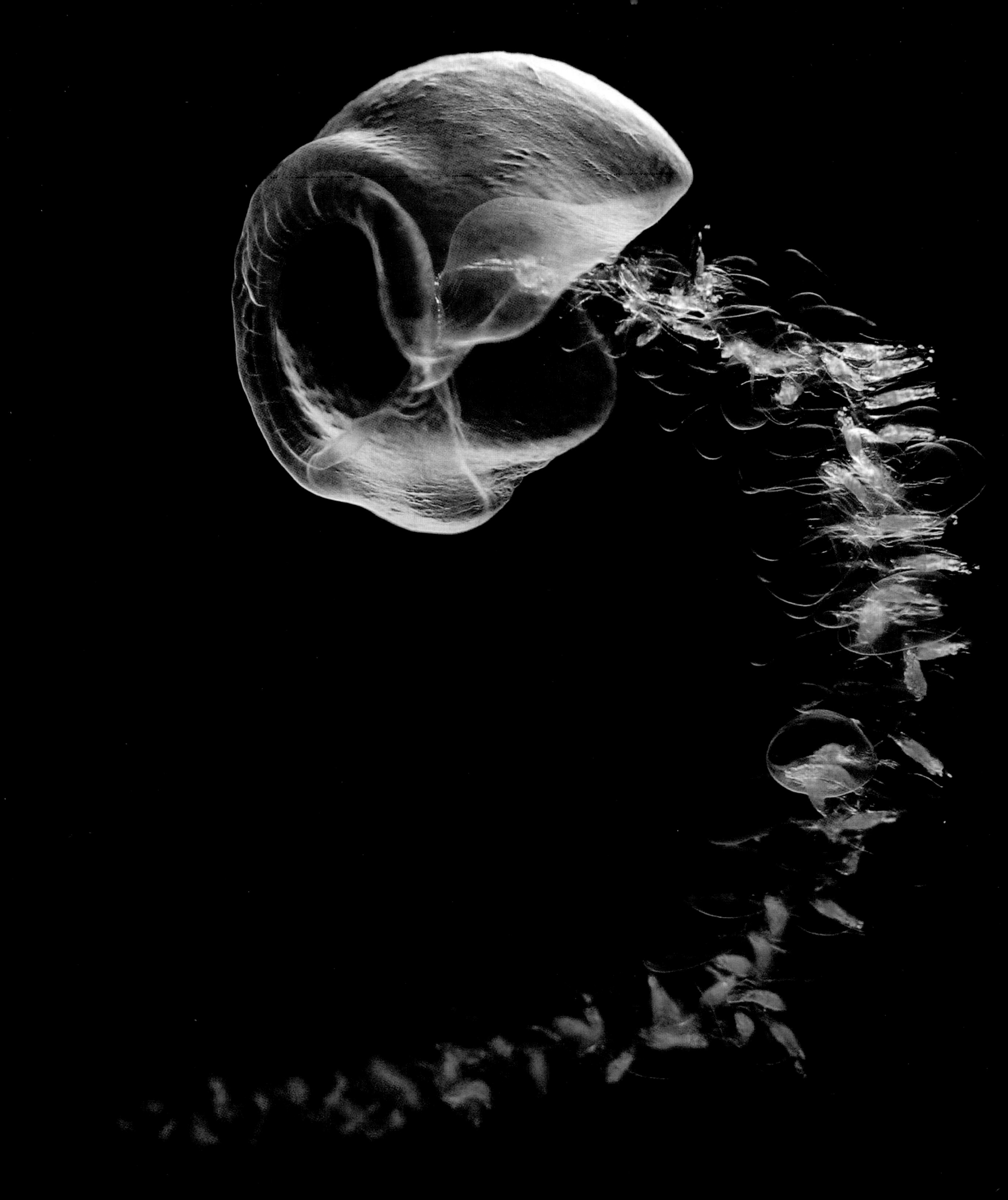

however, they generally do also feed on passing microscopic animals.

Since cnidarians are diverse and often very abundant, they have a considerable impact on aquatic ecosystems [35–37]. With their efficient means of catching prey they undoubtedly make a considerable dent on the populations of the animals that spend all or part of their lives among the plankton. On the other hand, cnidarians in turn are consumed by lots of other marine animals (regardless of their cnidocysts and toxins). Some species, such as the leatherback turtle, are very fond of pelagic cnidarians and other, similarly soft-bodied animals. Being fixed to the spot, some of the sessile species, such as coral polyps, fall prey to a wide range of predators, including fish, molluscs and echinoderms. In recent decades climate change and our continued exploitation of the oceans have had a huge effect on the populations of cnidarians. Numbers of some species have exploded alarmingly, while others have declined equally alarmingly.

One thing that makes the cnidarians so important in marine ecosystems is their remarkable ability to form elaborate structures, which is especially apparent in the corals [BOX 4].

The driving force behind the growth and productivity of tropical reefs is the symbiotic relationship between coral polyps and algae, hundreds of thousands of which are embedded in the polyp's tissues. As well as giving them their distinctive colours, these symbionts provide the polyps with simple carbohydrates, via photosynthesis, in exchange for certain nutrients they cannot synthesize, carbon dioxide and a good position in which to catch the sun's rays. As much as 90 per cent of everything the algae produce is taken up by the polyps, enabling them to thrive in waters where animal prey is lacking. The relationship between these two organisms is complex and lots of things can disturb it. For example, if the polyps become physiologically stressed by increasing sea temperatures, they eject their algal symbionts (a phenomenon known as bleaching). This is not a problem for all coral

BOX 3 ***Cnidocytes***

The continued success of the cnidarians long after their first appearance hundreds of millions of years ago probably owes a great deal to the tiny cnidocyte, or nettle-cell (from the original Greek for a nettle). It is a hugely modified epithelial cell that contains a tiny membranous capsule called the cnidocyst (or nematocyst) – a miniature harpoon and probably the most complex structure found within any single cell on earth (many cellular structures are in fact far more complex, but those comprise multiple cells).

Inside the cnidocyst is a tightly coiled tubule that is actually a hugely elongated extension of the capsule wall itself. For the cnidocysts to be discharged the cnidocytes of many cnidarians have a trigger – a fixed cilium on their outer surface. Typically, to prevent accidental discharge, both mechanical and chemical stimuli are required. For example, as the body of an animal brushes against a cnidarian, distinctive substances on and around it may increase the sensitivity of the trigger.

Once triggered, the cnidocyst erupts with explosive force, everting the tubule. The uncoiling structure achieves a velocity of around 2 m (6 ft) a second, which at such a small scale means its tip is accelerating at 40,000 Gs (by comparison you might experience around 4–5 G on the most vicious of rollercoasters). In many cnidarians, the tubule is intended to pierce the victim, but in others it is simply designed to snag the limbs of prey.

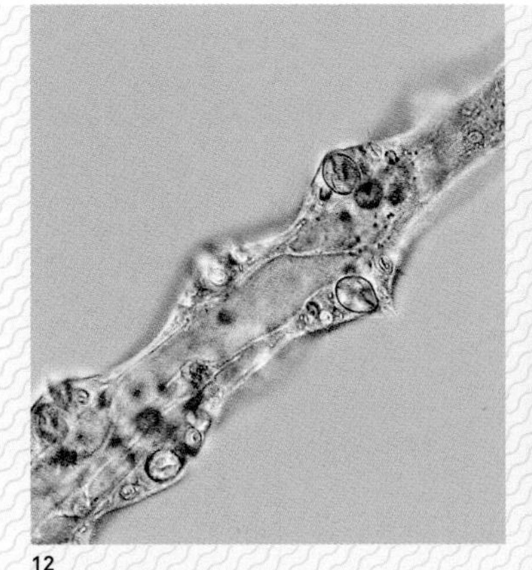

12

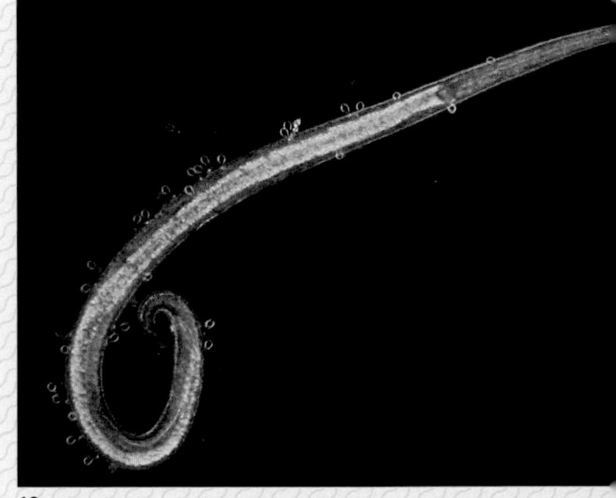

13

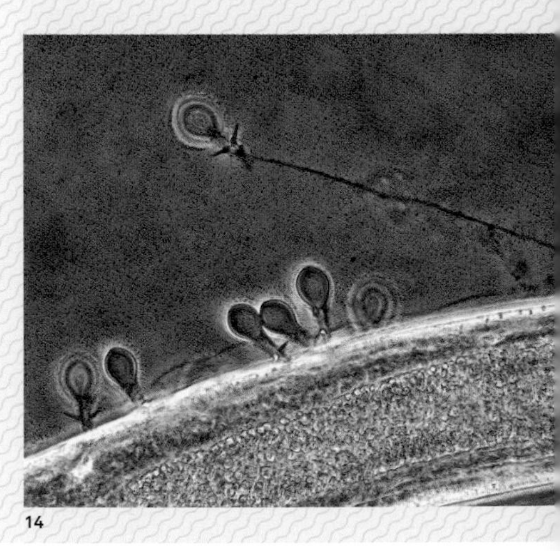

14

12 Studding the tentacle of this hydra are many cnidocysts, visible as ovoid capsules.

13 Numerous discharged cnidocysts piercing the body an unfortunate nematode.

14 Discharged cnidocysts embedded in the cuticle of a nematode. The cnidocysts are equipped with spines so they stick fast in the prey.

BOX 4 ***Reef-forming corals***

Many types of cnidarian, including some stony corals, octocorals and fire corals, produce massive stony skeletons that, en masse, form the structures we know as reefs. However, it is the stony corals that are the predominant reef-building animals alive today.

If you examine a small piece of reef produced by one species of stony coral, each of the tiny polyps you will see are morphologically identical zooids asexually derived from the original polyp that founded the colony. These zooids are all intimately linked to one another via extensions of their bodies, forming a veneer of living tissue around a secreted skeleton of calcium carbonate, essentially rock. Over decades, centuries and even millennia, successive generations of polyps lay down layer after layer of calcium carbonate until the thin living layer of polyps surrounds a massive accumulation of rock.

Considering that they are secreting rock, the growth rate of some coral species is extraordinary. The branching colonies formed by some corals can grow in height or length by as much as 10 cm (4 in.) per year (about the same rate at which human hair grows). Dome-or plate-like colonies are more bulky and may only grow by 0.3 to 2 cm (0.1 to 0.8 in.) a year. These rates can be sustained for thousands of years, forming huge and complex reefs. The Great Barrier Reef, off the coast of Australia, is the largest in the world and is composed of a multitude of colonies, many of which are formed by at least 70 species of stony, reef-forming corals. Stretching for over 2000 km (1250 miles) and visible from space, the Great Barrier Reef is thought to be in the region of 6000–8000 years old.

Tropical, shallow-water reefs are an incredibly biodiverse habitat, possibly the most biodiverse on the planet. Not only that, but the sheer size of coral reef structures alters the direction and flow of ocean currents, directly affecting weather patterns. All of this from the tireless deposition of rock by tiny polyps.

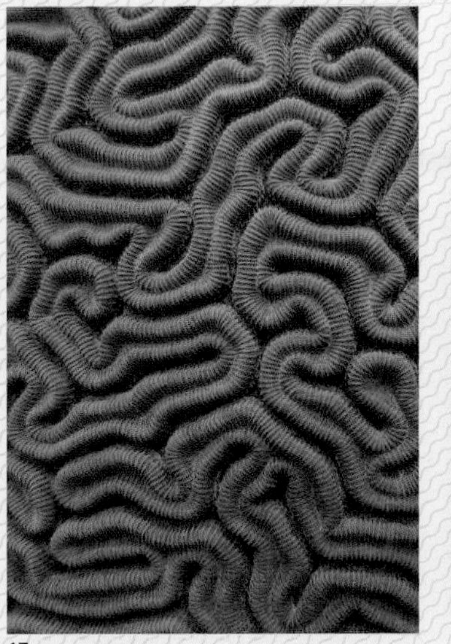

15 A stony coral colony (*Favia* sp.).

16 Soft corals, such as these *Dendronephthya* sp., are an important element of the reef community.

17 The surface of a brain coral (*Diploria labyrinthiformis*). The polyps live in the grooves between the ridges and typically extend their feeding tentacles at night. This species can form hemispherical 'heads' that can be more than 1 m (3 ft) in diameter.

18 Coral reefs are home to a bewildering variety of life forms, many of them small or hidden out of sight in the countless refuges offered by the structurally complex environment.

species, as they can live with or without their symbionts, but some do eventually die without their algal associates. Coral bleaching and death on a large scale has obvious repercussions for marine biodiversity since so many other marine creatures depend on corals for food and the structurally complex habitats created by the reef-forming activities of many species.

Reproduction in the cnidarians is a rather complex affair as many species go through both asexual and sexual stages. Asexual reproduction takes various guises, including 'budding', where the animal buds off miniature versions of itself [19], and 'fission', where splitting gives rise to two new individuals. The product of sexual reproduction is a fertilized egg that develops into a larva, the so called planula. Clothed in cilia, the planula is a planktonic creature that, after a short while drifting in the open water, settles on the bottom and metamorphoses into a juvenile polyp. Depending on the species in question, this polyp develops into a solitary form (such as an anemone or hydra), asexually buds off miniature medusae that grow into the familiar jellyfish, or asexually generates zooids to form a colony (corals, sea pens, sea fans, siphonophores). Polyps of one species, *Turritopsis nutricula*, give rise to medusae, which in turn can revert back to the polyp stage; a bizarre process that can go on indefinitely.

ORIGINS AND AFFINITIES

Some of the oldest fossils thought to be animals – the enigmatic Ediacaran organisms – have been interpreted as very early cnidarians, but we will never know exactly how they are related to the animals we know today. If we fast forward a little to around 400–500 million years ago, there are cnidarian fossils in profusion, showing that reef-building cnidarians have been a prominent part of the oceans for hundreds of millions of years, even though in some periods other reef-building organisms were dominant.

Apart from this rich fossil heritage there is little preserved in stone that gives us any idea of how cnidarians fit on the animal 'tree of life'. Once natural historians began to understand that the cnidarians were not plants, they realized that these creatures must be near the base of the tree, yet they were confused by the superficial similarity to other animals, such as comb jellies (see Ctenophora, pp. 26–35). Comparing DNA sequences has revealed that the cnidarians are indeed near the tree's base, but also that they are a distinct lineage without any close relatives.

‹ 19 A hydra (*Hydra* sp.) reproducing asexually by budding. Note the new individuals forming near the base of the adult. These cnidarians can move across the substrate inchworm fashion or by somersaulting.

20 A stalked jellyfish (*Lucernaria quadricornis*). These cnidarians can be regarded as an overgrown polyp, one end of which has partly differentiated into a medusa. The 'tufts' of the feeding tentacles are clearly visible.

Overleaf

21 The life cycle of some cnidarians consists of both polyp and medusa stages. The medusa in this image budded off from a small polyp attached to the seabed (*Gonionemus vertens*).

22 Wire coral (*Cirrhipathes* sp.). The yellow 'tufts' are the tentacles of individual polyps.

20

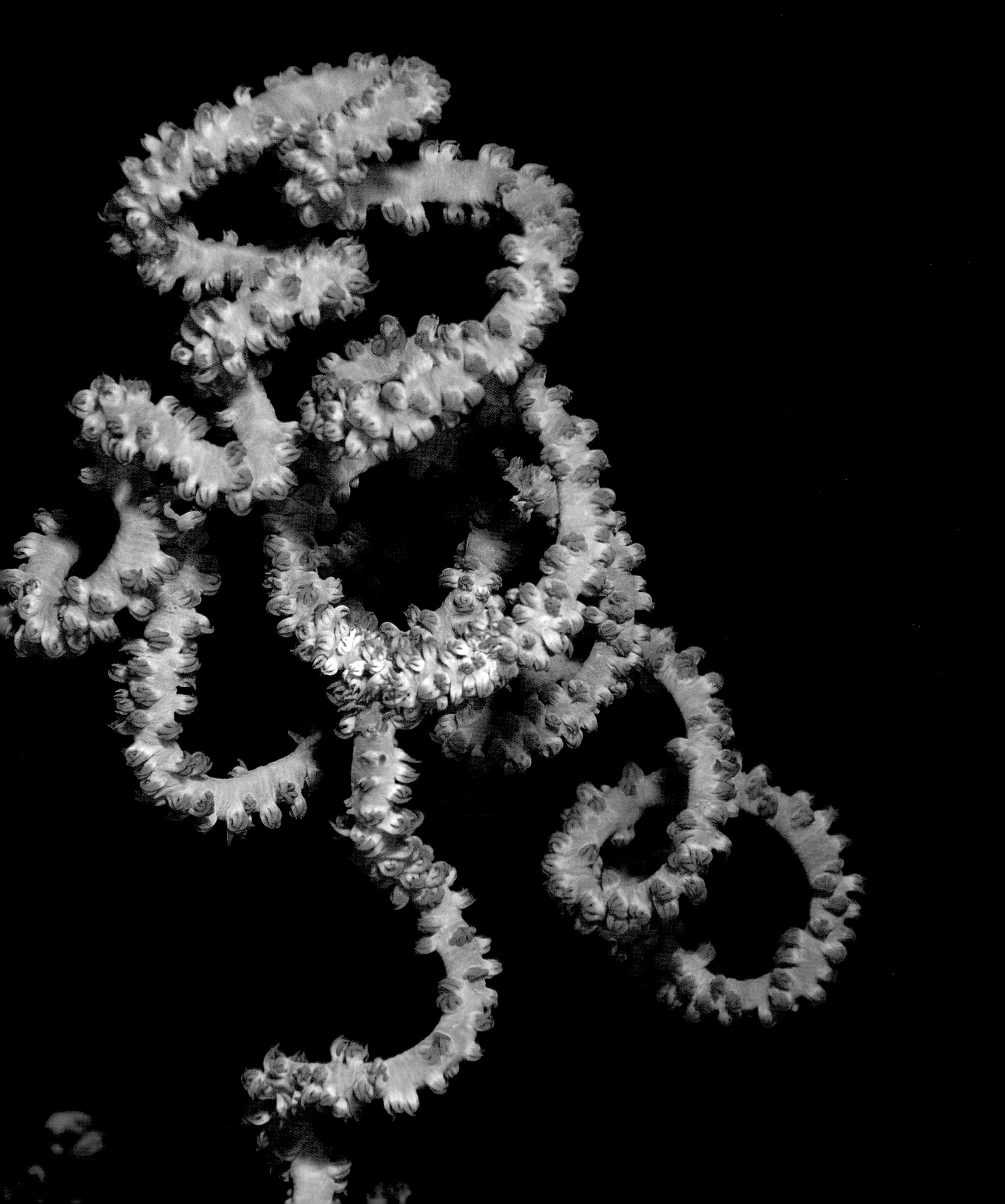

23

24

23 Individual polyp of the hydrozoan *Ectopleura larynx*.

24 Feeding tentacles arranged around the mouth of an anemone, *Metridium senile*.

25 An anemone, *Stomphia coccinea*, showing the tentacles encircling the mouth.

26 Some anemones can burrow into sediments, leaving only their feeding tentacles visible (unidentified edwardsiid).

› 27 A sea pen. For a long time, biologists did not know what to make of these colonial cnidarians. The central stalk, 'rooted' in the sediment, is the primary polyp, up to 3 m (10 ft) long in some species. The lateral 'branches' are secondary polyps, equipped with feeding tentacles. They are now known to be octocorals, close relatives of the soft corals (*Ptilosarcus gurneyi*).

25

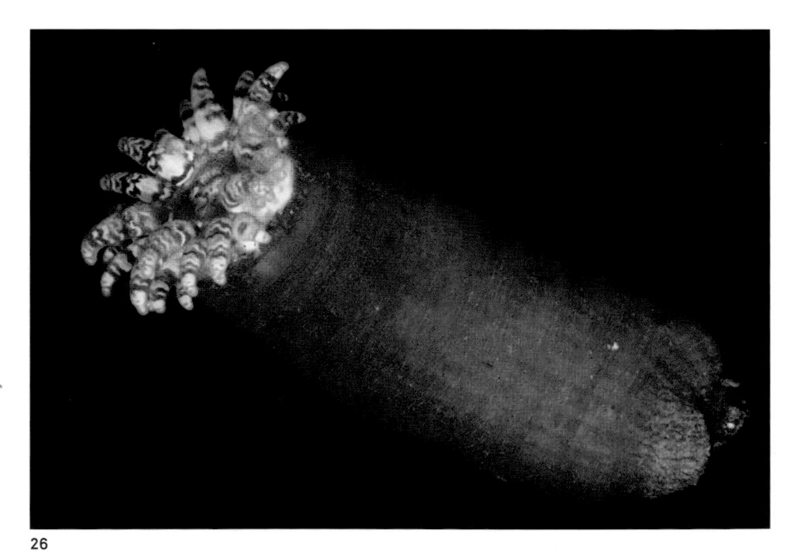

26

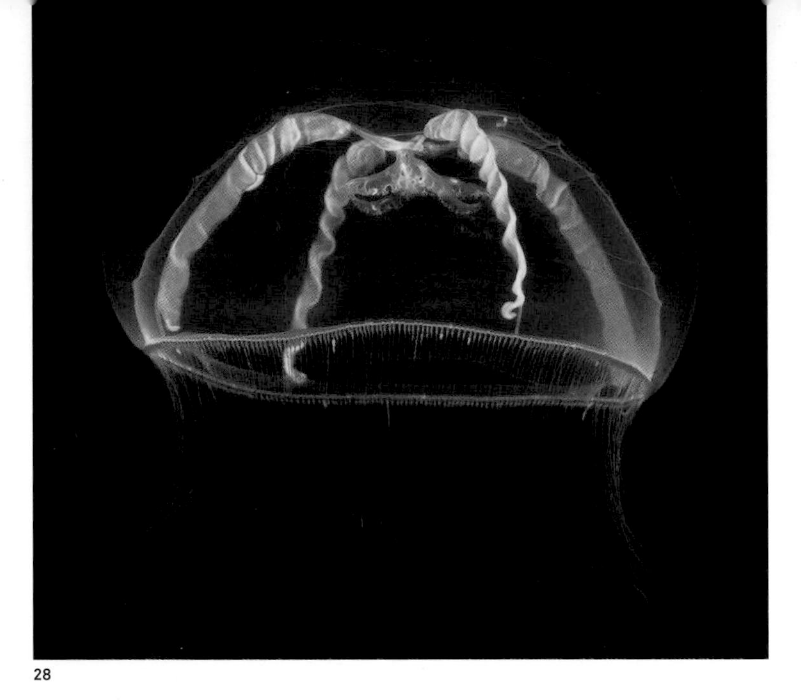

28

28 The anatomy of a jellyfish. The four, short, leaf-like structures in the bell (top centre of image) are the oral 'arms', at the centre of which is the mouth. The long, pale, arching structures are the four radiating arms of the gastric cavity (unidentified species).

29 The extremely long tentacles of *Cyanea capillata*. The bell of this jellyfish can sometimes be 2 m (6 ft) across.

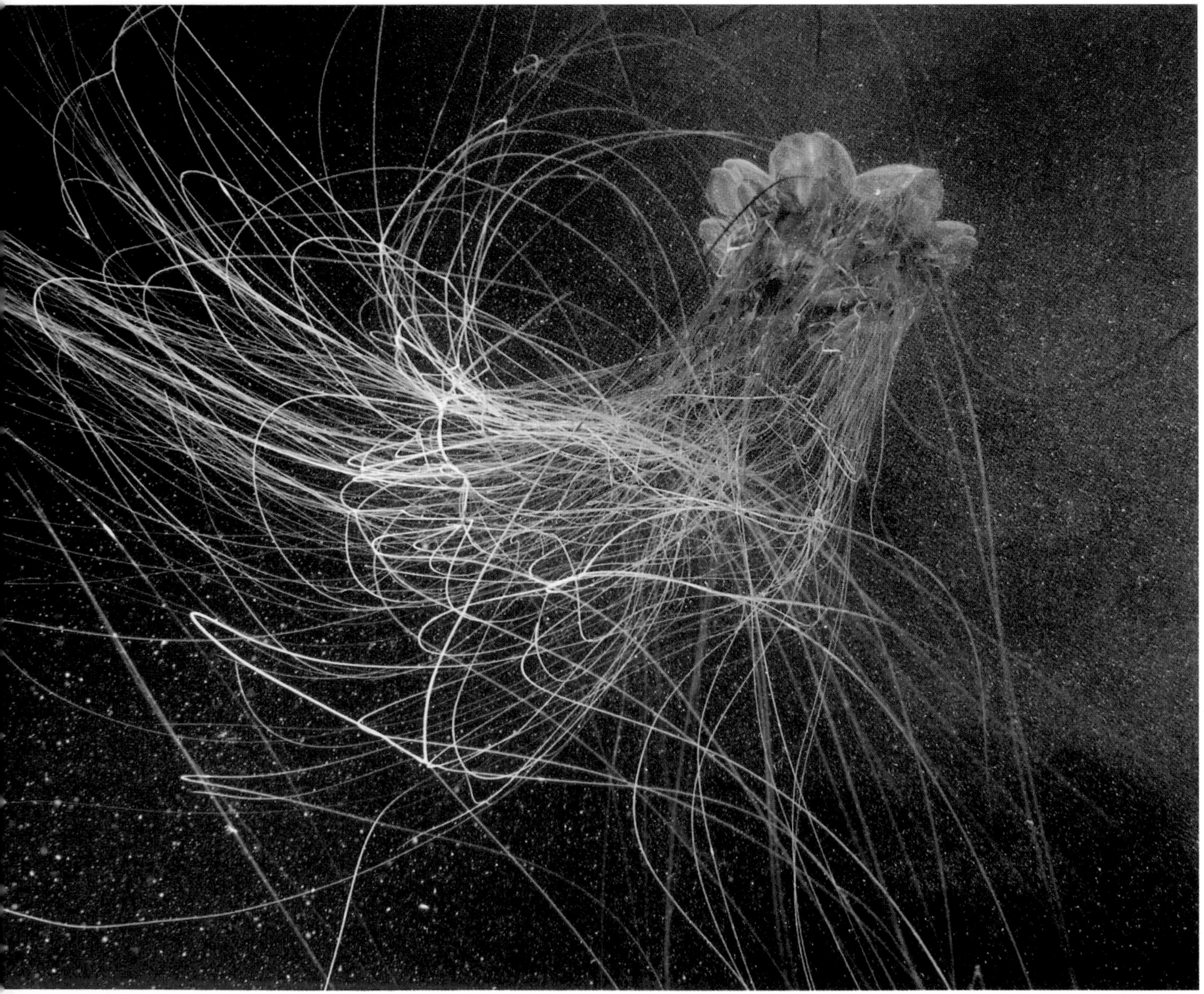

29

30

30 The sausage-shaped structures sprouting from the radiating arms of the gastric cavity of this jellyfish are its gonads (unidentified species).

31 *Carybdea marsupialis*, an extremely venomous box jellyfish. The small, white dots (centre and bottom centre of bell) are sensory structures known as rhopalia. Each of these is composed of two eyes (one facing upward and one facing inward) and a statocyst.

32 The jellyfish *Amphinema turrida*. Note the two long tentacles, sensory rhopalia (small, knob-like structures on the margin of the bell, each of which has a red eye spot), oral arms and the highly folded gonads (yellow).

Overleaf

33 The frilly oral arms of this jellyfish (*Phacellophora camtschatica*) are visible within the veil of tentacles.

34 A jellyfish (*Bougainvillia superciliaris*) with a hitchhiking amphipod (*Hyperia galba*).

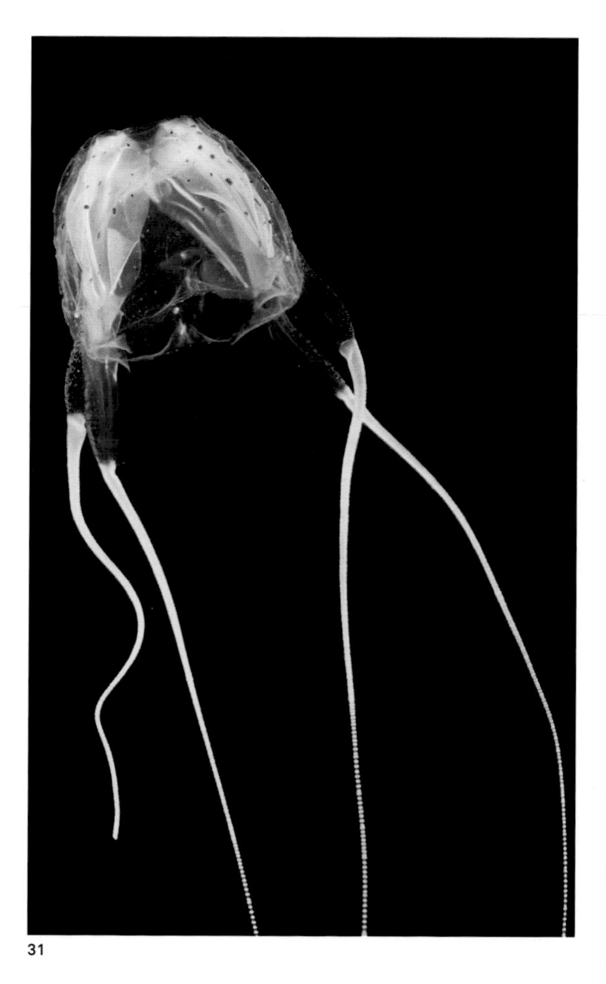

31

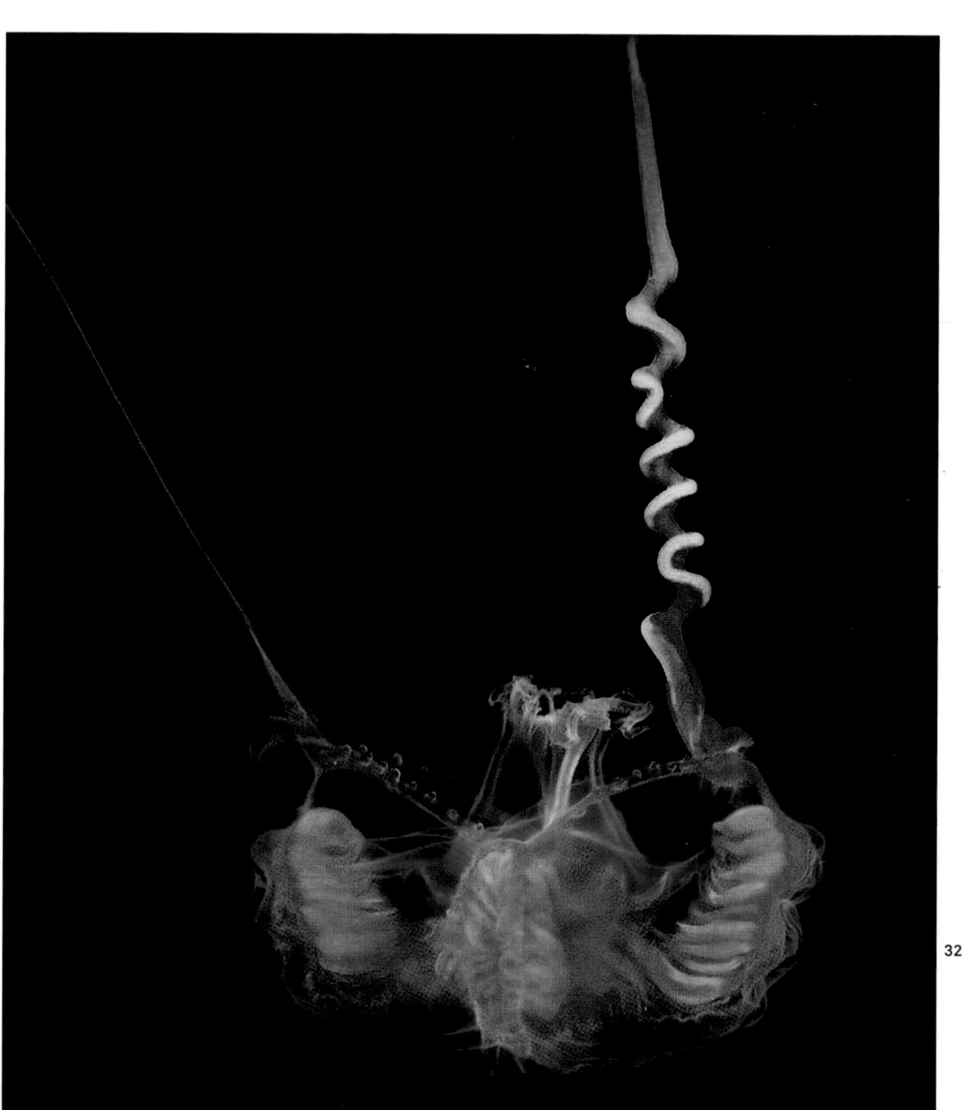

32

35 A hydrozoan colony (*Ectopleura larynx*). Hydrozoans are very diverse cnidarians and depending on the species their life cycle can include polyps, medusae or both.

36 Sea anemones on the back of a hermit crab. This is a mutually beneficial arrangement: the anemone gets a mobile, competition-free place to live as well as scraps of food, while the crab gets camouflage and defence courtesy of the cnidarian's cnidocytes.

37 Cnidarians are often the dominant animals of the seabed. These large anemones, *Metridium senile*, form a veritable thicket.

38 The feeding tentacles and swimming bell of *Cyanea capillata*, one of the largest jellyfish.

› 39 A benthic hydrozoan. Note the feeding tentacles and the spherical structures, known as gonophores. The latter are incompletely differentiated medusae that produce gametes (*Acaulis primarius*).

Cephalochordata

(lancelets, amphioxus)
(Greek *cephalo* = head;
chorda = cord)

Diversity
c. 30 species

Size range
4 to 8 cm
(1.6 to 3.2 in.)

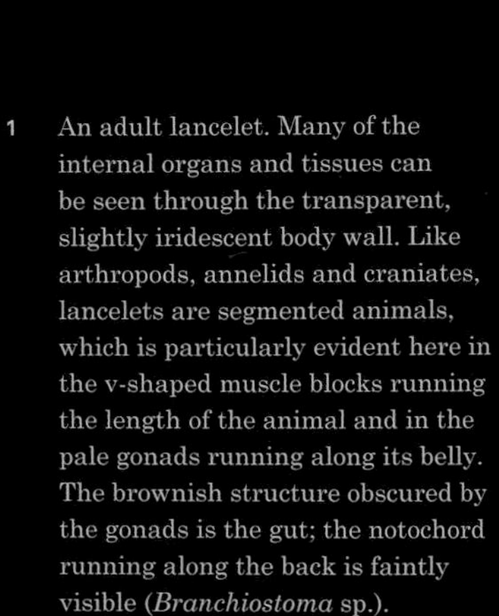

1 An adult lancelet. Many of the internal organs and tissues can be seen through the transparent, slightly iridescent body wall. Like arthropods, annelids and craniates, lancelets are segmented animals, which is particularly evident here in the v-shaped muscle blocks running the length of the animal and in the pale gonads running along its belly. The brownish structure obscured by the gonads is the gut; the notochord running along the back is faintly visible (*Branchiostoma* sp.).

The lancelets are a small group of marine animals, and though they are common they are rarely seen. With a transparent sliver of a body they bear an uncanny resemblance to young fish (and they are in fact a sister group of the lineage that includes fishes). Their interesting characteristics have long captured the imagination of zoologists, who consider them to be a tantalizing glimpse of how the first chordates may have looked and lived.

FORM AND FUNCTION

Put simply, lancelets look like small fish. Their slim body is flattened from side to side, tapered at each end and possesses a body cavity – the coelom – housing the internal organs. Lancelets have fins, and beneath their thin, translucent skin they have 50 to 75 well-developed muscle segments that they put to good use in swimming and burrowing into sediment [1].

Running along the back is the so-called notochord, a fibrous, flexible rod that acts like a spinal column. It gives the body rigidity and offers a good anchor for the muscles. Above the notochord is a hollow nerve cord, the front end of which ends in a swelling – the lancelet's rudimentary brain. Their light sensors are nothing less than surreal; like many other animals they have simple eye spots, but some lancelet species have more than 1000, and they are scattered along the front part of the nerve cord. That is unusual in itself, but there is more: the eye spots on the left side face upwards and those on the right and ventral side of the cord face downwards – a peculiar arrangement that is yet to be explained, although it may be a relic of the asymmetry seen in the larval lancelet (see below).

Other sense organs are thin on the ground. The nerve cord opens to the surface near the front end of the animal to form a ciliated pore, and other sensory structures are located in and around the mouth.

Surrounding the mouth are a number of finger-like projections known as cirri that help to screen out larger particles when the animal is filtering water for edible matter [2, 3]. At the inner end of the mouth is a structure known as the wheel organ [2]; it filters very fine particles from the water for food. Further along still, towards the belly of the beast, is the pharynx – a capacious tubular structure with numerous paired gill slits along its length, the significance of which will be examined below. Amassed food particles are digested in a well-developed through gut.

Lancelets also have a rather elaborate circulatory system of closed vessels, but they are heartless and the blood that courses through their veins contains no oxygen-binding pigment. For such small animals that is not important and most of the oxygen they need diffuses directly across the body wall into the tissues that need it. Their blood instead mainly transports nutrients and dissolved waste products. The main blood vessel does the work of a heart, contracting to squeeze the blood around the body.

With such a complex circulatory system supplying the tissues with nutrients and spiriting away waste there is a need for a specialized excretory system. This is to be found near the front of the animal where three organs filter the blood to remove metabolic waste, which is then excreted in much the same form as vertebrate urine.

LIFESTYLE

Lancelets are bottom-dwelling filter-feeders that spend most of their time buried obliquely in the sand, tail first, with their heads poking into the water. They swim well, but long-distance athletes they are not, and they tire easily. If they are disturbed by a predator or simply fancy a change of location they make a short rapid dart to a fresh patch of the seabed where they wriggle into the sand to start filtering again.

The capacious pharynx comes into its own in this filter-feeding lifestyle. A ciliated groove, which is actually the evolutionary forerunner of the thyroid gland in your neck, runs along the floor of the pharynx and secretes a mucus mesh in the form of two sheets. These latch onto each side of the pharynx and are steadily pulled up by more cilia. The constant motion of the cilia draws water through the numerous gill slits, and the mucus traps edible matter swept in on the current. Eventually these sheets of mucus reach another groove on the roof of the pharynx where they are rolled into a cord and passed along into the animal's digestive tract. The mucus is impressively efficient: its mesh is fine enough to trap phytoplankton, bacteria and even protein molecules suspended into the water.

Lancelets have separate sexes and the males release their sperm en masse to fertilize the mature eggs of the females. The fertilized eggs develop into planktonic, tadpole-like larvae

2

with bizarre asymmetries – a large mouth on the left side of the head and gill slits on the opposite side. Like the adults, these juveniles are filter-feeders that strain edible matter from the water.

Buried up to their necks in sand, straining food from seawater does not take a great deal of energy or space, which allows lancelets to live in huge population densities where suitable conditions prevail – most often shallow temperate and tropical waters. In such places it is not unheard of to find 5000 lancelets in every sq. m of sand (or more than 4000 per sq. yard). Not only do they exist in vast numbers but they are also eminently edible and their nutritious, boneless bodies are gobbled up by a great variety of predators, including humans. Where their population densities are high their ecological role must be significant since they process large amounts of suspended matter and convert it into biomass, which is relished by other, larger animals further up the food chain.

ORIGINS AND AFFINITIES

From an evolutionary perspective lancelets are special because they clearly exhibit the four key features that unite all the chordates (see Introduction, p. 13), a grouping on the tree of the life that includes, themselves, all the animals we are most familiar with (see Craniata, pp. 84–97) and some we are not so familiar with (see Tunicata, pp. 72–83). These features are a notochord, a hollow nerve cord running along their back, gill slits and a tail that extends beyond the anus. In fact these hallmarks are at their most evident in the lancelets, which prompted early zoologists to propose them as the ancestor of all the other chordates. Once again, DNA has shone a light on the murky parts of the animal tree of life to reveal that the lancelets, the craniates and the tunicates did indeed evolve from a common ancestor but that their evolutionary trajectories diverged long ago.

If the lancelets are just a sister group of the other living chordates, then what did the common ancestor of these animals look like? Since the principal chordate features are so stark in the lancelets, they appear to have diverged little in form and function from this common ancestor, leaving us with the possibility that they are very similar to it, at least in the general organization of their body. Intriguingly, some Cambrian fossils of an animal from the Burgess Shale, some 500 million years old, bear a striking resemblance to lancelets. Known as *Pikaia,* these 5-cm- (2-in.-) long creatures may represent the common ancestor, or very close to it, of all living chordates, including humans and all the more familiar animals.

2 In this image the oral cirri (finger-like projections) around the mouth are clearly visible. Inside the animal the wheel organ is also clearly visible (thicker, horizontal projections to the right of oral cirri) (*Branchiostoma caribaeum*).

› 3 Cirri help to screen out larger particles when the animal is filtering water for edible matter. These animals spend most of their time buried in the sediment with only the front of their body visible. In both appearance and lifestyle lancelets are probably very similar to the first chordates (*Branchiostoma* sp.).

Tunicata

(sea squirts, appendicularians, salps)
(Latin: *tunica* = *shirt*)

Diversity
c. 2860 species

Size range
~1 mm to >20 m
(colonial forms)
(~0.04 in. to >65 ft)

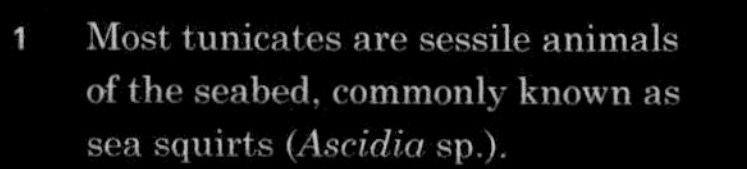

1 Most tunicates are sessile animals of the seabed, commonly known as sea squirts (*Ascidia* sp.).

2

2 A colony of brightly coloured sea squirts (*Rhopalaea* sp.). The vertical brown/white lines are sperm ducts. *Rhopalaea* sea squirts are predated by *Nembrotha* nudibranchs (see Mollusca, p. 239).

The tunicates are nothing short of captivating, bizarre in both appearance and lifestyle. Exclusively marine and often extremely numerous, some have taken on a sponge-like form, living out their entire adult lives glued to the spot and encased in little fibrous jackets; others, akin to jet engines, band together to form huge floating colonies, and still more have evolved an elaborate way of filter feeding.

FORM AND FUNCTION

The vast majority of tunicates (some 2000 species) are sessile: these are the archetypal sea squirts [1, 2]. They have a well-developed pharynx, gut, musculature and circulatory system (complete with heart), as well as reproductive and excretory organs and a rudimentary central nervous system. An individual is encapsulated within a tough, living jacket known as a tunic, unique amongst animals because it contains a form of cellulose as a major constituent. Exactly how these marine animals developed the capability to synthesize this substance is something of a mystery, but recent evidence suggests that their ancient ancestors acquired the necessary genes from bacteria.

Apart from its protective function, the tunic forms the base of the attachment of benthic species to their anchorage; extensions grow from its lower surface and cling so firmly to rock that the animal cannot be pulled from its attachment without killing it. Sessile species commonly enhance the protective qualities of their jacket by including inorganic spicules or sand, in some cases so much so that it becomes decidedly rock-like.

The most obvious outward feature of an adult sea squirt is a pair of siphons, one of which – essentially the mouth – is the inlet for water while the smaller siphon is the outlet [13, 16]. Peeling away the tunic reveals a translucent body, the most notable feature of which is the more-or-less U-shaped gut, which begins and ends at the siphons [14]. Extending from the base of the incurrent siphon to the anterior end of the gut is a large pharynx, a capacious ciliated bag with gill slits in its wall [15].

Situated at the base of the gut is one of many sea squirt oddities: the heart. Every few minutes this pumping organ reverses its action to drive blood first in one direction and then in the other; this adaptation ensures that all the animal's organs and its tunic receive adequate oxygen and nutrients. Another enigmatic feature is the ability of sea squirts to accumulate heavy metals, most notably vanadium, in their blood and tissues, stabilized by the presence of sulphuric acid. Exactly how and why they do this is not fully understood, but it could be to deter predators. By-products of metabolism are also thought to accumulate in their tissues and tunic as another line of defence.

The sea squirt nervous system is composed of a number of nerves emanating from a ganglion – the brain. This is embedded in the tissue between the two siphons along with the so-called neural gland, which is thought to be the evolutionary precursor of the pituitary

3

3 Some of the colonial sea squirts form flower-shaped colonies. Each petal-shaped zooid has its own inhalant siphon, but all in the colony share an exhalant siphon (at centre of each colony) and a common tunic (*Botryllus* sp.).

gland deep inside the human brain. It is thought to keep a sea squirt's fluids in balance with its environment.

Sea squirts often band together in colonies that come in a vast array of shapes and sizes. Depending on the species, an individual squirt – a zooid – in such a colony may be an almost independent animal or intimately linked to the other zooids. Some highly specialized colonial species exist as flower-like arrangements of individuals embedded in a common tunic [3]. Each zooid within a 'flower' has its own inhalant siphon, but they all share a single central exhalant siphon.

Around 140 species of tunicates are pelagic animals that have evolved what must be among the most bizarre of all animal forms [5, 8, 12, 18]. The salps have barrel-like bodies with bands of muscle, an inhalant siphon at the front and an exhalant one at the back – a form reminiscent of a jet engine [8]. The muscle bands contract to squeeze water through the animal, propelling it forward. With a body that is little more than a stubby tube, there is precious little room for internal organs and these are all accommodated in the thin body wall of the animal [8, 18].

Many of these pelagic tunicates exist as colonies, with tunics meshing together to form incredible structures [9–12, 17, 20, 21]. Some exist as huge, vase-like structures that can be 20 m (65 ft) long and capacious enough to allow a diver to venture inside [20, 21]. Others form long chains [12], while some have specialized individuals [11].

LIFESTYLE

With a few exceptions the tunicates are filter-feeders that pump water through their pharynx and trap edible matter in a mesh screen in the same way as lancelets (see Cephalochordata, pp. 68–71). Some species, especially the sea squirts, can pump prodigious volumes of water, as much as 170 l (45 gal.) over a 24-hour period – impressive for an animal just a few centimetres long [16]. As in the lancelets, the mucus net is sufficient to trap edible particles as small as suspended protein molecules.

Perhaps the most peculiar of all the tunicates are the appendicularians; they have evolved a singular way of filtering edible morsels from seawater [BOX 1]. In the deep sea there are also aberrant sea squirts that have given up the benign existence of filtering seawater for a predatory way of life. Their inhalant siphon is enlarged to snare small animals inhabiting the dark, benthic zones.

Tunicates have male and female sexual organs. In most cases sperm and eggs from individuals of the same species are shed into the water where they meet to begin the arduous journey that will take the new generation into adulthood. There are some species that retain their eggs within the confines of the tunic-clad

BOX 1 ***Appendicularians***

The appendicularians, looking a bit like mutant tadpoles, have an extraordinary way of collecting food. From specialized epithelial cells they secrete an elaborate mucus house, little bigger than a walnut, which has intricate structures for channelling and filtering water. Wriggling its tail constantly the animal, only about 10 mm (0.4 in.) long, draws water through the filters, concentrating any edible matter into a broth that can be sucked up.

Despite the fact that creating the house takes time and precious resources, it is completely disposable and these industrious little animals think nothing of building 4 to 16 of them every single day. When the filters are clogged up with matter, the house is simply abandoned and another one secreted. The larvacean inflates the nascent house and in about one minute it can carry on with the pressing task of collecting food.

And why do they go to all this effort? Unlike the other adult tunicates, their pharynx is too small to filter adequate amounts of food; the mucus house gets around the problem by acting like a complex extension of the pharynx.

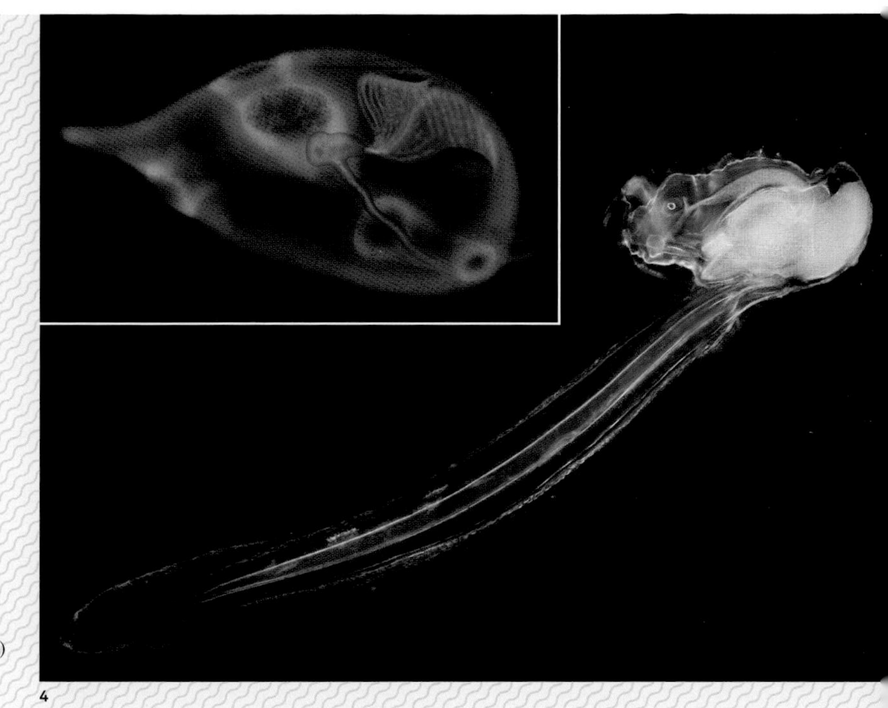

4 An appendicularian (*Oikopleura* sp.) and its mucus house (inset).

body, and when these are fertilized by drifting sperm they develop inside the adult until they hatch into larvae. The tunicate larva, which looks like a microscopic tadpole, is a non-feeding creature equipped with a nerve cord, stiffening notochord and well-developed tail [7, 19]. With its thrashing tail, the larva is mobile, but this freedom to roam is short-lived and before long the little creature gets the urge to stick itself headfirst to a suitable bit of seabed. It then begins the transformation into adulthood. Metamorphosis completely strips the larva of its distinctive features until what was once a mobile creature is little more than a feeding bag.

Tunicates that live as floating colonies have truly mind-bending life cycles and figuring out what develops from what and how it does so has been something of a zoological detective story. Alas, that story is still incomplete. Some do not have the distinctive tadpole larva, but instead go through sequences of asexual and sexual reproduction to build up their huge colonies. In some species eggs and sperm from a mature colony produce a mother zooid that will, if it is lucky, go on to form a big colony by budding off replicas of itself. In some of the salps, what looks like a single organism – a small transparent barrel with a short tail – is actually an asexual colony made up of vastly different, genetically identical individuals. There are zooids specially adapted for feeding, some for swimming and others for reproducing sexually to spawn new colonies via the intermediary of the tadpole larva [11]. Other salps exist as colonies composed of a single, barrel-shaped zooid that sprouts a long chain of identical reproductive zooids that nurture their embryos – baby barrel-shaped zooids – with a structure akin to a placenta [9, 10]. The life cycle of the appendicularians is rather mundane by comparison. Sexual reproduction yields a tadpole larva, but instead of a radical metamorphosis the larva simply grows and retains all of its characteristics into adulthood, except for the placement of the tail, which shifts 90 degrees and ends up dangling down [4].

Tunicates are hugely significant in the marine ecosystem. Collectively, they filter vast amounts of seawater, feeding on tiny organisms and suspended matter. Few other filter-feeding animals are as effective at pumping water and collecting food. Their very fine mucus filters enable them to trap particles that simply slip through the nets of other filter-feeders. As larvae, juveniles and adults they are food for a range of other marine animals [6]. As we have seen, however, the tough tunic of the adult sea squirts is often laden with acid, toxic heavy metals and noxious by-products of metabolism, enough to deter all but a few specialist predators [2].

The complex life cycles of the colonial pelagic species give them a considerable advantage when it comes to exploiting patches of abundant

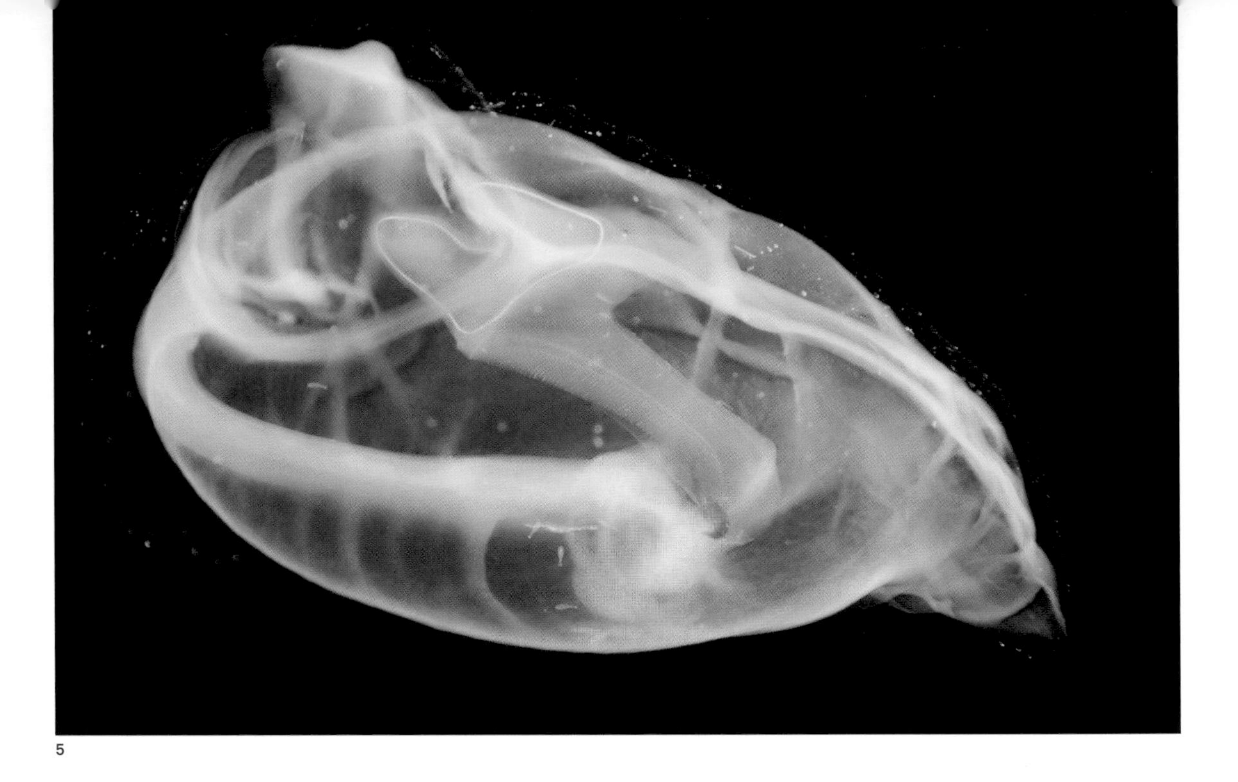

5

suspended food. With their ability to reproduce asexually, the populations of these floating tunicates can quickly reach enormous densities. In natural systems, larger animals such as fish control these population explosions. But in recent decades commercial fishing has so depleted predator numbers that the prevailing balance of the marine ecosystem has been disturbed and huge swarms of salps and other pelagic tunicates are now seen for longer periods and occur regularly in areas where they were once rare.

Another, less obvious contribution to marine ecosystems is the copious waste that the tunicates produce. High population densities, short life cycles, ceaseless feeding and the frequent secretion of disposable structures all contribute to the production of vast amounts of floating debris, all of which inexorably sinks to the seabed. This debris makes up a significant proportion of 'marine snow', which showers the seabed and acts as an energy and nutrient link between the sunlit and productive surface waters and the relatively depauperate and sunless ocean depths.

Some tropical colonial sea squirts have even forged a symbiotic association with microscopic photosynthesizing organisms. These reside in the zooid's body and tunic converting sunlight into food, some of which is shared with the host. Some of these colonies slowly creep over the substrate as they grow, perhaps as a way of controlling the amount and intensity of light reaching their symbionts, to make them as comfortable as possible.

ORIGINS AND AFFINITIES

The soft bodies of the tunicates mean that the remains of long-dead members of the lineage are few and far between. But fossils from China, at least 540 million years old, have been interpreted as ancient stalked sea squirts a few centimetres long. It seems clear that this body plan, along with all the other templates for future animal diversity, was established more than 500 million years ago.

Back in the late 19th and early 20th centuries, zoologists struggled to work out where the tunicates belonged on the tree of life. For a long time the sea squirts and the pelagic forms, very poorly known at the time, were assumed to be completely different animals. The former were thought to be relatives of the sponges, since the two groups have certain superficial similarities. The different types of pelagic tunicates were also thought to represent distinct lineages.

It was not until embryology took off as a science that zoologists started to put the puzzle together. The key was the 'tadpole' larva: although small, it links these creatures – wildly different as adults – in a single lineage and it reveals roughly where they fit on the tree of life. Its features [7, 19] show that the tunicates are an offshoot of the lineage that gave rise to mammals and all the other chordates. It is both a sobering thought and a reminder of the beauty of evolution that the lowly sea squirt, eking out an existence on the seabed and seemingly a million miles from a mammal, is actually quite a close relative of ours.

5 The solitary, asexual stage of a salp. What looks like a comb running through the animal is the gill bar, which is important in filter feeding since it rolls up the food-laden mucus net (unidentified species).

6 Several species of crustacean depend on pelagic tunicates for refuge and food. There are even some amphipods (*Phronima* spp.) that kill the tunicate, hollow it out to leave only the tunic, and then lay their eggs inside. They push this impromptu egg case through the water. Here, one of these 'pram bugs' and its eggs can be seen inside a hollowed-out tunicate.

7 Tunicate larvae are proof of the evolutionary relationship between these animals and the other chordates (craniates and lancelets), since they have a nerve cord, a stiffening notochord running along their back and gill slits. These first two traits are lost as the larva metamorphoses into an adult (unidentified species).

8 The pelagic tunicate, *Thalia democratica*. Note the tubular form, long, comb-like gill bar and the faintly visible muscle bands.

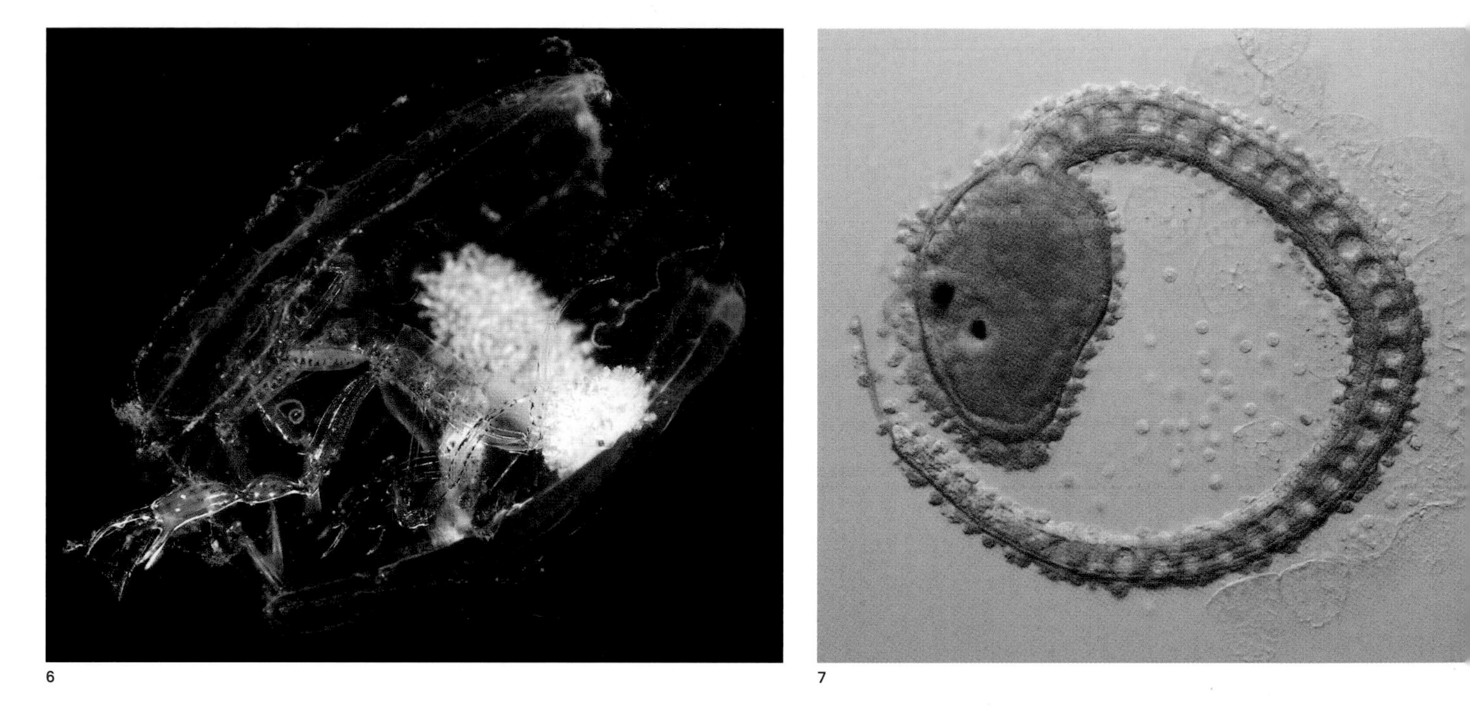

6

7

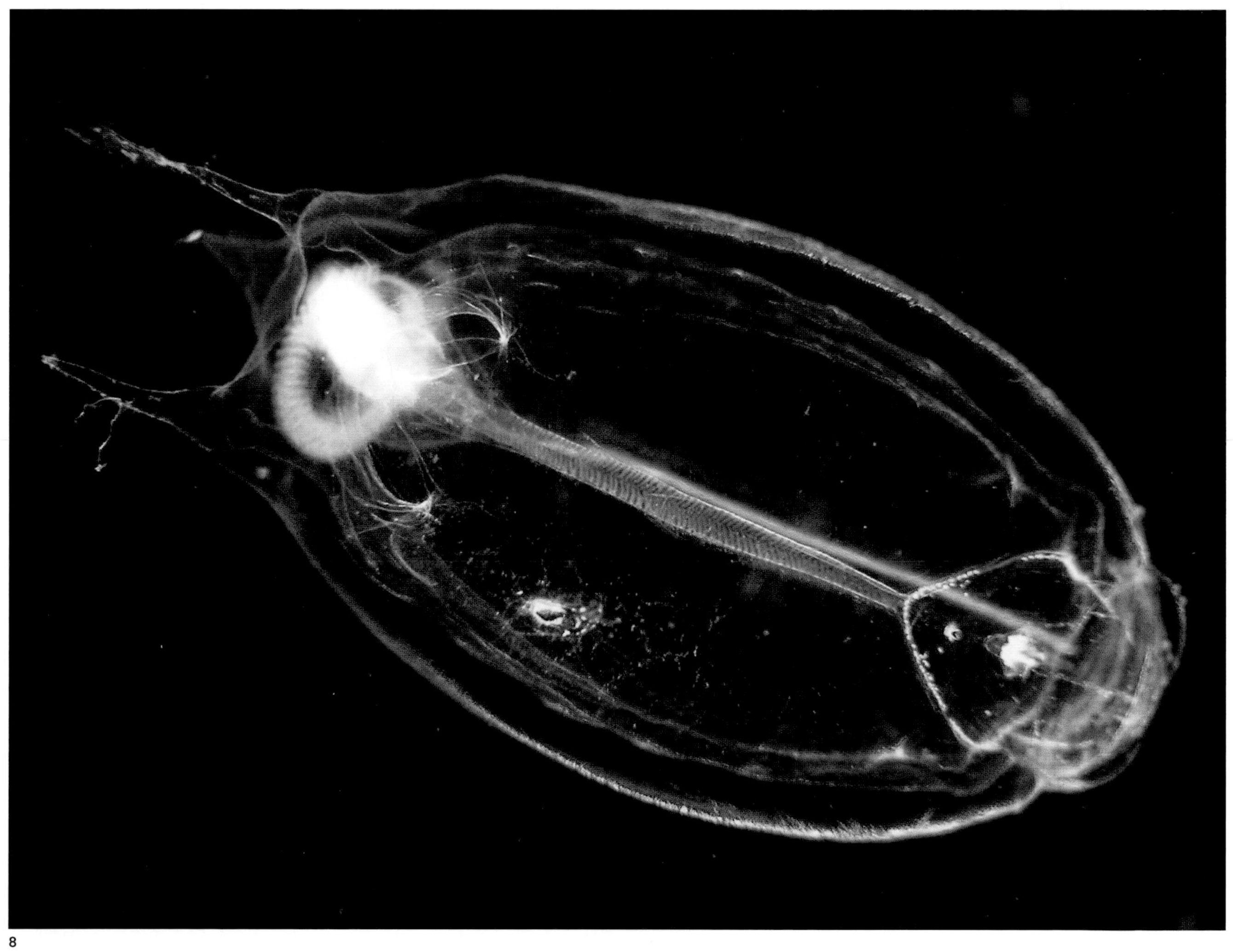

8

9

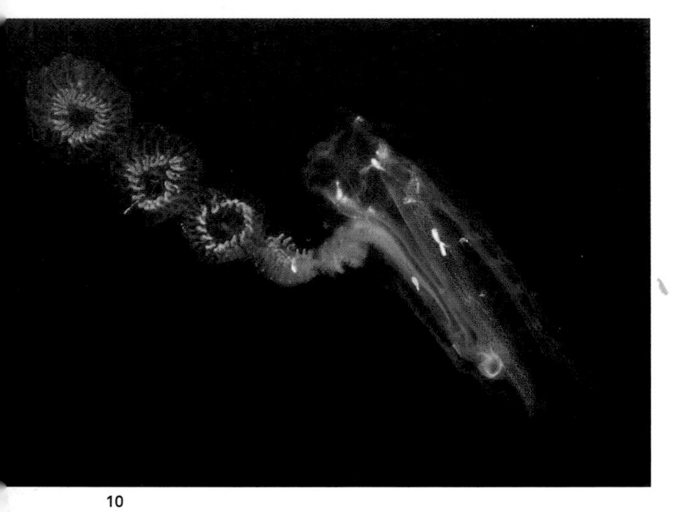

10

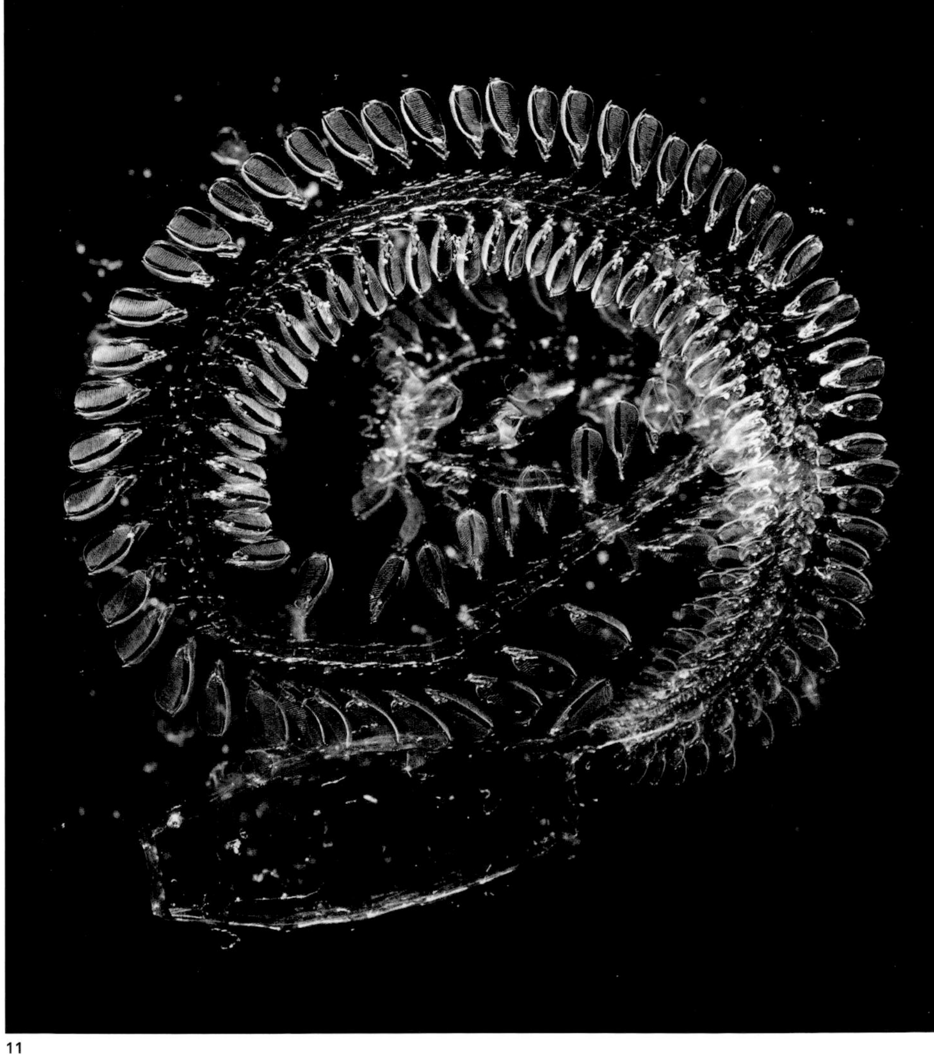

11

9 Colonial sexual zooids of *Cyclosalpa affinis*, a salp. The orange, U-shaped gut in each of these zooids can be clearly seen.

10 The solitary phase of *Cyclosalpa affinis* (tubular organism, right) asexually producing a chain of colonial, sexual zooids in tight whorls. These chains will detach from the solitary phase and each zooid will reproduce sexually, eventually nurturing a new solitary phase individual via a placenta-like structure.

11 The tubular, asexual solitary phase of a doliolid, *Dolioletta* sp., with a long chain of sexual zooids.

› 12 Colonial sexual zooids of *Pegea confoederata*, another salp.

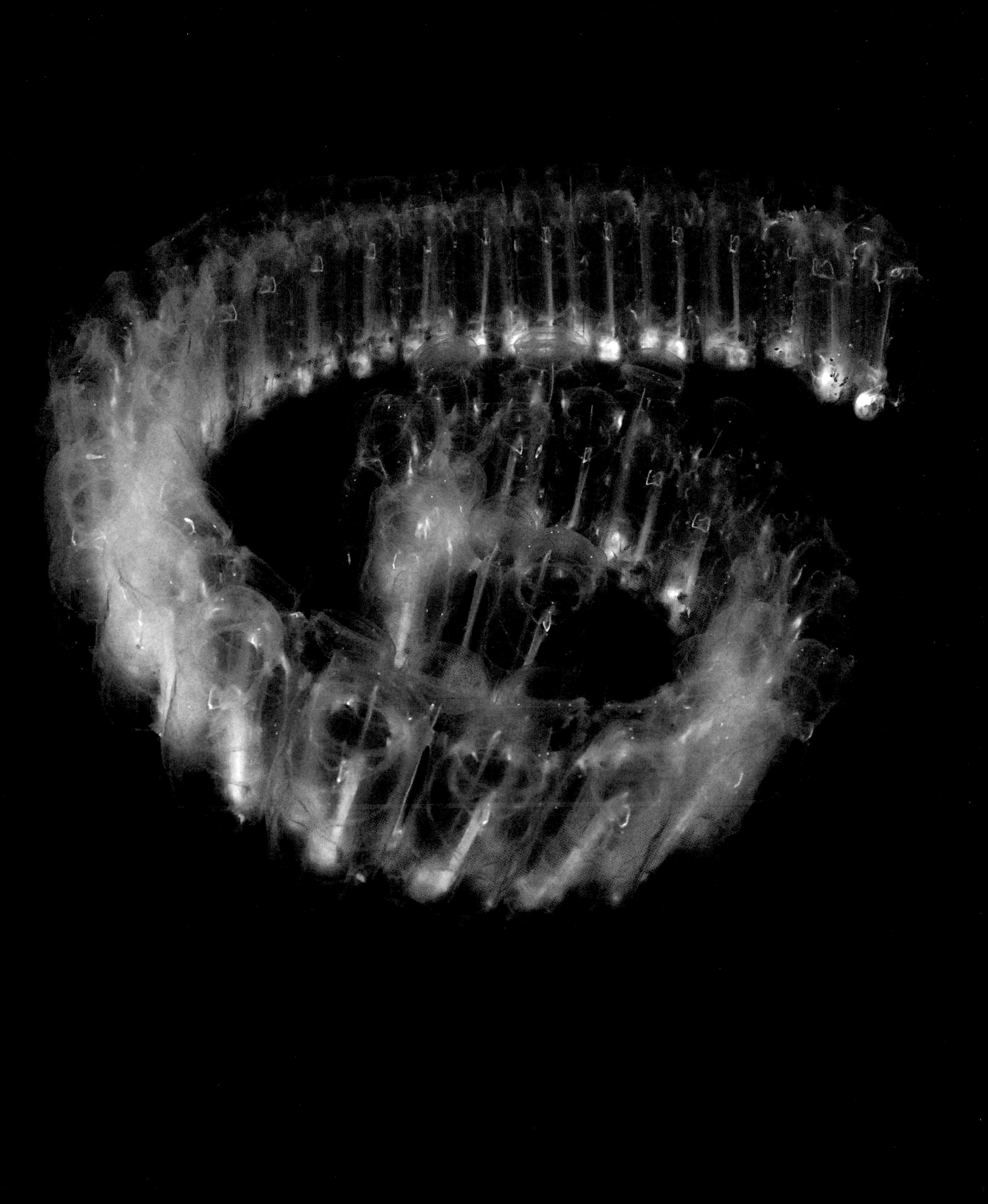

13

14

13 The inhalant siphon (opening upwards) and the exhalant siphon of a sessile tunicate, *Styela montereyensis*. Siphons pointing in different directions may be an adaptation to avoid flow disruption.

14 The U-shaped gut extending from the inhalant siphon (top) to the exhalant siphon (right) can be clearly seen in this transparent sea squirt, the aptly named *Ciona intestinalis*.

15 The colonial lightbulb sea squirt (*Clavelina huntsmani*), so called because of its bright pink 'filaments'. Forming a ribbed, tubular bag between these filaments is the pharynx with its numerous gill slits. One of the pink filaments is the endostyle, which secretes the mucus net that lines the pharynx to trap food. The other is the dorsal lamina, a ciliated gutter that rolls the net mesh and its cargo of food particles into a cord and conveys it further down the digestive tract.

15

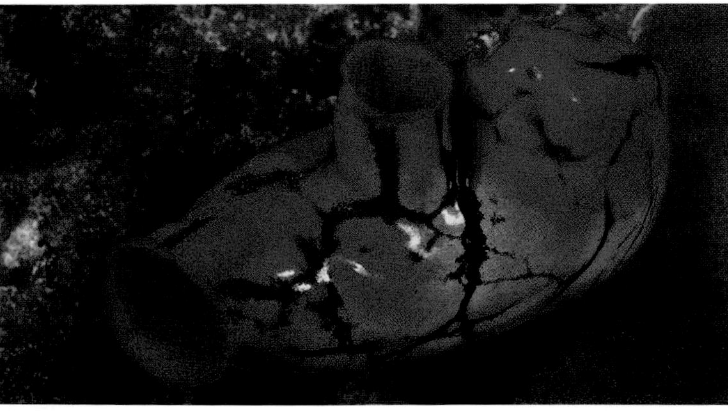

16

16 Sessile tunicates can pump huge volumes of water to collect particles of edible matter. They are also chemically defended, which is sometimes broadcast to potential predators with bold colours (*Polycarpa* sp.)

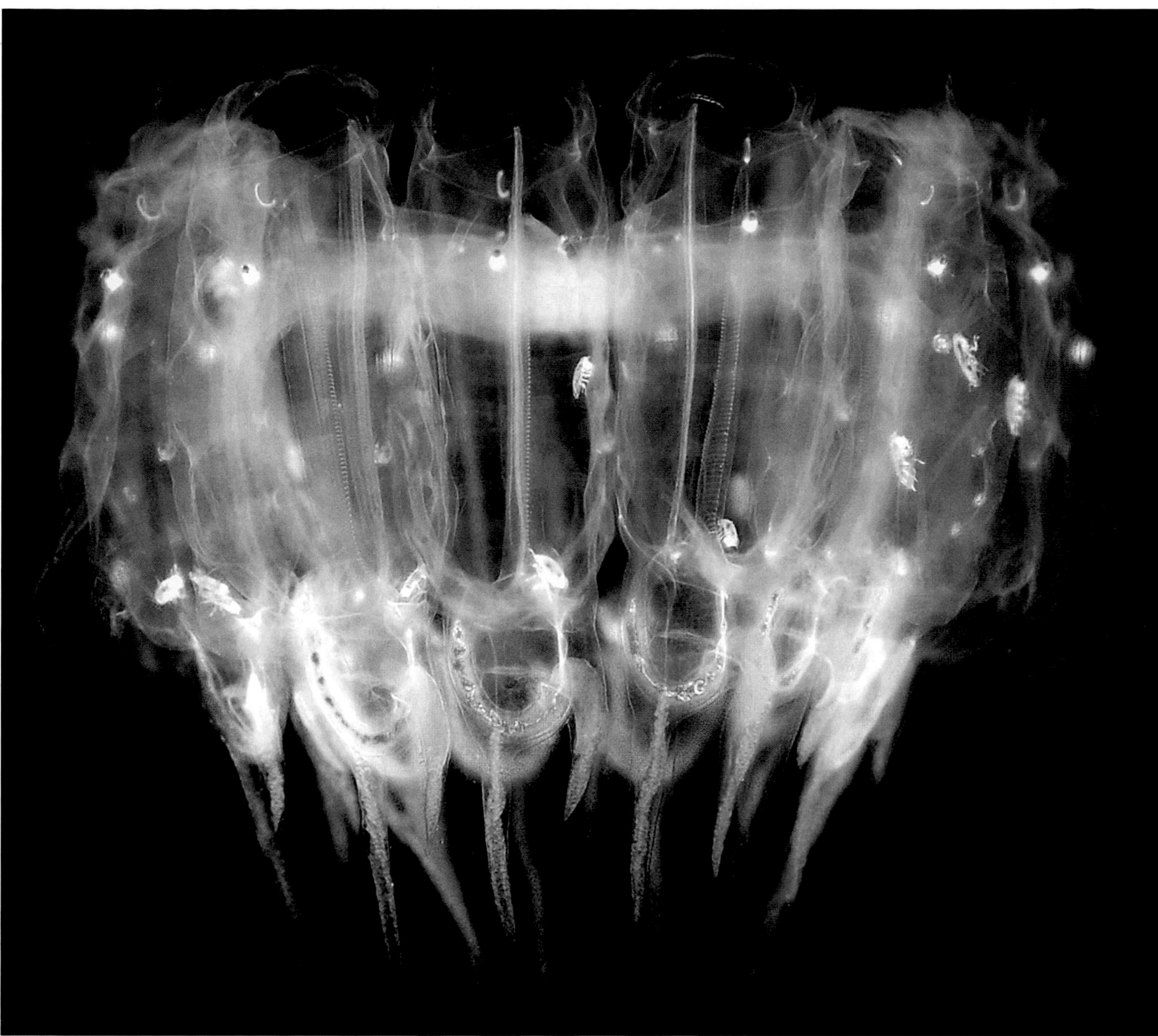

17

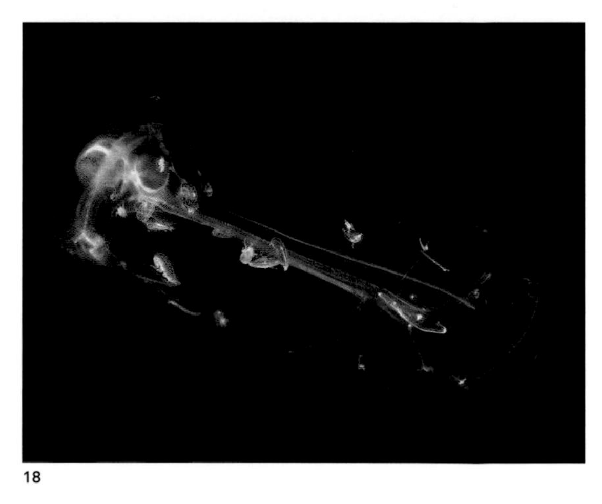

18

17 Colonial sexual zooids of a salp, *Cyclosalpa bakeri*. Note the numerous, small amphipods within the individual zooids.

18 The solitary, asexual stage of a salp. In these pelagic tunicates the siphons are at opposite ends of the body. Muscle bands squeeze water through the animal for filter feeding and for propulsion (unidentified species).

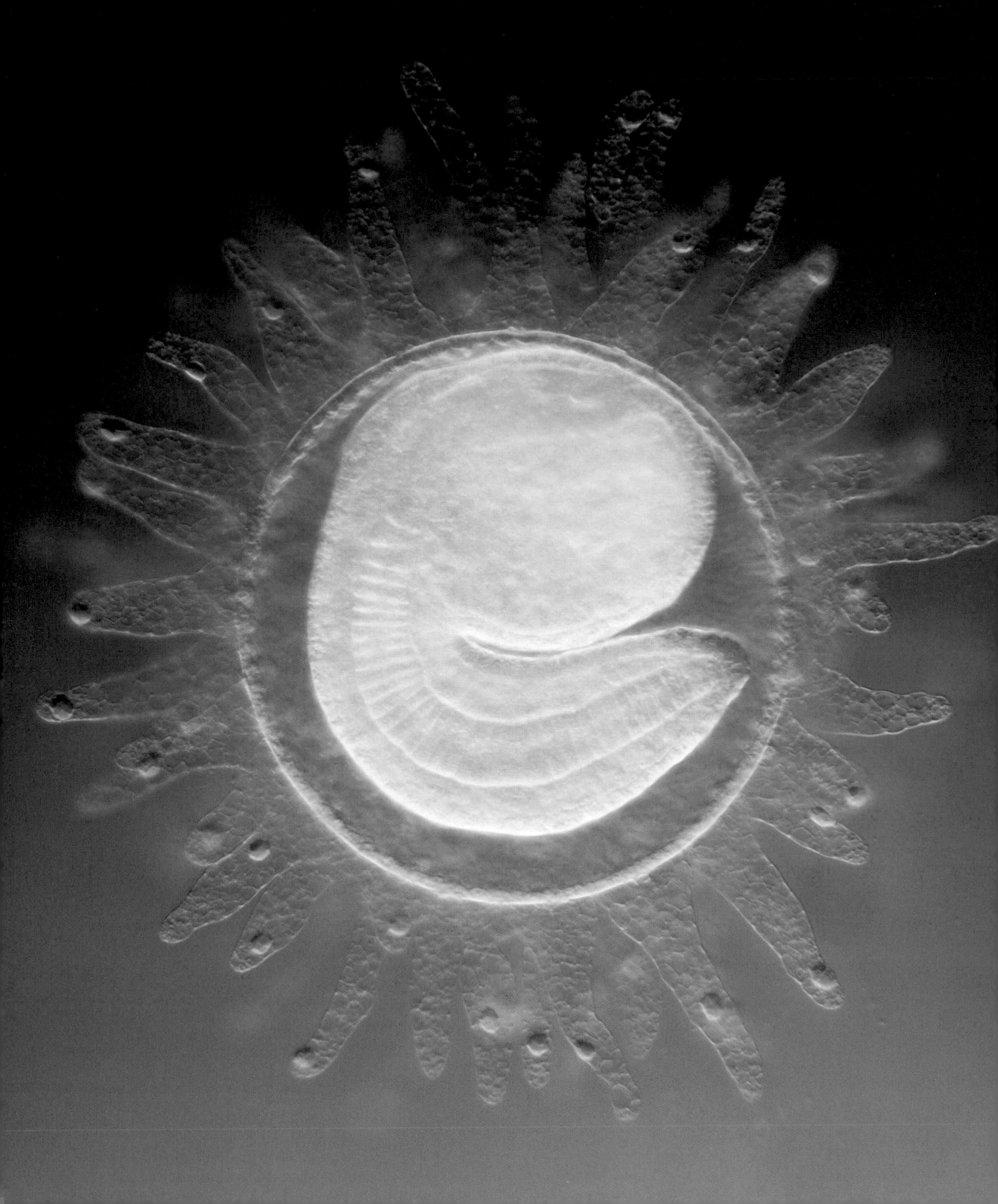

‹ 19 A developing tunicate larva inside its elaborate egg. For the purposes of dispersal the eggs must float for as long as possible, so they are furnished with finger-like projections and oil droplets (apices of projections), both of which improve buoyancy (*Ciona* sp.).

20 The colonial pelagic tunicate, *Pyrosoma* sp. The zooids of these tunicates are relatively small, but they can form huge colonies.

21 Close-up of the zooids forming a *Pyrosoma* sp. colony.

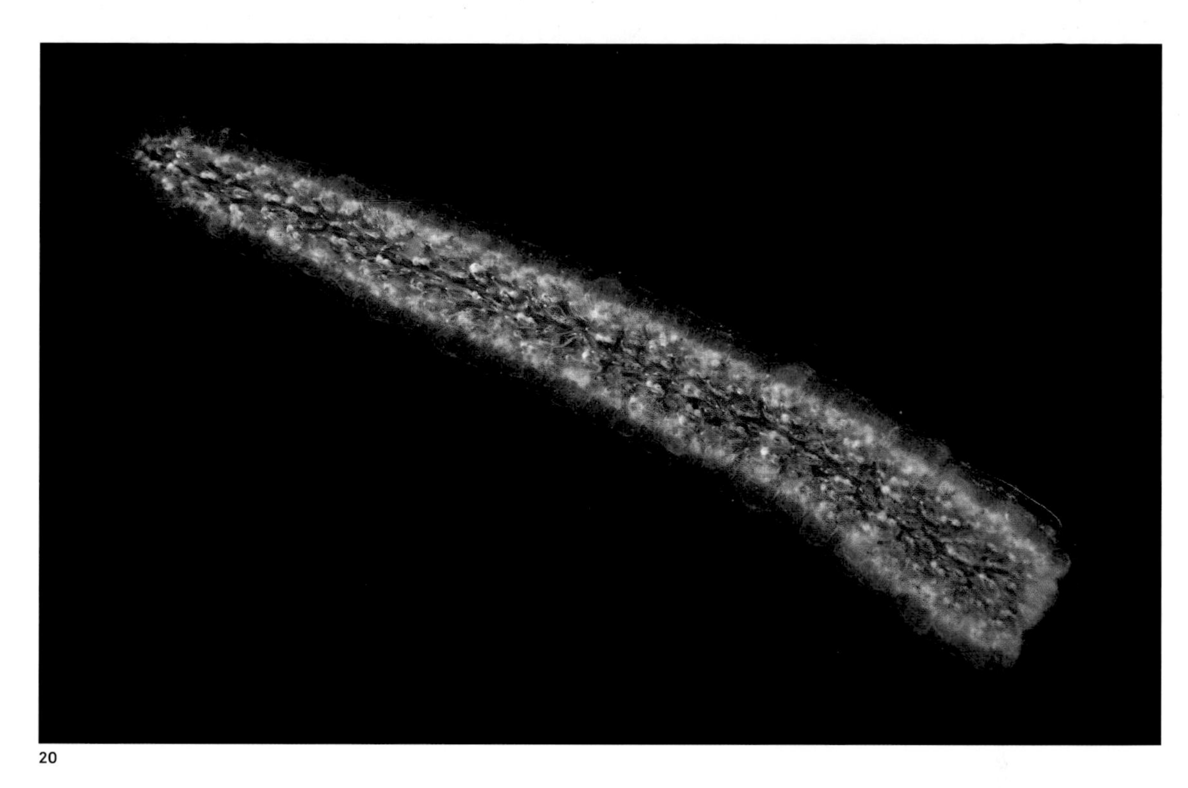

20

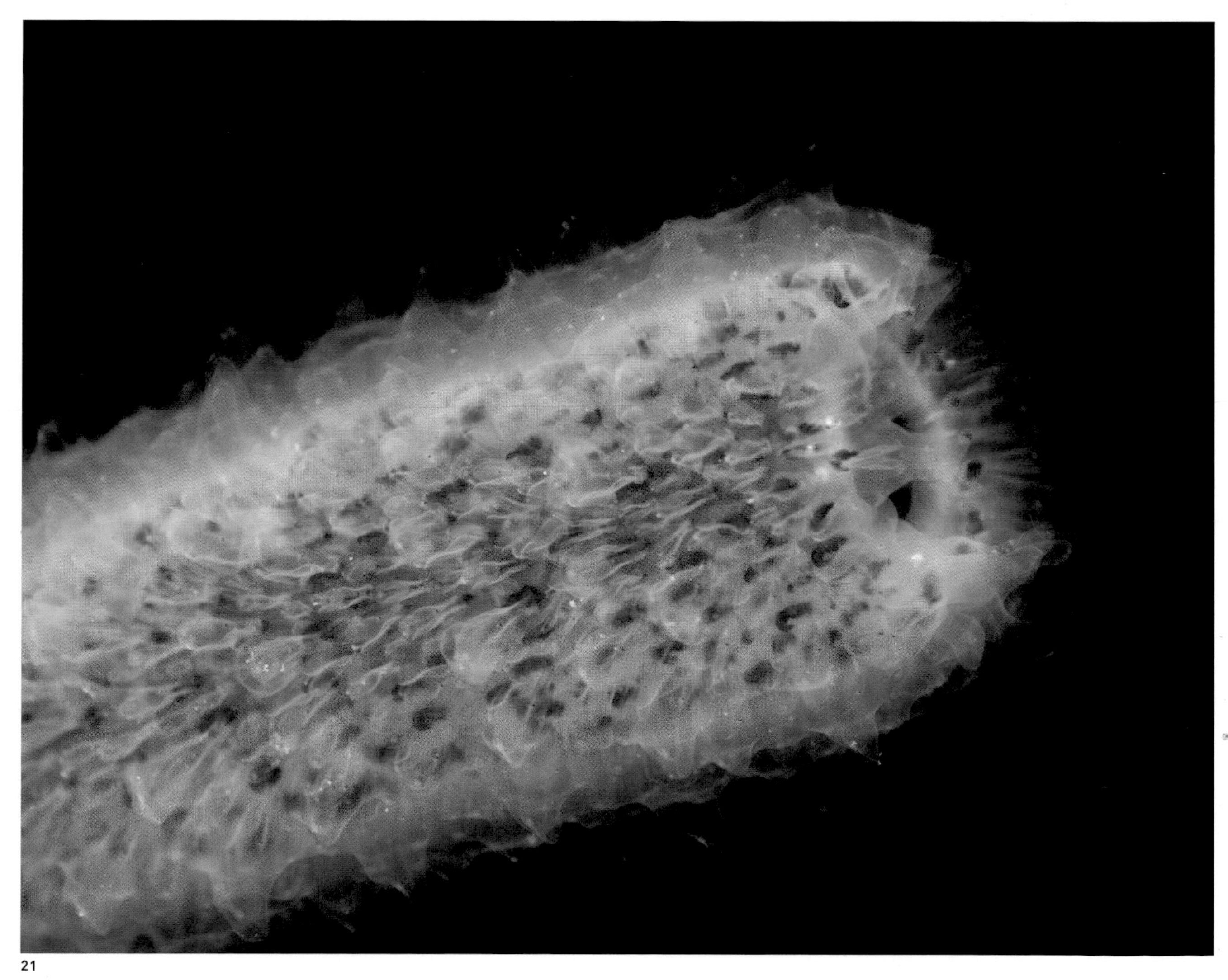

21

Craniata

(hagfish, lampreys, cartilaginous fishes, bony fishes, lobe-finned fishes, amphibians, lizards and relatives, mammals, crocodilians, birds)
(Greek *kranion* = brain)

Diversity
c. 64,830 species

Size range
~10 mm to ~30 m
(~0.4 in. to ~100 ft)

1 The evolution of limbs in a branch of the craniates was one of the important stepping stones in the colonization of the terrestrial domain by these animals. Here, a young nile crocodile (*Crocodylus niloticus*) hatches from its egg. The basic crocodilian body plan is a very successful one; modern species closely resemble their Cretaceous ancestors of around 80 million years ago.

2

Birds represent the only group of dinosaurs to have survived the extinction event at the end of the Cretaceous period. The wings of these small, feathery dinosaurs are modified forelimbs that have allowed them to exploit the air with considerable success. Hummingbird aerial abilities are on a par with the most accomplished flying insects (*Archilochus colubris*).

Without question, the craniates are the most familiar and best studied of all the animal lineages, since both humans and the creatures we directly interact with most often are members. But because they are so familiar we often fail to appreciate their zoological significance. Their basic body plan is extremely flexible and has been adapted to myriad lifestyles, ranging from aquatic filter-feeding to tree-top fruit picking and aerial predation. Some of the craniate organs and organ systems have become very elaborate, fine-tuning internal environments and allowing for complex behaviours, intelligence and even self-awareness. These traits have enabled colonization by the craniates of all the habitable places on earth, an achievement that can only be matched by one other lineage – the arthropods.

FORM AND FUNCTION

A number of features unify the craniates, though some are only evident when these animals are nothing more than tiny embryos. Perhaps most well known is the cranium, a cartilaginous or bony case that protects the brain. Running along the back is an internal fluid-filled nerve cord, a structure that gives rise to the brain and the spinal cord. Below this tube of nervous tissue is the notochord, a stiffening rod that in most species is replaced by a bony, articulating structure as they develop – the vertebral column. The cranium and vertebral column form part of the endoskeleton, one trait that literally underpins the success of the craniates [BOX 1].

The pharynx, equipped with gill slits, is retained into adulthood by aquatic species, and in some of those forms it has undergone extensive remodelling into extremely efficient gills and associated structures. In most of the terrestrial forms and in those that have returned to the water after an evolutionary history on land, only the embryos have gill slits.

Inside the body of all craniates there is a capacious body cavity – the coelom – lined with mesoderm, sheets of which support and confine all the organs packed into this space. Craniate bodies do not, in general, appear to exhibit the same degree of segmentation (see Introduction, p. 16) as, say, the arthropods do, but they are in fact segmented in fundamental ways, and there are clear signs of segmentation in the notochord or vertebral column, the nerve cord, and the muscles surrounding these structures [4, 13]. The digestive tract extends from a well-developed mouth and has pouches for storing food and initiating digestion, a process during which the food passes through lots of sinuous intestine and eventually exits at the anus.

With the exception of the hagfish and the lampreys, eel-like animals of the ocean and fresh water, all the craniates have well-developed jaws. The evolution of biting jaws was a real boon for the further diversification of this lineage, since a stout pair of jaws can make short work of prey that is too large or too well defended for a jawless animal to tackle, and the physical process of breaking food up into small morsels improves the efficiency of digestion.

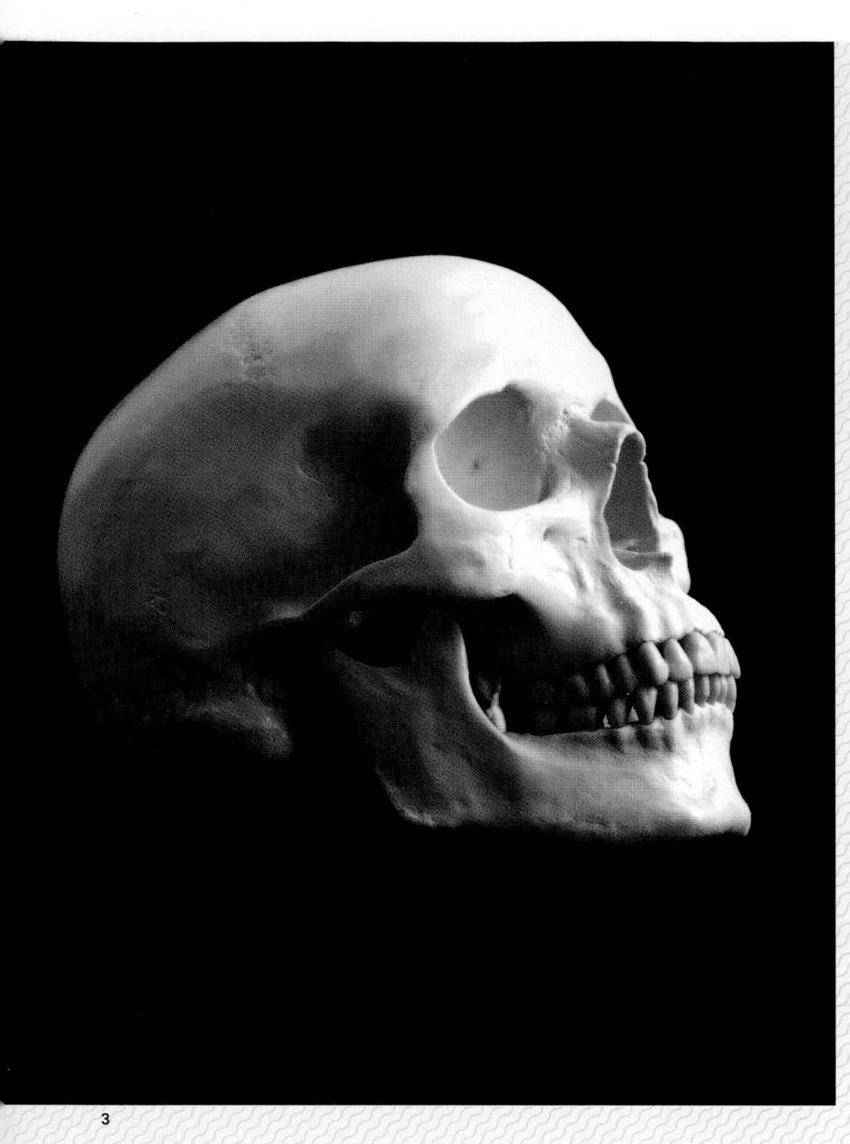

3

4

BOX 1 ***The endoskeleton***

In some craniates, the endoskeleton is nothing more than a cartilaginous reinforcement of the notochord and a rigid box surrounding the brain, but in others the cartilaginous template has been suffused with calcium phosphate to form a bony, articulated scaffold, an incredible evolutionary marvel.

Although bone is admirably suited to mechanical and protective roles, it probably evolved as a solution to a metabolic problem. The cells of the hypothesized ancestral craniate, very likely an animal capable of short bursts of frenetic activity, were hungry for calcium and phosphorous; therefore, a repository of these minerals would have been advantageous. This was crucially important because it meant that the lineage had to develop both the physiological means to store and to deplete the repository. As a result, the organisms that employed the material in a structural instead of just a physiological role were able to remodel themselves both to change their shape and size as they grew, and as their functions and demands changed, and they were able to do so without sloughing their skeletons (like the arthropods), and without relying on fixed forms such as the shells of molluscs. Over time, these mineralized tissues conferred other, mechanical benefits, including protection of soft organs, a solid anchor for the action of muscles, and support of the body.

Continual refinements and adaptations of this flexible, multifunctional scaffold produced the complex skeletons we see in the craniates, particularly the tetrapods. These skeletons, with their hip and pelvic girdles, have sprouted limbs that have evolved into an amazing variety of forms – just think of a bird's wing, your own hand or the shovel-like paw of a mole.

3 The name 'craniate' relates to the possession of a cranium, a protective case of cartilage or bone surrounding the brain. In addition to the cranium this image of a human skull also shows another distinctive feature of most craniates – the jaw.

4 The skeleton of a rock pigeon (*Columba livia*). The large keel beneath the neck is the hugely enlarged sternum that serves as a point of attachment for the flight muscles. Segmentation is a key structural principle in all craniates, but it is typically hidden from view; it is revealed here in the repeating units of the vertebral column, the processes growing from the vertebrae and the bones of the limbs.

An array of senses hooked up to a very sophisticated central nervous system has also been central to the success of the craniates. There are sense organs to detect light, pressure waves, gravity, touch, magnetic fields and the electrical energy discharged by other organisms. Some birds have among the sharpest eyes of any animal, while many of the cartilaginous and bony fishes can pick up the faint electrical fields emanating from their burrowing prey. Some of the mammals, however, such as the micro-bats and the toothed whales, have become creatures of the night, the deep sea, or murky water, at times or in places where eyes are useless. Instead, they 'see' with sound waves, emitting pulses that they bounce off their surroundings, building up a visual picture from the echoes.

To deal with this sensory challenge the brain has become extremely elaborate, with specific regions devoted to making sense of the continuous streams of information and enabling the animals to react accordingly. This radical development of the front part of the central nervous system is most marked in primates, toothed whales and birds, creatures where something altogether peculiar has happened – the evolution of intelligence and self-awareness to a degree unrivalled in other lineages. These brains, with their billions of neurones and trillions of connections, are fiendishly complex, so much so that a complete understanding of their inner workings is still way beyond our reach. What seems crucial in the evolution of these higher mental functions is not the absolute size of the brain, but the way in which the individual neurones are connected. But while undoubtedly the single greatest product of evolution, the running expenses of these brains are enormous. Your brain, for example, accounts for roughly 2 per cent of your body mass yet it receives 15 per cent of all the blood pumped by your heart, 20 per cent of all your oxygen and 25 per cent of all your glucose.

Supplying all the organs and tissues with oxygen and nutrients is a sophisticated circulatory system consisting of closed vessels and an efficient, muscular pump – the heart. Oxygen from the water or air diffuses across the surface of gills or lungs and is bound up by the haemoglobin packed into special blood cells. The blood is also extremely important in mobilizing the defences if any pathogens manage to get inside the body. A variety of specialized cells make up this defence – the remarkably efficient immune system that can 'learn' from a threat to improve its response next time it is encountered. Miraculously compact kidneys filter the blood to remove waste as well as to regulate its acidity and water content.

LIFESTYLE

A skeleton coupled with powerful muscles, an array of senses and a sophisticated nervous system was a recipe for success and today the craniates live a huge range of lifestyles. There are the hagfish and lampreys, which typically, as adults, scavenge on the bodies of dead and dying animals or rasp at the flesh of the living [8–11]. The cartilaginous fish, the well-known sharks and rays and the lesser-known chimaeras are animals of the seabed and open water [16, 17]. Many of the pelagic species are beautifully adapted predators with finely tuned senses, powerful muscles for rapid bursts of speed, and fearsome jaws. Some of the largest forms (whale sharks, basking sharks, and manta rays) have opted for a filter-feeding existence, swimming vast distances searching for algal blooms and the associated swarms of zooplankton that they need to fuel their massive bulk.

The majority of the craniates – about 29,000 known species – are bony fishes, a hugely successful and diverse group that abounds in marine and freshwater habitats [12–14, 18–22, 26]. Some of them, such as the tuna and their kin, are very close to hydrodynamic perfection; they have extremely slippery bodies propelled by powerful flicks of a scimitar tail driven by huge amounts of muscle [15]. The fastest have been estimated to reach speeds of 110 km/h (70 mph). Bony fish occupy a range of aquatic niches from benthic herbivory to pelagic predation, while their abundance and cosmopolitan distribution makes them integral to the diet of larger animals, including humans who consume huge, unsustainable quantities every year.

Today, the lobe-finned fish are a rather small craniate offshoot, represented by six known species of lungfish and two species of coelacanth. The ability of lungfish to breathe air enables them to survive in ephemeral pools or bodies of water that are very low in oxygen. Some species can even survive seasonal drying out of their habitat by forming cells below the surface of the mud. The coelacanths (see illustration on p. 22) are large, slow-moving predators of deep water, but their greatest claim to fame is that they are very closely related to the aquatic animal that gave rise to all the terrestrial craniates, which are collectively known as tetrapods.

5 Two groups of mammals are capable of powered flight: the megachiroptera or fruit bats (pictured; *Leptonycteris yerbabuenae*) and the microchiroptera, mostly smaller, insectivorous species. These two groups are not closely related, but it is the bones of the tetrapod forelimb, some of which are hugely elongated, that form the structural framework for the wing in both.

The fleshy fins of the extinct relatives of the coelacanth formed the basis for adaptation into limbs. The coelacanth of today even moves its fins in a sequence like that of a walking amphibian, lizard or mammal. The development of limbs brought to a close the arthropod monopoly of the land, since the ancestral tetrapod had a means of supporting its body whilst out of the water [29]. Not only were these animals equipped with nascent legs of muscle, bone and sinew, but their tough skins prevented too much water from being lost in this new, very arid frontier. With only the arthropods to compete with, there was nothing stopping the tetrapods and they quickly (in geological terms) diverged into an amazing array of forms.

It is among the tetrapods that the basic craniate body plan has been tweaked the most [1, 2, 5–7, 28, 31]. The simple limbs of the ancestral form have been moulded into legs for running, arms for climbing and wings for flying, not to mention flippers in animals that have gone full circle and returned to a semi- or fully aquatic existence (such as ichthyosaurs, penguins and whales). There even are those tetrapods (the limbless squamates and caecilians) that have forsaken their limbs for a worm-like form that comes into its own for a life spent at ground level or even below ground [23–25, 27]. The snakes are an amazing product of these evolutionary experiments in limblessness, since they have had to come up with other ways of capturing and subduing prey; most notably highly modified saliva – venom – that plays havoc with the cells and tissues of their unfortunate victims.

The living tetrapods range in size from a tiny frog, 7 mm (0.28 in.) long, to the gargantuan blue whale, which at 30 m (100 ft) and around 200 tonnes is probably the largest animal ever to have lived. This huge range in size and appearance is a consequence of adapting to different ecological niches, perhaps at its most apparent in the mammals [BOX 2]. Key to the adaptability of the mammals is their ability to maintain a regular body temperature regardless of the ambient conditions, a trait they share with the birds. Crucially, metabolic heat, insulation and evaporative cooling allow mammals and birds to survive in the hottest places on earth as well as the coldest, areas that are strictly off limits to many other animals.

The finer details of reproduction in the craniates are extremely varied, but there are a few general rules. In the vast majority of species there are separate sexes, with the female producing eggs and the male producing large numbers of comparatively tiny sperm. In some aquatic and all truly terrestrial forms, internal fertilization makes the task of the sperm a whole lot easier. In those species that use this method,

6

BOX 2 *Mammal adaptability*

Two creatures that together beautifully illustrate the amazing plasticity of the mammal form and the power of evolution are the sperm whale and the camel. Both evolved from a common ancestor, a small, scurrying creature that dodged the dinosaurs and survived the cataclysm that brought about the latter's demise. In no time at all, geologically speaking, this ancestral form gave rise to one creature perfectly at home in the ocean, and, at the other extreme, an animal able to thrive in habitats where there is very little water.

Both are extreme in every sense of the word and perfectly tuned to their respective habitats. The sperm whale hunts at the bottom of the ocean at depths of as much as 3000 m (10,000 ft). These long, deep dives are made possible by muscles that soak up oxygen like a sponge, thick blubber and a collapsible ribcage. Once at the bottom, they 'see' and subdue their prey using pulses of sound focused through a huge reservoir of wax in their head.

The camel, a study in desert survival, is incredibly frugal in its use of water to the extent that (on good forage at any rate) it can go for six months without drinking. It can also subsist on meagre vegetation and has an air cooling system in its nose that prevents its brain being baked by the desert heat.

5

an embryo can be encased in an egg and laid or retained inside the body of the female. If it is retained, it is either nourished by secretions from the female, such as yolk, or by a direct link to the mother's circulatory system, via some form of placenta. Parental care is also at its most elaborate in the craniates [30].

Along with the insects and pterosaurs, only the birds and mammals have successfully managed to colonize the air. Birds are the only group of craniates where flight is the rule rather than the exception. Descended from feathered dinosaurs, birds [2] are among the most aerially accomplished of all the animals – some species of swifts do not set foot on solid ground from the time they fledge until the time they mature and raise their own young. The swift is to the air what the tuna is to the water: a near perfect exercise in aerodynamic efficiency. Avian mastery of the air is made possible thanks to a very light but strong skeleton, huge muscles around the ribcage, incredibly efficient gas exchange and acute senses, particularly sight.

Bats are the only mammals capable of powered flight. In seeking vacant niches they have become creatures of the night using sound rather than light to 'see'. Exploiting insects and plants when other animals like birds are fast asleep, they have become extremely successful and widespread, especially in the humid tropics where the fruit bats (megachiroptera) play a crucial role in pollinating plants and dispersing their seeds [5].

Collectively, the craniates are of considerable ecological significance. Many, especially the mammals, are large animals, since a larger body can be more efficient than a smaller one in many circumstances, particularly when it comes to maintaining an optimal body temperature. Large animals are typically at or near the top of a food chain, so they regulate the populations of the organisms they feed on and in doing so can modify ecosystems. Spectacular examples of this are easy to see in terrestrial habitats where large herbivores can completely alter the vegetation structure.

But we must be careful not to equate size with ecological significance. The canopy of a tropical forest, for example, may support large numbers of monkeys, sloths and birds, all of which eat leaves and fruit, but the collective population of these animals (their biomass) and therefore their impact on the vegetation of the forest canopy is a lot less than that of the arthropods sharing the same habitat.

ORIGINS AND AFFINITIES

The enigmatic creature from the Burgess Shale known as *Pikaia* (see Cephalochordata, pp. 68–71) is thought by many to represent a

7

6 The flexibility of the craniate body plan has enabled some tetrapods to go full circle and return to the ocean with some style. The sperm whale, with a number of incredible adaptations, is capable of diving to great depths in order to hunt (*Physeter catodon*).

7 The dromedary camel has adapted to extremely arid environments, almost as far as it is possible to get from the watery beginnings of life (*Camelus dromedarius*).

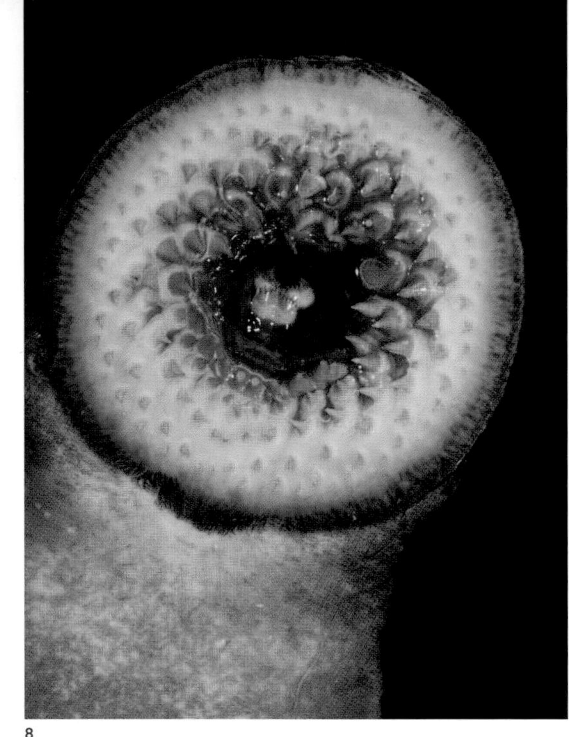

8

9

10

11

8 Lampreys are jawless craniates. As adults, some are parasites of other aquatic animals, often bony fish. To cling on to their host they have an elaborate sucker equipped with numerous, recurved teeth. The tongue is also tipped with teeth, used to rasp the flesh of the host (*Petromyzon marinus*).
9 The head of a hagfish, showing the mouth and sensory appendages (*Eptatretus stoutii*).
10 Hagfish can be very abundant animals. Here, a huge number have been trawled up from the seabed. The pale spots are gill slits (*Eptatretus stoutii*).
11 A lamprey; note the seven gill slits behind the eye (*Petromyzon* sp.).

Opposite
12 Mimicry is a common defensive/aggressive strategy in the craniates, particularly among the bony fish. The sargassum fish (*Histrio histrio*) is excellently camouflaged amid sargassum seaweed.
13 Bony fish are by far the most speciose group of craniates (*Brotulotaenia* sp.).
14 A blenny (*Ecsenius ops*) surrounded by the tentacles of anemones.

very early point in the evolution of the craniates, but their very early diversification will forever be shrouded in mystery by the scarcity of fossils and problems with interpreting those that do come to light. Unless and until dramatic new finds come to light, our best prospects for elucidating early craniate evolution lie in DNA, but even that has its limitations. Some of the later evolution and diversification in the craniates does have a better fossil record, with ancient jawed fish more than 400 million years old and ancient amphibians from 360 million years ago heralding the colonization of the land by the craniates, and remnants of the first mammals at around 200 million years ago.

The morphological plasticity of the craniates especially confused early zoologists; purely on outward appearances, one might venture that a mole-rat, a swift and a porpoise belong on separate branches of the tree of life. But on looking more closely at the embryos and the genes you find these all belong on the same branch, other offshoots of which are lineages that superficially appear to be about as far from the craniate body plan as it is possible to get (see Cephalochordata and Tunicata, pp. 68–83).

12

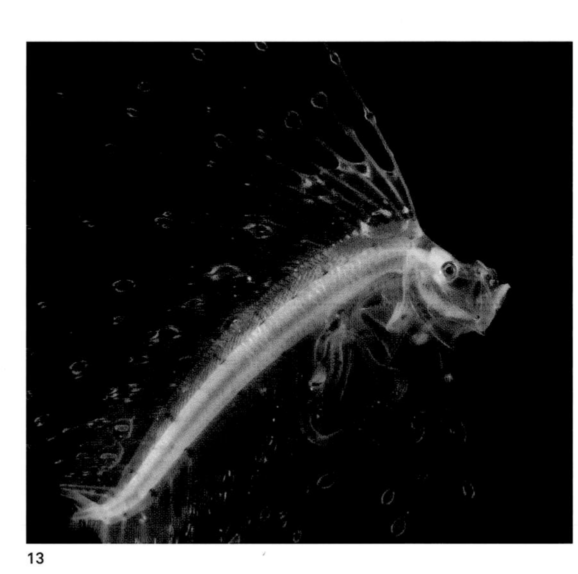

13

14

Overleaf

15 Some of the bony fish are near the pinnacle of hydrodynamic efficiency (juvenile *Scomber* sp.).

16 Sharks and their relatives are beautifully adapted carnivores and typically apex predators in the habitats in which they are found. Their skeleton is cartilaginous rather than bony (*Carcharhinus amblyrhynchos*).

17 Rays are very close relatives of sharks that have adapted to a benthic existence, evolving flattened bodies and hugely enlarged pectoral fins. However, some species, such as this eagle ray (*Aetobatus narinari*), have used these adaptations for a life in open water.

16

17

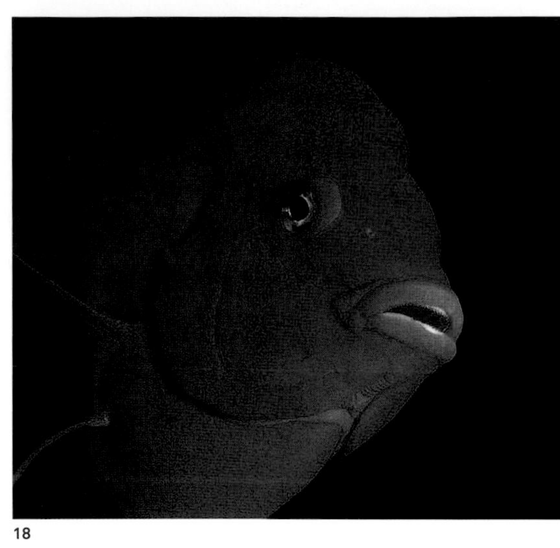
18

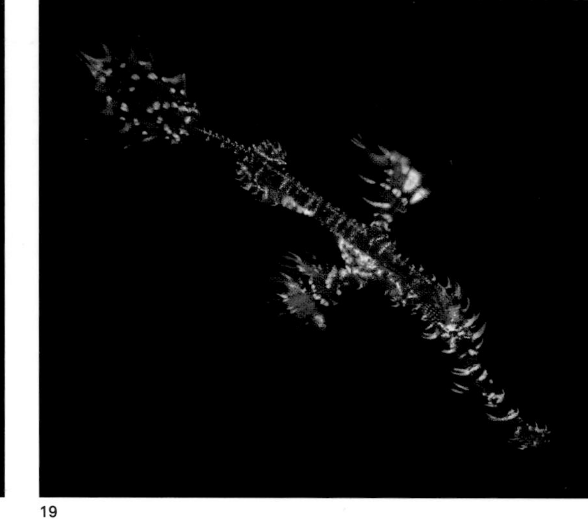
19

20

21

22

18 A bright orange garibaldi (*Hypsypops rubicundus*).

19 An ornate ghost pipefish (*Solenostomus paradoxus*). This species floats almost motionless in the water, head downward and against a background that it almost seamlessly blends into. Females brood their eggs in a pouch derived from the pelvic fins.

20 Some of the bony fish, such as this moray eel, are ambush predators of the seabed (*Gymnothorax mordus*).

21 A frogfish (*Antennarius striatus*). In some benthic, bony fish the foremost fin ray has evolved into an elaborate lure that is used to attract prey.

22 With incredible camouflage this pygmy seahorse blends almost seamlessly with the fan corals on which it lives (*Hippocampus bargibanti*).

23 Snakes have secondarily lost their tetrapod limbs, although in some species tiny vestiges remain. Many, such as this *Bothrops* sp., subdue their prey, begin the process of digestion and defend themselves with venom – highly modified saliva.

24 Caecilians are elusive, limbless amphibians of the soil and fresh water. Little is known about their diversity and biology, but in this species the newly hatched young feed on the top layer of their mother's skin for two months (*Boulengerula taitana*).

25 Amphisbaenids are limbless lizards supremely adapted to a fossorial way of life. Their eyes are mere vestiges covered by a transparent scale, their skulls are heavily reinforced for ramming through the soil, and the skin is only loosely attached to the body, allowing them to move through their tunnels backwards just as easily as forwards (*Amphisbaena alba*).

23

24

26 Many types of fish have evolved a worm-like form as an adaptation to burrowing in sediment and negotiating small spaces (*Moringua edwardsi*).

27 Even without limbs, snakes have adapted to arboreal habitats. This is a so-called vine snake (*Oxybelis brevirostris*).

25

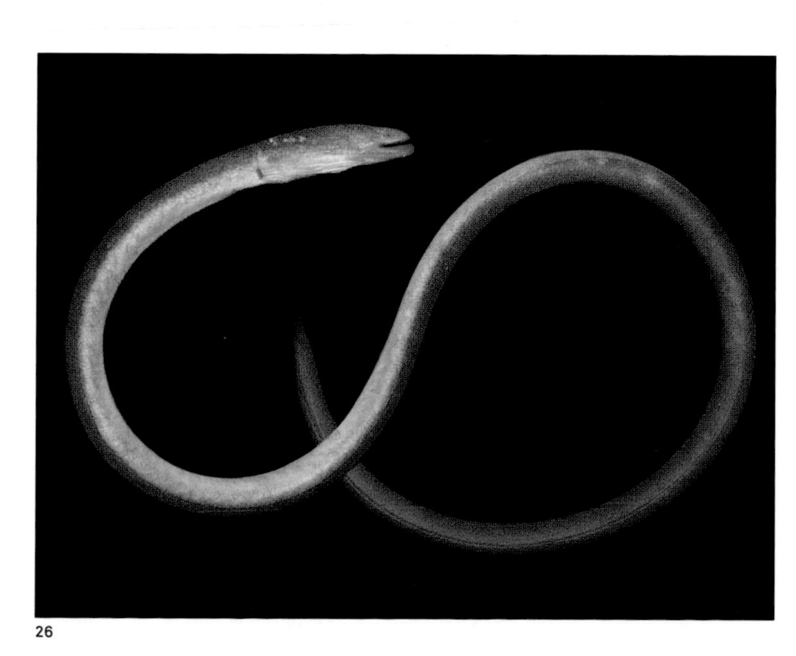

26

27

29

28 Lizards are very successful craniates, particularly in tropical habitats (*Gonocephalus borneensis*).

29 Some of the amphibians are superficially similar to the early tetrapods that gave rise to all the terrestrial craniates living today (*Ambystoma mexicanum*).

30 Some craniates go to great lengths to give their offspring the very best chance in life. Males of certain poison arrow frogs wait for the eggs to hatch and then carry the tadpoles into the upper reaches of rainforest trees where small pools of water accumulate in epiphytic plants, such as bromeliads. These pools are where the tadpoles will develop, relatively free from competition (*Ranitomeya* sp.).

31 Some arboreal frogs of southeast Asian forests are capable of gliding considerable distances using webs of skin between their long digits (*Rhacophorus reinwardtii*).

30

31

Hemichordata

(acorn worms (enteropneusts),
sea angels (pterobranchs))
(Greek *hemi* = half;
Latin *chorda* = cord)

Diversity
c. 120 species

Size range
~1 mm to ~2.5 m
(~0.04 in. to 8 ft)

1 In this acorn worm the proboscis is extended and finger-like projections along the body are clear to see. These are hepatic caeca – extensions of the body-wall and gut that run along the back of the animal. These may increase the surface area of the gut available for the absorption of nutrients (*Ptychodera flava*).

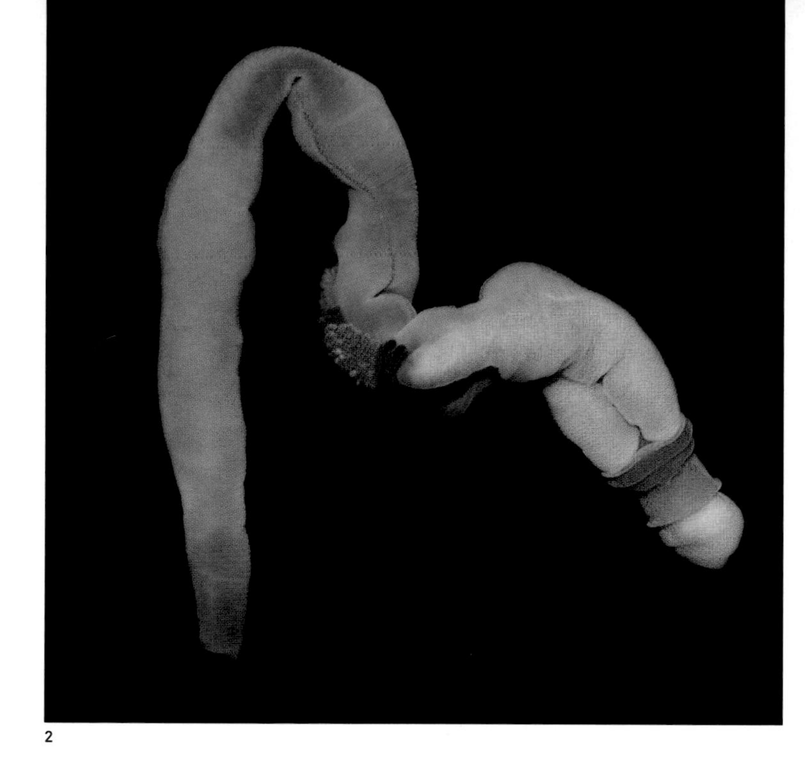

2

3

Rarely seen and poorly known, the acorn worms and sea angels are enigmatic animals. The former are burrowing creatures of the sediment while the latter are small colonial animals that live in tubes. For a long time their odd appearance had zoologists scratching their heads trying to work out how they were related to other animals.

FORM AND FUNCTION

The acorn worms (enteropneusts) are so called because of their proboscis and collar, which are more than a little reminiscent of an acorn nestled in its snug cup. The longest species may grow to as much as 2.5 m (8 ft), but all are very fragile and the sinuous body often breaks when the animals are handled. At the front end of the body are numerous gill slits. Hemichordates have a well-developed gut, muscles and circulatory system, but that is about it: there are no obvious sense organs and nothing that could be called a brain [1–3, 7].

The sea angels (pterobranchs) are decidedly odd little animals. Their internal workings are very similar to those of the acorn worms, but on the outside they look quite different. At their front end they have two or more arms bristling with tentacles and a disc-shaped oral shield. Like many other small marine animals, the sea angels are not a single being, but rather they exist as an intimately linked community of clones known as zooids, each of which is rarely more than 7 mm (0.3 in.) long. At the rear end of a zooid there is a stalk that gets progressively thinner, eventually forming thin tubes of tissue linking the zooids together and enabling the colony to function as a single organism. Small and soft, the sea angels have evolved a defence in the shape of a branching network of collagen tubes secreted by their oral disc – effectively an exoskeleton [4–6].

LIFESTYLE

In general, acorn worms are burrowing animals, although one species can drift above the seabed, possibly buoyed by mucus. They drive their muscular proboscis through the sediment with waves of muscular contractions along their body. Many species project their proboscis out of their burrow and simply flop it out on the seabed where edible bits and pieces get trapped by mucus and are conveyed by cilia to the mouth. Others build a burrow with a small feeding funnel at one end. As they ingest sediment from the bottom of their burrow, more mud and sand tumbles into the feeding funnel and down towards the hungry acorn worm. This is an interesting example of the exploitation of a pit with walls at a critical angle (certain terrestrial insects also take advantage of a similar set up to trap their prey).

Sea angels creep to the end of their protective tubes on their oral shield, extend their arms into the water and wait for organic matter and tiny organisms to get trapped in the sticky mucus coating their tentacles, a slow process hastened a little by the occasional flick of the arms. Wafting cilia transport these particles to the mouth.

2 Note the collar and acorn-like proboscis at the anterior end of this acorn worm (*Saccoglossus* sp.).

3 Acorn worms are rarely seen burrowing animals. Their bodies are very fragile and break easily when handled (unidentified species).

4 A sea angel (pterobranch) colony. Each of the individual zooids has feeding arms bearing tentacles for collecting edible particles suspended in the water.

5 Illustration showing part of a sea angel colony (*Cephalodiscus* sp.). A common adhesive disc anchors the zooids, via a stalk, to the interior of the tube.

6 Illustration showing a sea angel zooid. To feed, the zooid creeps to the aperture of its collagen tube on an adhesive structure known as the oral shield.

7 The hepatic caeca are clearly visible on this acorn worm (unidentified species).

8 The early life of an acorn worm is spent as a hat-shaped, planktonic stage, known as a tornaria larva (*Ptychodera flava*).

Acorn worms and sea angels that have obtained adequate food to mature turn their attention to reproduction. In the former there are typically separate sexes, although a recently described deep-sea species (*Allapasus aurantiacus*) is hermaphroditic and is the only known animal with external ovaries. The adult acorn worms release their gametes into the water and in most cases fertilization yields a strange planktonic larva called a tornaria [8]. This short-lived immature stage uses cilia to propel itself through the water, albeit weakly, before settling on the seabed and undergoing a radical transformation that gives rise to a juvenile acorn worm.

The sea angels can reproduce asexually as well as sexually. New zooids in a sea angel colony can 'bud' from the thin stalks that link all the individuals or the adhesive disc that anchors the entire colony to the substrate. In some cases this adhesive disc can even fragment or divide and each part will go onto form a new colony. In sexual sea angels there is an interesting division of labour as female zooids in some species have rudimentary tentacles of little use for feeding. These females concentrate on producing eggs, but the males have to produce sperm as well as collecting enough food for the entire colony. In other species this situation is reversed and the males get an easy ride. In most cases a fertilized sea angel egg develops into a short-lived planktonic larva that settles after a day or two and secretes a cocoon, firmly attaching itself to the substrate. The larva metamorphoses into a zooid, which breaks out of the upper part of its cocoon to found a new colony.

Acorn worms, like their close relatives the echinoderms, have exceptional powers of regeneration, which is a good thing considering just how fragile they are. They can regenerate lost parts over a period of weeks and this is how some species deal with predation. For example, the auger snail (*Terebra dislocata*) skulks around the burrow of an acorn worm waiting for the occupant to reveal its rear, which it does when it needs to defecate. Taking its cue the snail lunges and bites a chunk from the rump of the acorn worm. The snail, satiated, retreats and the acorn worm regenerates its ragged rear.

In addition to their powers of regeneration the acorn worms are also able to secrete nasty, bromine-containing compounds from the gland cells of their proboscis. It is thought these chemicals deter predators and inhibit the growth of microorganisms.

The hemichordates may be rarely seen and poorly known, but they are important components of marine ecosystems, which is particularly true of the acorn worms. Some acorn worms are relatively large animals and they can exist in high densities. Collectively, their burrowing behaviour and consumption of deposited and suspended matter plays an important role in churning sediments and maintaining the flow of energy and nutrients through the marine ecosystem. The sea angels are perhaps of more limited ecological

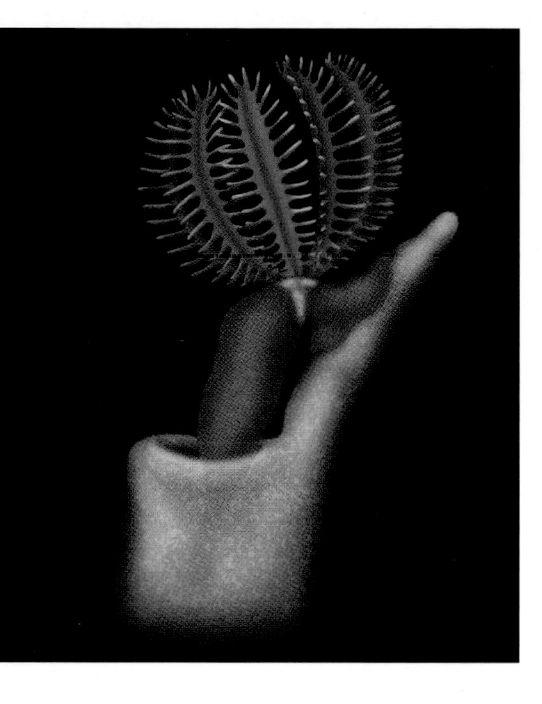

6

7

8

significance, but there is still a great deal to learn about these animals and their abundance in often poorly studied marine habitats.

ORIGINS AND AFFINITIES

Exceptional Cambrian fossils some 525 million years old with well-preserved soft parts appear to be long-dead acorn worms and sea angels. The remains of other creatures interpreted as ancient members of this lineage are also known from deposits of around the same age, but generally speaking fossils of such soft creatures are rare. However, it is thought that the extinct graptolites, known from very common fossils anywhere between 320 million and 510 million years old, also belong to this lineage.

For well over a century zoologists have been aware of the affinities between the hemichordates, the echinoderms (see pp. 102–15) and the craniates (see pp. 84–97). The early stages in the development of these animals are very similar and even as adults they retain features that demonstrate a close evolutionary relationship. More recently, analyses of DNA sequences have shown these animals are indeed closely related, but questions still remain and it will some time before we have a detailed understanding of the evolutionary history of the hemichordates and their relatives.

Echinodermata

(sea stars, sea urchins, sand dollars, sea biscuits, brittle stars, basket stars, sea cucumbers, sea lilies, feather stars, sea daisies) (Greek *echinos* = spiny; *derma* = skin)

Diversity
c. 7500 species

Size range
~10 mm to ~2 m
(~0.4 in. to ~6 ft)

1 The spherical test and impressive spines of a sea urchin, *Coelopleurus floridanus*. The mobile spines offer protection from predators. Since this species lives in relatively deep water, the purpose of the bright pigments in the skin and underlying skeleton is unknown.

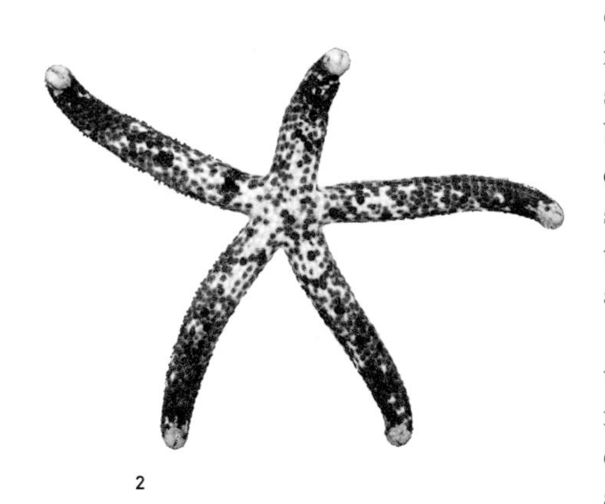

2

Outlandish in appearance, but with a symmetry that delights the eye, the echinoderms have in equal measure perplexed and enchanted people for centuries. Aside from the many variations on star-shaped forms, some of the sedentary species bear a superficial resemblance to plants, while other small burrowing echinoderms look like something that belongs in a purse. But despite their oddness, they are animals nonetheless and actually quite closely related to humans.

Echinoderms are exclusively marine, and they often can be found in huge numbers even in the unforgiving environment of the abyssal depths. The sedentary nature of many species and the durability of their remains mean that we often encounter them in shallow seas and on the strandlines of beaches. However, even though we see them frequently, they are not to be taken for granted; much of their biology is as unusual as their appearance.

FORM AND FUNCTION

One of the most striking aspects of a typical echinoderm is its unusual symmetry – the inspiration for a good number of common names, since many are flower- or star-like [2, 8–12]. The technical term for this layout is pentaradial symmetry, meaning that there are five similar parts arranged around a central axis in such a way that each rotation of one fifth of a full turn leaves the animal looking much as it did at the start. Depending on the echinoderm in question, such symmetry is not always obvious, particularly in the case of the sea cucumbers. This pentaradial layout is also reflected in much of the internal anatomy. There is no front or a back in the conventional sense, just an oral surface bearing the mouth and an aboral surface through which the anus issues. This weird layout consigns most species to spending their entire adult life mouth-down on the seabed, although the sea lilies and feather stars (crinoids) are turned over with the mouth facing up [13–15]. With no eyes, no head, nor even a rudimentary brain, it is unlikely they will be lamenting what seems to us a rather unfortunate existence.

The echinoderms all possess a tough endoskeleton reinforced with ossicles, often ornate structures composed of calcium carbonate [19, 24]. In the sea stars these ossicles articulate with one another, which lends a degree of flexibility to the body, but in the urchins they are fused to form a rigid skeleton known as a test [1, 20, 21, 25, 26]. It is these urchin tests that can be found washed up on the strandline, after all the soft bits have decomposed. The ossicles are embedded in so-called mutable connective tissue (such tissue occurs in other lineages including the arthropods, but this is a special kind unique to the echinoderms). The rigidity of echinoderm mutable connective tissue can be varied at will, sometimes in less than a second; this ability is at its most extreme in the sea cucumbers, whose bodies can take on a fluid-like flexibility.

Ossicles have another function beyond simply providing a skeletal framework, namely keeping the animal's body free of other organisms [22]. The upper surface of a sea star represents a nice patch of real estate for the juveniles of sessile animals that are looking for somewhere to settle. Fortunately, hugely modified ossicles that look a bit like miniature forceps are on hand to grab any of these trespassers and cast them back out into the water.

Inside the body there is a well-developed coelom (body cavity) that has been partitioned into smaller zones, one of which is the water vascular system, unknown elsewhere among the animals. This complex internal plumbing is filled with liquid very similar to seawater, but it contains some cells, proteins and high levels of potassium. This fluid is shunted around by muscular pressure and into offshoots of the vessels known as tube feet [17, 18], which protrude out through the underside of the animal and are used in locomotion, feeding and gas exchange. The undersides of sea stars, brittle stars and urchins bristle with these hydraulically controlled tube feet. They are a cunning, multifunctional adaptation, but they are a bit leaky so the water in the system is continually topped up via a calcareous sieve that opens to the outside through a distinctive pore. A direct opening to the outside seems like a sure-fire way to pick up harmful microorganisms, but specialized parts of the water vascular system act like an immune system to remove and eliminate any microscopic interlopers.

In addition to the unique water vascular system, echinoderms also have a network of blood vessels. This is at its most complex in the sea cucumbers, where it is thought to function in nutrient transport.

LIFESTYLE

As adults, most echinoderms spend their time creeping around on the seabed using their tube feet, spines and arms. It is hard to see them as anything more than benthic decorations, but

2 The sea stars represent the most familiar echinoderm body form (*Linckia multifora*). Calcium carbonate ossicles beneath the skin of sea stars and their relatives form a tough, flexible endoskeleton.

3

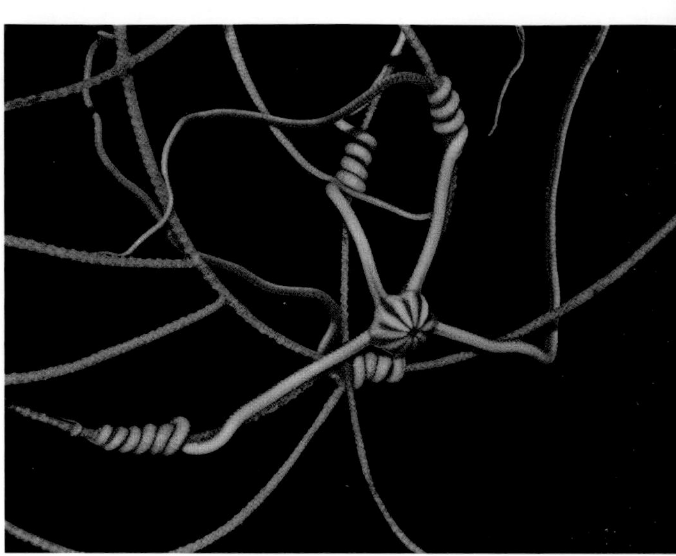

4

they are actually very effective scavengers and carnivores, more than capable of dealing with a range of marine fare. Some species, sea stars in particular, prey on sessile marine animals such as bivalve molluscs by employing a unique strategy. Arching over the victim and locking its body into place with its mutable connective tissue, the sea star grips onto the bivalve's shell with the suckers of its tube feet and begins to heave [3]. Eventually, the shell of the prey yields a little – even a gap of 0.1 mm (0.004 in.) is enough – and the echinoderm slips its stomach inside and secretes digestive enzymes all over the soft parts of the bivalve. After some time these enzymes have done their work and the prey has been digested in its own shell, ready to be slurped up by the sea star. Some other species have a similar approach, but they evert their stomach over corals or even patches of seabed, consuming coral polyps in their rocky refuges, plus whatever edible matter lies on the surface.

Brittle stars are equally benign-looking animals, but they too are efficient predators and scavengers. Some species simply use their tube feet to mop up edible detritus from the sea floor, while others extend their long arms into the water to trap suspended morsels of food [6]. The carnivorous brittle stars cannot evert their stomach out of their mouth at will like the sea stars, but they do have heavy-duty jaws. They also have long, mobiles arms [4] that some species actually use to capture live food, including small prey and even squid. Sea urchins have jaws too, but theirs are part of a singular structure known as Aristotle's lantern – an assemblage of ossicles and muscles that can be poked out of the mouth to graze algae on the seabed [23]. Most sea cucumbers have opted for an extremely economical existence by lying on or in the seabed, using modified tube feet to mop up or shovel sediment and whatever edible matter it contains into their mouth [28, 29, 31]. Some of the deep-sea species have needed to be a little more proactive in this harsh environment, where food sources are very patchy; these trundle around on massively elongated tube feet, sniffing the water for the tell-tale signs of a feast, sometimes a dead marine mammal that has finally come to rest on the seabed. The plumes of decay draw these walking oddities from far and wide to feast on the corpse. In a strange quirk of evolution some of the sea cucumbers, quintessential bottom dwellers, have taken to a pelagic existence, swimming near the seabed [30]; some have fins, sails and even bell-like structures, akin to those seen in the jellyfish.

The elegant form of the sea lilies and feather stars has evolved to collect suspended particles in the water. Their outstretched arms bear long tube feet studded with protuberances that secrete sticky mucus, which traps edible morsels. Periodically the tube feet then convey the catch to the waiting mouth.

Perhaps the most remarkable feature of the echinoderms is their ability to regenerate. If a sea star is torn asunder until nothing is left but a bit of the central disc and a single arm, the tattered, sorry-looking fragment can heal and completely regrow the missing parts [16]. The process is slow, and complete regeneration may take a year, but it is still an extremely impressive feat. These regenerative powers also allow sea stars to reproduce asexually by fission

3 By stiffening its mutable connective tissue an echinoderm's body becomes a rigid scaffold against which the tube feet can act to pry open the shells of prey, such as this bivalve mollusc. Even a tiny gap allows the sea star to digest the unfortunate victim within its own shell.

4 Some brittle stars have extremely long, very mobile arms that can be used, like here, for anchorage, as well as collecting food (unidentified species clinging to coral).

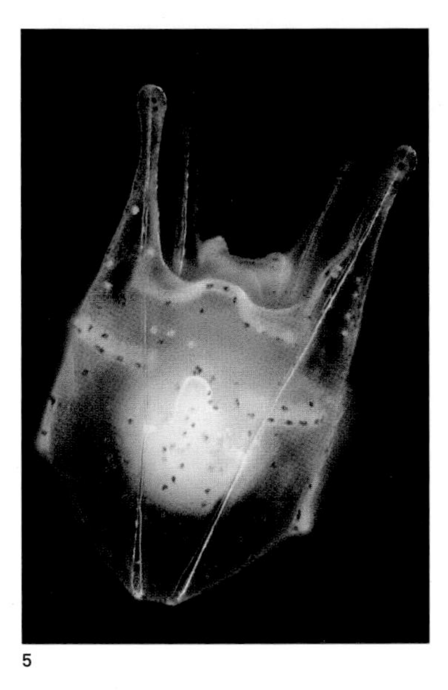

5

5 Larva of the purple sea urchin (*Paracentrotus lividus*).

Overleaf

6 Some brittle stars (commonly known as basket stars) have branching, plant-like arms. These are the real monsters of the echinoderm world with an arm spread that can reach more than 1 m (3 ft), especially in the cold water species, such as this *Gorgonocephalus arcticus*.

7 Echinoderm larvae are among the most outlandish planktonic organisms (unidentified sea urchin larva).

and budding. Typically this involves the animal tearing itself in two, each half regenerating the missing sections, but some species push the boundaries even further by simply casting off an arm, which then regenerates into a complete new individual.

The sea cucumbers have evolved some novel defences based on their ability to regenerate. If suitably irritated, some can direct their anus at the threat and disembogue a bundle of very sticky, toxin laden structures, known as Cuvierian tubules, which extend by as much as 20 times their original length and quickly ensnare the target [27]. For smaller predators, the embrace of these tubules can be fatal, and for larger animals it is annoying enough to make them search elsewhere for an easy meal. The Cuvierian tubules grow from the sea cucumbers' gills – known as respiratory trees – and they regenerate once discharged. Evisceration is another sea cucumber trait that draws heavily on regeneration: if attacked or stressed, some species can burst their body at will, giving their attacker a face full of unpleasant-tasting guts. The predator turns tail, while the sea cucumber regenerates its lost parts.

They may be lacking in brains, but echinoderms have enormous gonads capable of producing prodigious numbers of gametes. Most species are composed of separate males and females, which tend to spawn en masse during a short breeding season each year, giving the sperm a fighting chance of bumping into an egg. Normally, the fertilized eggs float around in the sea and develop into very peculiar pelagic larvae, of which some kinds are reminiscent of Sputnik 1 [5, 7]. The larvae eventually settle on the seabed, at which point the body goes through a huge overhaul with everything being shifted around into the adult form, which first becomes evident as a juvenile rudiment on the side of the larva. When metamorphosis is complete, the juvenile trundles off on its podia to begin life as a benthic animal.

Echinoderms are diverse and abundant animals in all oceans, and they play a crucial role in the ecology of these ecosystems. During their early life, they are integral components of the zooplankton, feeding on other organisms and getting consumed by creatures higher up the food chain. Many are important as scavengers, patrolling the seabed for dead and decaying matter, ultimately keeping the nutrients and energy moving through the tangle of the marine food web. Lots of species live on or in the sediment of the seabed, and by moving on and through it they help to stir it up and bring nutrients and organisms closer to the surface, thus providing food for other animals. Their role as predators is easily overlooked, since most echinoderms look about as threatening as a cushion, but collectively they consume huge quantities of animal life, and thereby profoundly influence marine population dynamics. The grazing species scour vast areas of seabed each year, rasping at algal mats that form wherever there is sufficient sunlight. In many places the rampant growth of these algae would choke the growth of corals and other sessile organisms, so grazing echinoderms, such as sea urchins, fulfil an important role in the maintenance of these marine habitats.

ORIGINS AND AFFINITIES

The echinoderms have a long and illustrious past, one borne out by a very rich and decidedly old fossil record. The diversity of these animals as we see them today, although still impressive, is but a shadow of their former glory. The oldest fossils interpreted as belonging to this lineage date from the Cambrian period, some 550 million years ago. The petrified remains of these animals bear features that putatively demonstrate the link between the echinoderms and the craniates and tunicates, most notably gill slits and bilateral symmetry. Today, the idea that these lineages are on a distinct branch of the animal family tree (the deuterostomes – see Introduction, p. 13) is beyond refute. There is clear evidence of this relationship in the way the eggs divide following fertilization, not to mention the similarity of their DNA.

But if the echinoderms are so closely related to the craniates and tunicates, why do they look so radically different? Based on their adult form, these creatures appear to be about as different from us as it is possible to be, but once again morphology deceives. Long ago, it seems that a lineage of mobile animals with bilateral symmetry and gill slits became more and more adept at suspension feeding. Millions of years and countless generations later, they had left their old ways behind and adopted radial symmetry, since, for suspension feeders, this body plan offers clear advantages. Most importantly, a suspension-feeding animal is surrounded by food, and since it no longer has to look for its meals, there is no need for a head. These animals were the forerunners of all the echinoderms alive today.

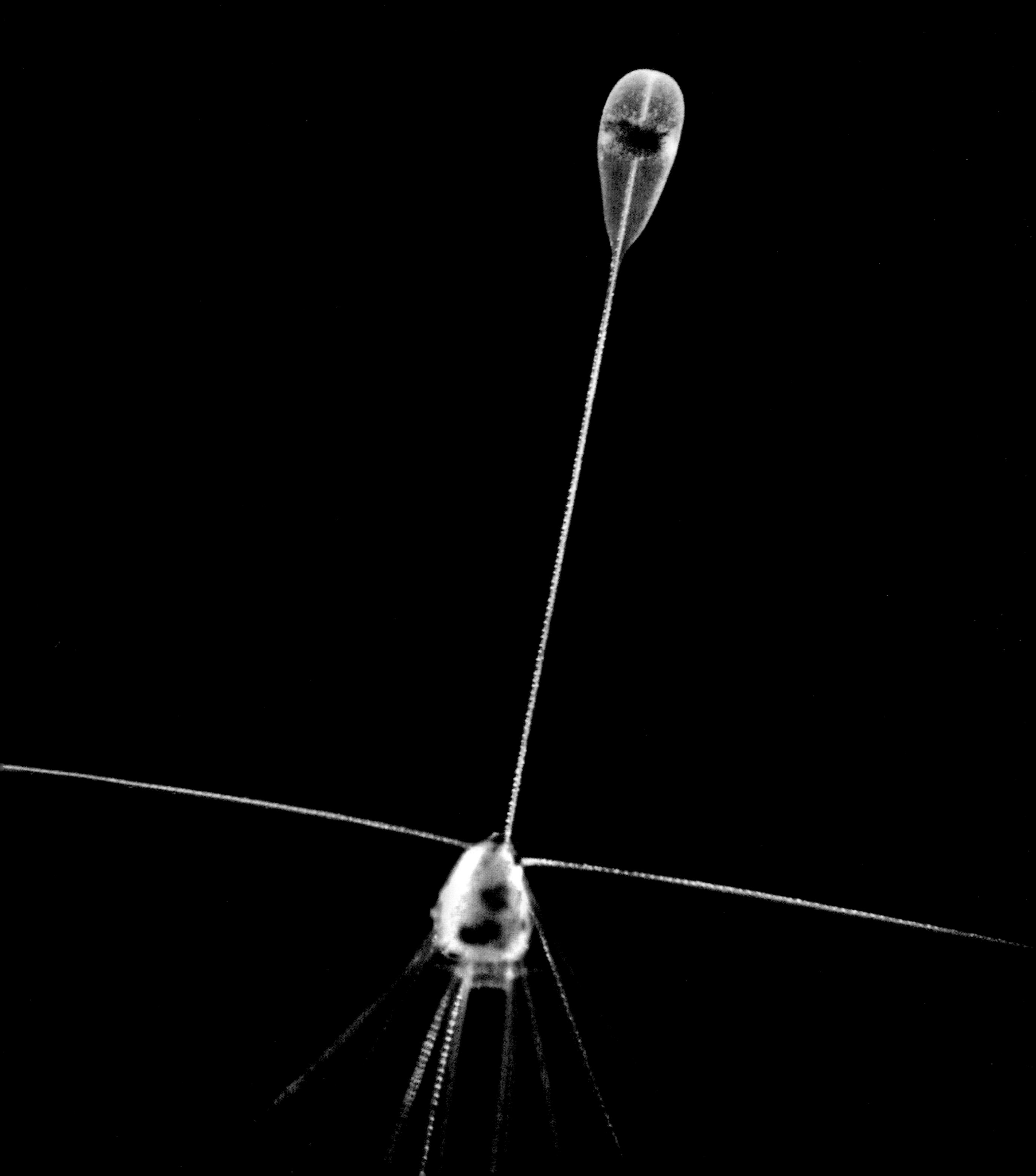

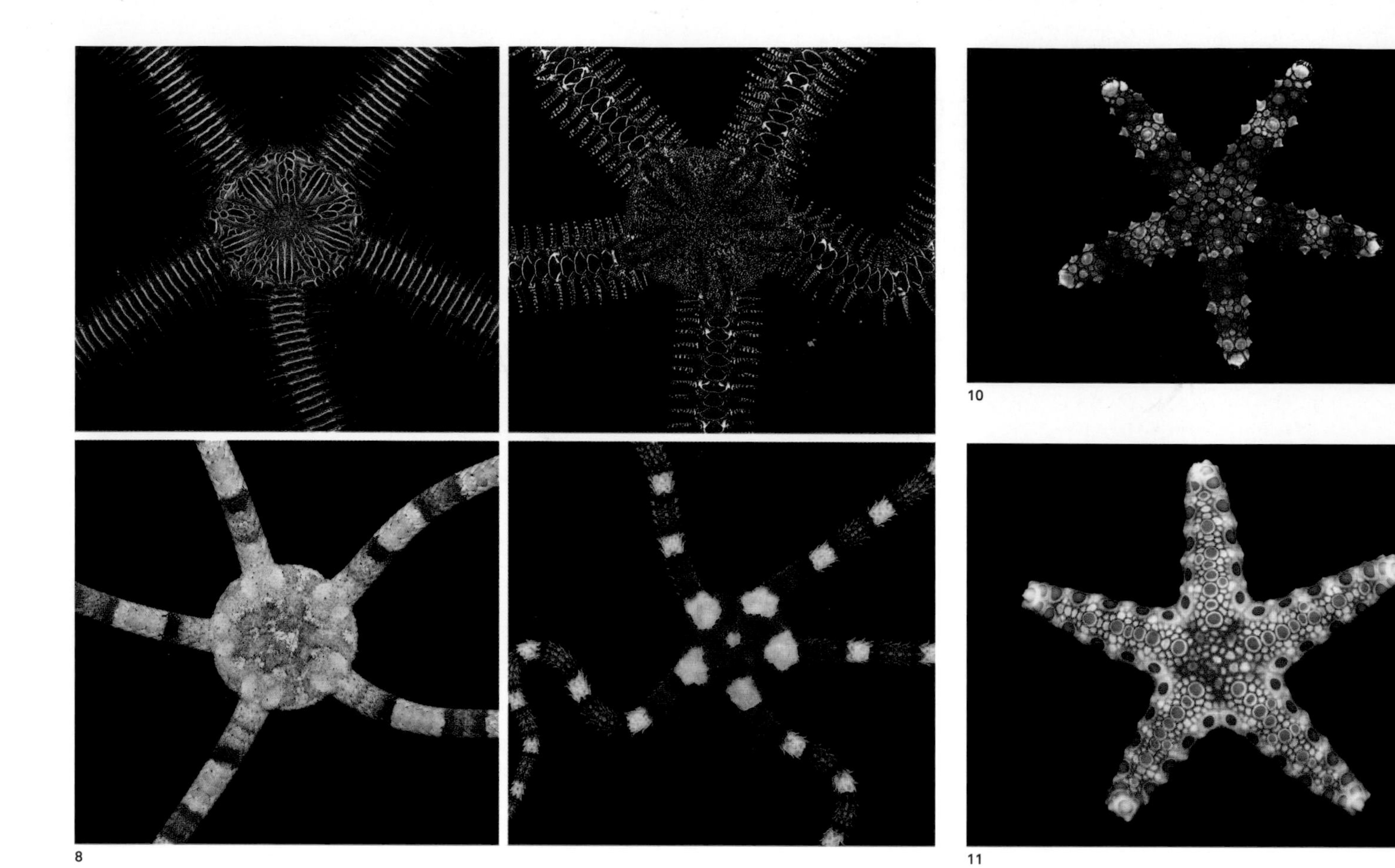

10

8

11

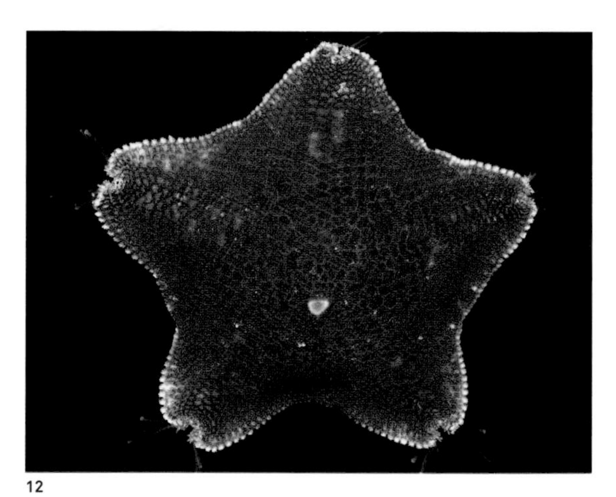

12

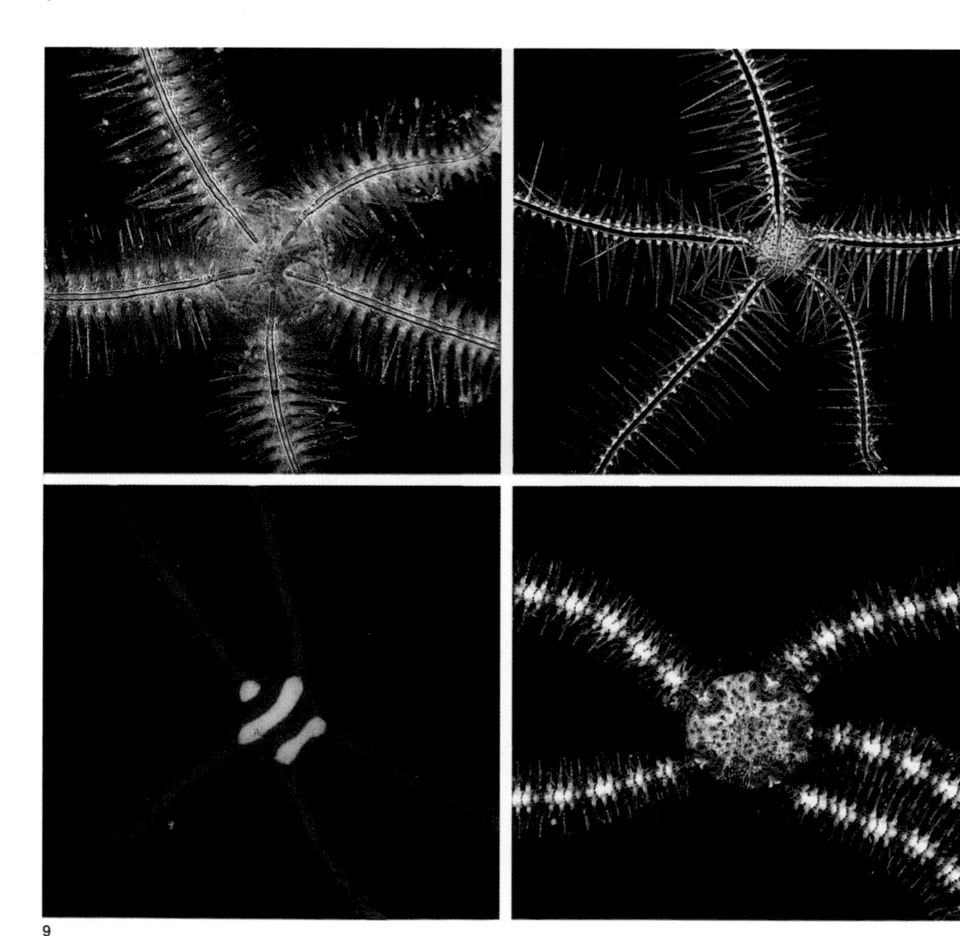

9

8, 9 Brittle stars are the most diverse group of echinoderms, but they have cryptic lifestyles and relatively few species are found in the shallow water habitats humans typically visit (various unidentified species).

Sea stars are instantly recognizable animals:

10 *Gomophia egyptiaca.*
11 *Neoferdina cumingi.*
12 *Patiriella regularis.* Note the long tube feet of this species.

13

14

13 In the crinoids, the oral surface of the animal faces up into the water. (unidentified species). Some of these echinoderms can even use their arms to swim short distances. Pentaradial symmetry is always obvious in the central disc of echinoderms, but sometimes the number of arms can be misleading (especially in many-armed species that are missing a few).

14 Feather stars are mobile animals that cling to the seabed using specialized structures known as cirri (unidentified species).

15 The arms, adapted for filter feeding, and the anchoring cirri of a feather star, *Heterometra savignyi*.

16 A sea star in the process of regenerating most of its body, probably after an attack by a predator (*Linckia columbiae*).

15

16

17

17 The tube feet of echinoderms are extensions of their water vascular system (*Asterias rubens*).

18 Multifunctional tube feet are able to attach to surfaces using suction, slowly move the animal over the seabed, pull apart the closed shells of prey to get at the soft tissues inside, and pass morsels of food to the mouth (*Asterias rubens*).

19

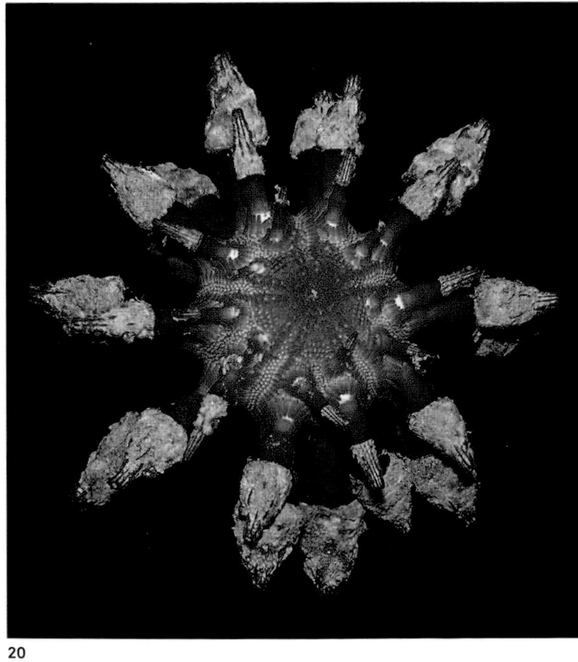

20

19 The calcareous skeleton of a sea star allows for the formation of defensive structures such as spines. In this species (*Acanthaster planci*), the fragile spines break off easily, remaining in the skin of an attacker and releasing irritants.

20 The ossicles of sea urchins form a more-or-less spherical, rigid test bearing a huge variety of spines and processes, such as those of the almost hallucinogenic *Chondrocidaris brevispina*.

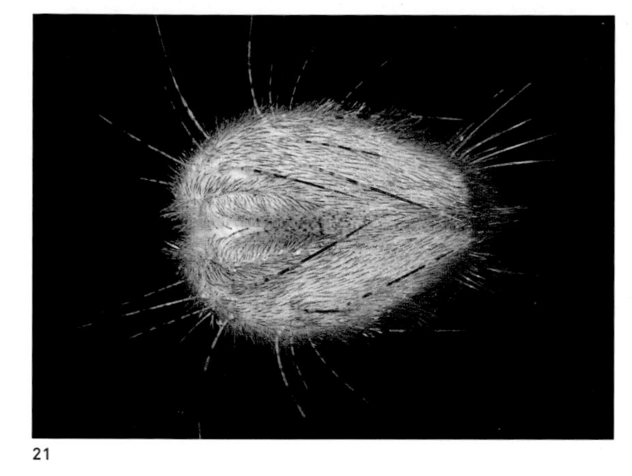

21

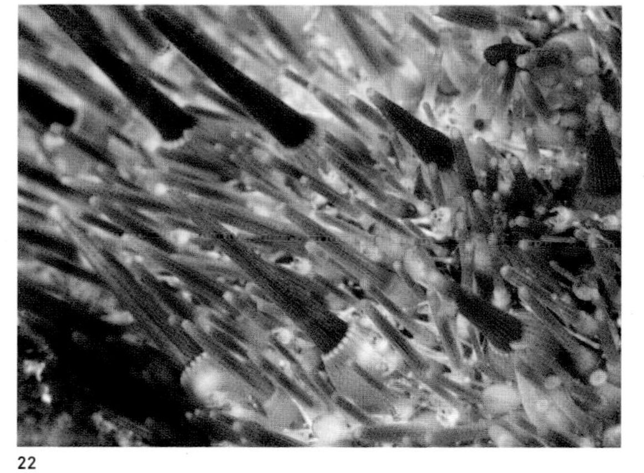

22

23

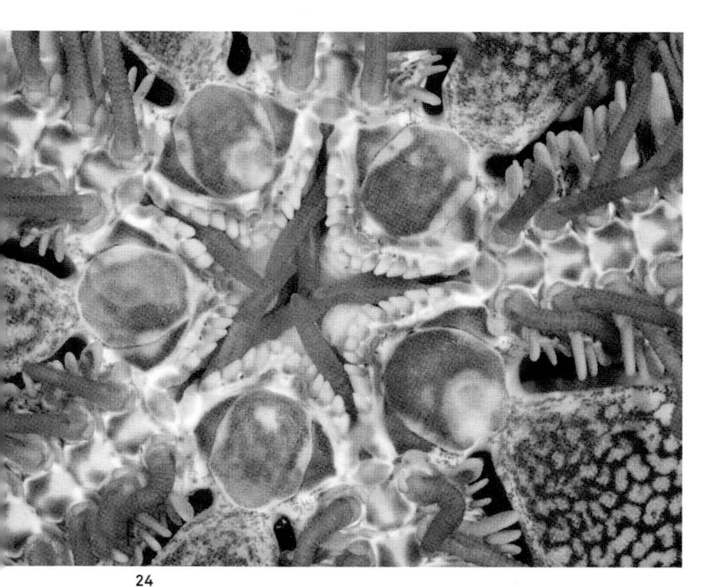

24

21 A sea urchin, *Lovenia* sp.

22 Spines work together to keep an urchin's body free of detritus and settling organisms. A number of spines on this urchin (*Strongylocentrotus droebachiensis*) are moving around several small fragments of debris. Also visible are specialized, forceps-like spines known as pedicellariae.

23 The oral surface of a sea urchin, showing the unique jaw apparatus known as Aristotle's lantern. The five 'teeth' (modified ossicles) of this structure are clearly visible. Note also the numerous tube feet (short stalks terminating in a circular pad) (*Strongylocentrotus droebachiensis*).

24 The underside of a brittle star, showing the ossicles that form five protective plates on the central disc (unidentified species).

25 Some of the urchins have evolved into flattened, burrowing forms with imaginative common names such as sand dollars and sea biscuits (*Clypeaster cf. humilis*).

26 Sand dollars (*Dendraster excentricus*). Only the remains of these animals, bare tests devoid of spines, are normally encountered, washed up on the strandline.

25

26

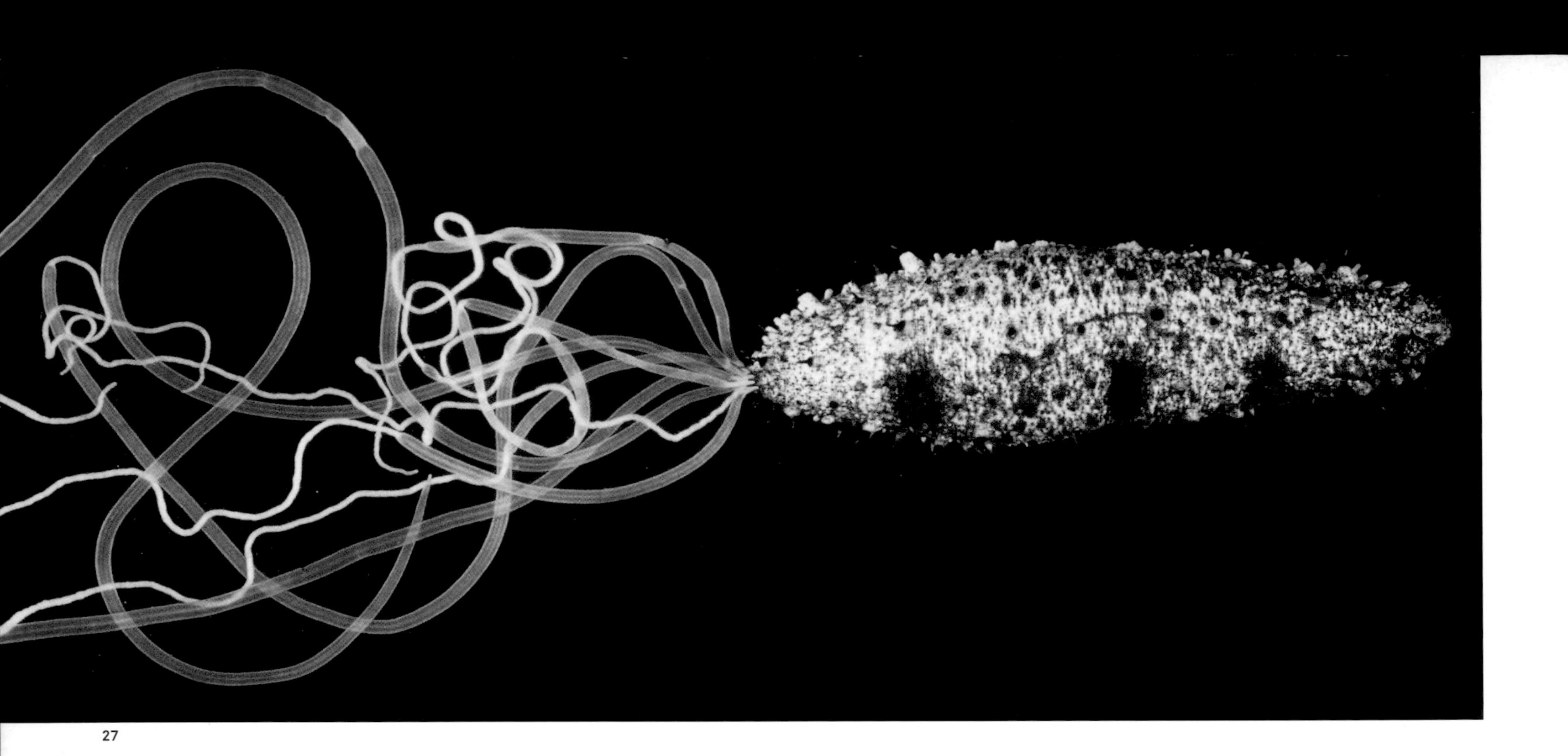

27

27 A sea cucumber releasing its Cuvierian tubules to defend itself (*Holothuria pervicax*).

28 The modified, branching oral tube feet of the sea cucumber, *Eupentacta fraudatrix*. Most sea cucumbers spend the majority of their life buried in the sediment or nestled in crevices. The rest of this sea cucumber's body is buried out of sight.

29 Sea cucumbers use their tube feet to collect edible particles from the seabed. In this image, a sea cucumber (*Apostichopus japonicus*) is joined by some colourful sea stars (*Asterina pectinifera*).

30 Sea cucumbers are typically benthic animals, but some species swim. This transparent, swimming sea cucumber was photographed at a depth of 500 m (1650 ft) (unidentified species).

31 A sea cucumber (*Chiridota* sp.). The modified, branching tube feet growing from the oral pole of the body are clearly visible here.

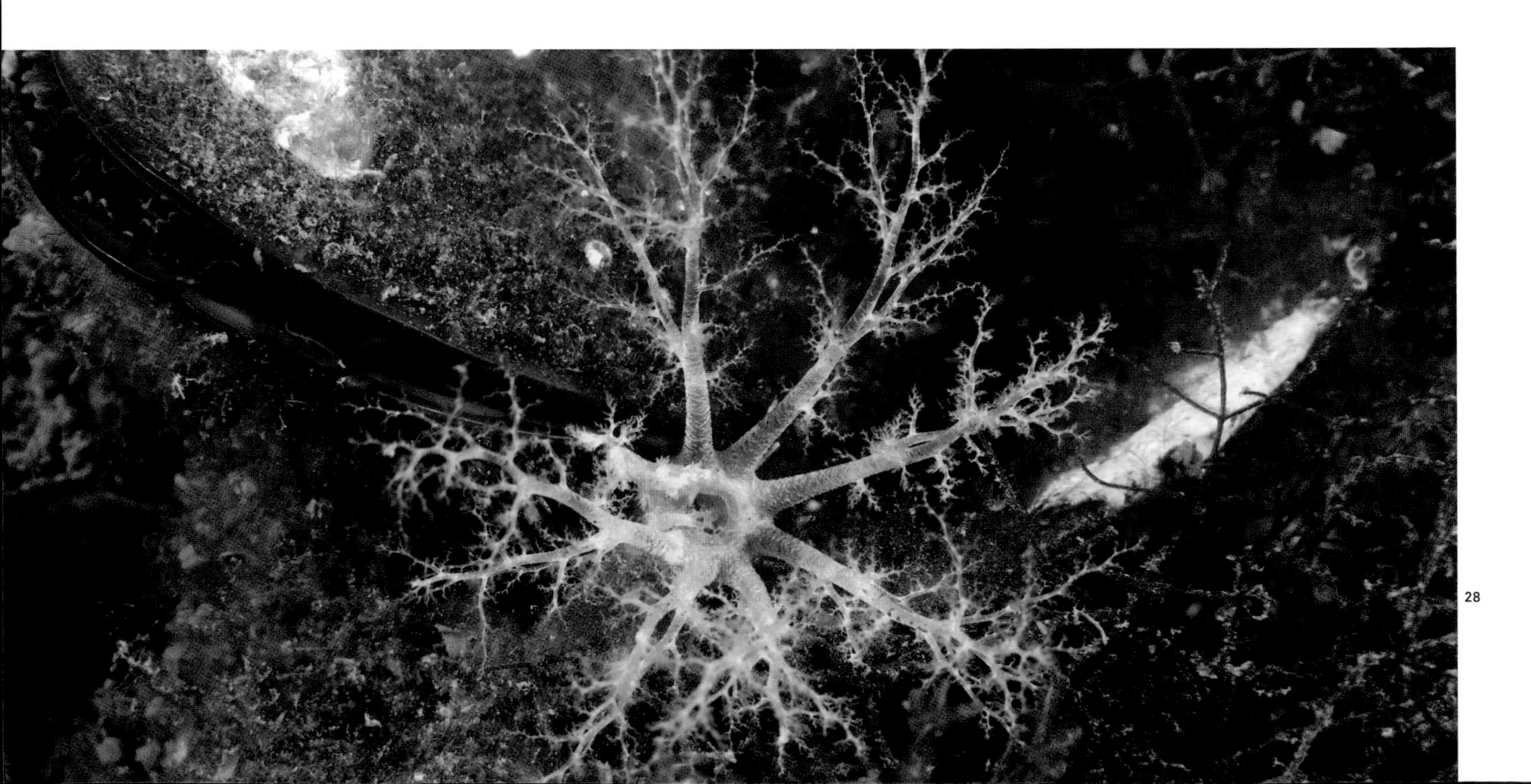

28

29

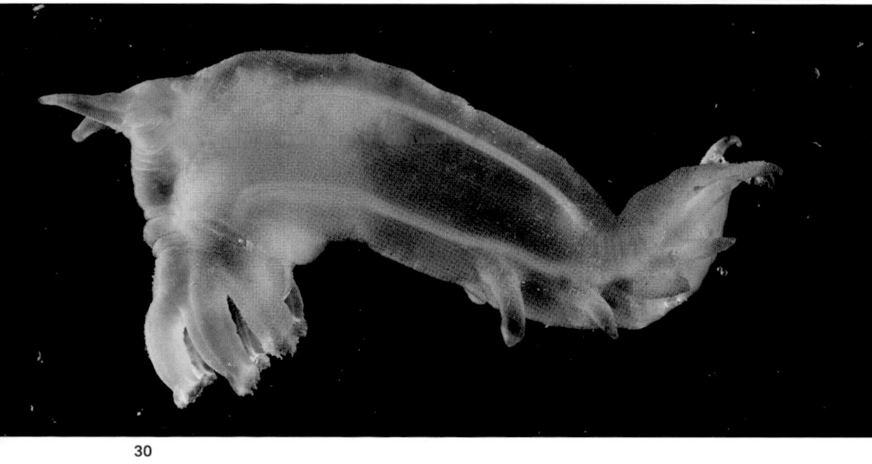

30

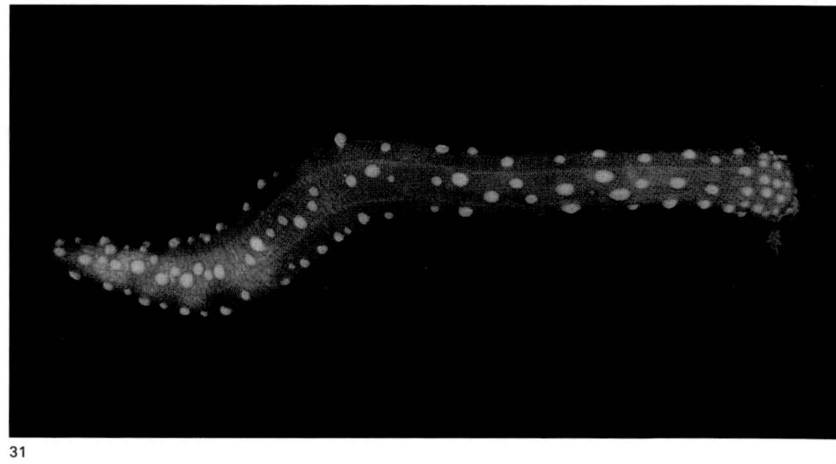

31

Xenoturbellida

(strange worms)
(Greek *xenos* = stranger;
turbellida, referring
to the flatworms from
Latin *turbella* = bustle,
turmoil)

Diversity
1 species known

Size range
3 to 4 cm
(1.2 to 1.6 in.)

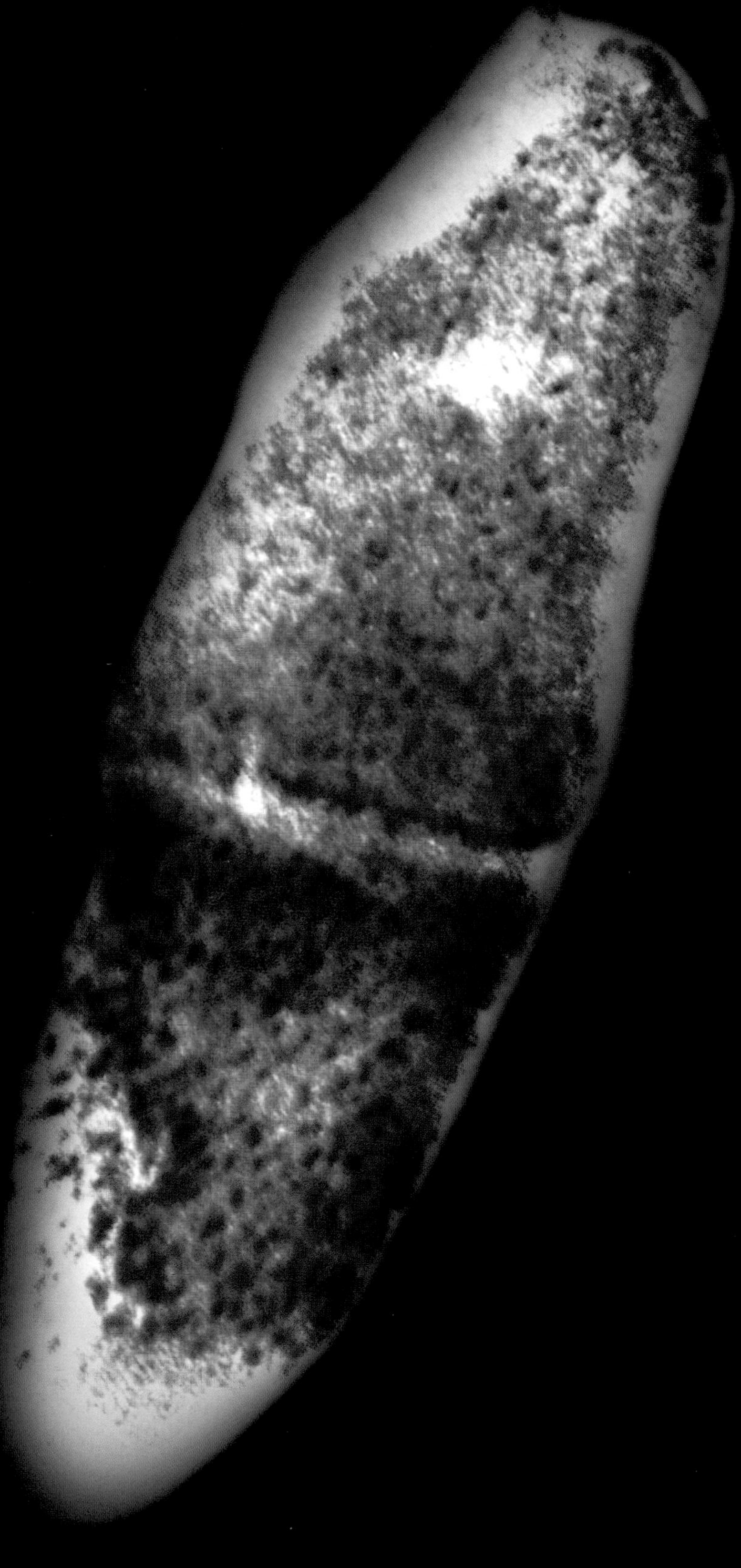

1 *Xenoturbella bocki*, currently the only agreed upon strange worm species. Note the midline furrow, the position of which corresponds to a thickening in the underlying nerve net.

These bizarre beasts have baffled scientists since they first were discovered in 1915 off the coast of Sweden. Now, after almost a hundred years of head scratching and argument, and a great deal of shifting around, we have a rough understanding of where the strange worms fit in the animal family tree. However, we still know next to nothing about how they live their lives in the murky ocean depths.

FORM AND FUNCTION

Some biologists have likened the strange worms to little more than ciliated bags, and it is true they offer morphologists very little to go on. They have a mouth, a sac-like gut and a network of intersecting nerves beneath or within their skin. Thickenings in this nerve net trace barely perceptible furrows running down their sides and around their midline that are thought to be sensory in function. At the front end there is a statocyst [3], which is distinct from the comparable organs of ctenophores and acoelomorphs (see pp. 26–35 and 118–21), but any other discernible structures are few and far between. There is little sign of any specialized organs, no nerve cords, ganglia or bundles of nerve cells, or anything that could be described as a brain. In fact these animals do not seem to have any internal nerves at all – their nerve fibres seem to be wholly within the epidermis. They do not even have any distinct reproductive organs to speak of.

All the same, xenoturbellids are reasonably active; they glide in and over the soft mud of the seabed using cilia, and in doing so they use their muscles to twist and steer [2].

LIFESTYLE

Strange worms have been collected from waters 40–100 m (130–330 ft) deep off the coasts of Sweden, Norway and Scotland. Some biologists who have studied them suggest that they simply absorb nutrients dissolved in the water, and no doubt they do to some extent; however, it is a good bet that organisms with sac-like guts will not rely solely on nutrient absorption as their main source of food. Also, the remnants of molluscs and worms have been found in those capacious guts – in fact some analyzed specimens contained so much genetic material from mollusc eggs or tissues that they confounded the first attempts to classify the xenoturbellids (see below).

Symbiotic bacteria are embedded in the tissue lining the gut. It is not certain what these bacteria are doing, but since a xenoturbellid has no obvious specialized excretory organs – nothing like kidneys, for example – some scientists suggest that they help their host process metabolic waste. However, considering the diverse and complex relationships that exist between endosymbionts and other animals, it is unlikely that this is the whole story. It is possible that the strange worm symbionts also assist in digestion and the production of essential nutrients.

What we know of xenoturbellid reproduction tells us that it is extremely unusual. These animals are hermaphrodites and produce both sperm and eggs, but they have no specialized organs for doing so. Sperm occur individually and in clumps throughout their bodies; so do eggs. How and where the young develop is not currently known, but it is possible the juvenile stages parasitize other marine animals.

ORIGINS AND AFFINITIES

What are the strange worms and where do they fit on the animal evolutionary tree? This question has divided biologists for the past century. Initially there did not seem to be much difficulty: the strange worms looked much like any other free-living flatworm, and flatworms are a weird bunch anyway, so their scientific name seemed to be adequate acknowledgement of their stranger features. More recently, and in light of DNA investigations, they were labelled as extremely simple molluscs, but it turned out that samples had been contaminated by the remains of mollusc prey within the strange worms' stomachs. Now, after more testing, we seem to be nearer the answer.

The strange worms, along with the acoelomorphs, appear to be closely related to the hemichordates and echinoderms (see pp. 98–107). The xenoturbellids bear little superficial resemblance to those lineages, but this is a good reminder that the outward appearance, ecology and anatomy of an animal means little when analysing such relationships. But DNA, although difficult to decipher, does not lie (as long as one has read enough of it to sift out any confounding factors).

The strange worms have lost all the features that characterize their close relatives, but until we find out more about how they live in the wild we can only speculate why this should be so.

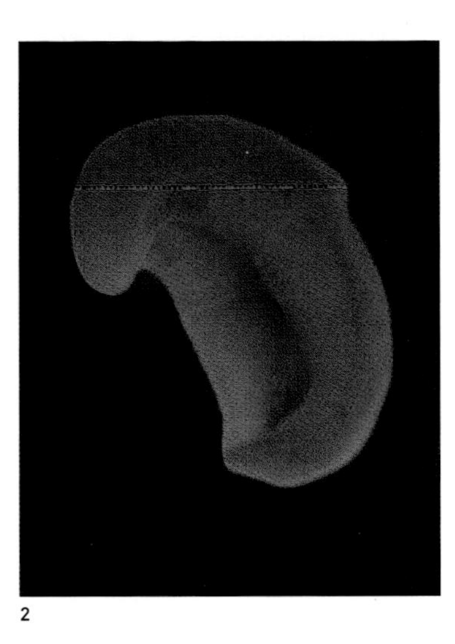

2

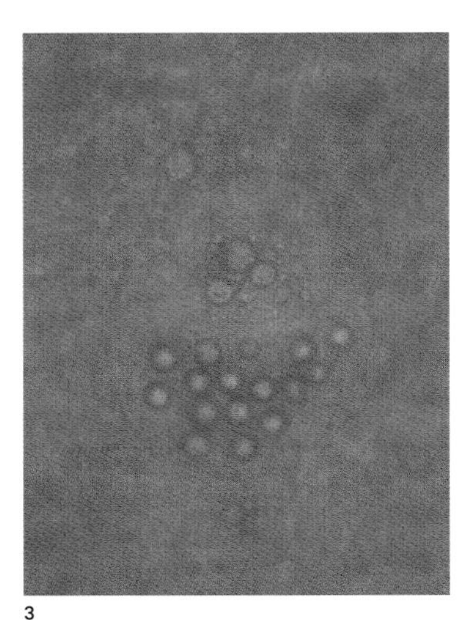

3

2 In addition to the beating cilia that strange worms use to move across the seabed, their well-developed musculature helps them twist and steer.

3 A xenoturbellid statocyst. This organ is a sensory receptor that detects gravity and is therefore crucial in orientation. Tiny spheres of dense material move around in the fluid-filled sac of the statocyst stimulating different parts of its lining, giving the animal a sense of orientation in relation to gravity.

Acoelomorpha

(acoels and nemertodermatids)
(Greek *A* = no;
coel = body cavity;
morph = shape)

Diversity
c. 400 species

Size range
<0.5 to 15 mm
(<0.02 to 0.6 in.)

1 Acoelomorphs bear a superficial resemblance to flatworms, but any similarities are shared ancestral features or evolutionary convergences (*Convolutriloba longifissura*).

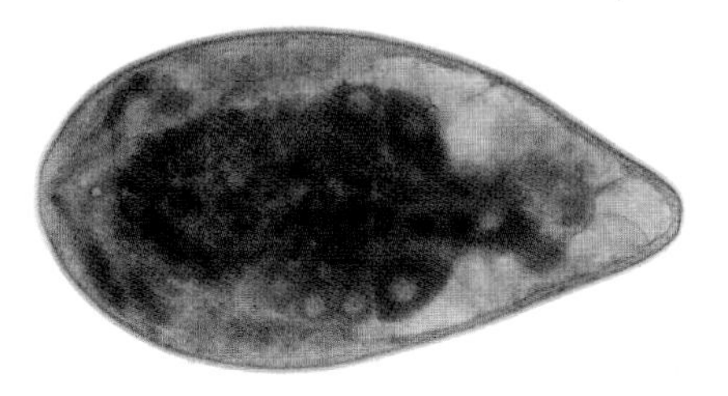

2

Generations of confused biologists have shunted these animals from one part of the 'tree of life' to another. Like the related xenoturbellids (see pp. 116–17), the acoelomoprhs are *really* strange, so much so that it is hard to know what to make of them. Most of them live in the ocean, but some have colonized freshwater habitats. We know next to nothing about how most of the acoelomorphs live.

FORM AND FUNCTION

These small, flattened animals have some very odd features, some of the most peculiar being those that are missing, often things you would hardly think an animal could do without [1, 2, 4, 5].

The body is solid; it does have a mouth, but in the majority of species this opens straight on to a unique structure known as the digestive syncytium [BOX 1]. The acoelomorphs also lack any specialized structures to get rid of waste, so solid bits of undigested food pass out of the mouth, while metabolic waste products simply diffuse out into the surrounding water, a process facilitated by their small size.

At their front end they have an aggregation of nerve cells, the commissural brain, from which a number of nerve fibres branch out to form a diffuse nerve net. Near this 'brain', such as it is, there is a statocyst organ [BOX 2] that is thought to help these animals orient themselves, although some adult acoelomorphs lack a statocyst. The cilia on their surface are their main means of locomotion.

Another oddity of these animals is their lack of discrete ovaries and testes, or any tubes leading to or from them. They do have male and female reproductive tissues, but not contained in their own little pouches. They do have structures for exchanging sperm though (see below).

LIFESTYLE

The acoelomorphs are animals of substrates in aquatic (typically marine) habitats. Using their cilia they glide around looking for morsels to eat and others of the same species to exchange sperm with [6]. A huge variety of tiny organisms, including crustaceans, other worms and algae are eaten by acoelomorphs. Some species feed on algal cells and diatoms when young before progressing on to small animal prey as they reach adulthood. A few species harbour photosynthesizing organisms that supply all the nourishment their host needs in exchange for a relatively safe and mobile place to live [7, 8].

Reproduction in many of these animals is a brutal business. With the ability to produce both eggs and sperm they have evolved some rather grisly means of exchanging sperm with others of their species. Many use a pointy penis to rupture the body wall of a sexual partner to deposit their sperm, while others deposit sperm on the epidermis of their partner that enzymatically digests its way into their body. The male gametes are a lively lot and they make a bee-line for the unfertilized eggs.

2 Acoelomorphs have a very simple stucture, but it seems they have evolved from a more complex ancestor. In adapting to their current way of life, many features were surplus to requirements and were progressively lost (*Isodiametra pulchra*).

BOX 1 ***Syncytia***

When we think of a living organism, we generally think of it as being made up of cells, each of which is formed of similar components. There is good reason for this view: it is largely correct. Cells are the building blocks of living things, and by putting cells together and adjusting their individual settings many different structures with all sorts of different functions are possible.

However, there are complications. Messages and supplies need to be passed from cell to cell, and cells have an amazing submicroscopic apparatus to do this, but there are some things for which direct communication or direct sharing is best. One example in our own bodies is in our voluntary muscles; each of our striated muscle fibres is contained in a single cell membrane, but this contains a number of nuclei. It is in effect a team of cells sharing the same membrane and the same internal apparatus. This gives better and faster control and mechanics than passing every message on from cell to cell.

There are numerous examples in other creatures. As a rule these shared structures are called syncytia (syncytium in the singular), meaning 'cells together'. It can be the product of repeated division of the nucleus (in this case it is called a coenocyte or a plasmodium), or of a number of cells fusing by breaking down the cell membranes separating them, either partly or completely. Sometimes formation of a syncytium is caused by a viral disease, but there are many ways in which syncytia are vitally important in healthy creatures. Among other roles, these super-cells play key roles in insect embryology, mammalian placentae, and in the formation of the glassy skeletons of sponges and the teeth of echinoderms.

BOX 2 ***Statocysts***

Over and above the traditional five, among the senses of animals that we often overlook are balance, orientation and acceleration. A number of animal lineages use a common mechanism for sensing these things, that of the statocyst and statolith (Greek for 'static pouch' and 'static stone').

The statocyst is a hollow lined with nerve receptors (sensory hairs known as setae), filled with mucus or another suitable fluid, and containing a mineralized lump, or statolith, that is free to move within the hollow. If the animal moves suddenly, the statolith touches the nerve receptors lining the inner sides of the hollow enabling the animal to react appropriately to turning, falling and sudden changes of direction or speed.

Humans and most other mammals are also able to sense balance, orientation and acceleration, but through a different mechanism (the vestibular system in the inner ears).

3

3 In this acoelomorph the double statocyst is very obvious. In addition there is a pouch-like digestive tract, the mouth of which closes after a meal to become a tiny pore in the skin (*Sterreria* sp.).

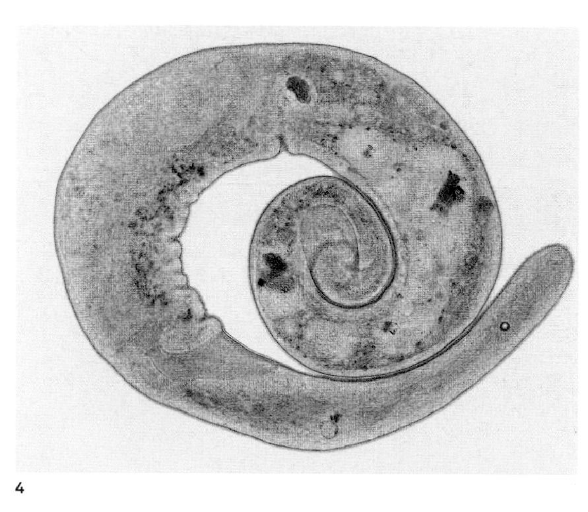

4

4 Most acoelomorphs are are animals of the oceans, where they glide around in and on the seabed. They consume other small animals as well as algal cells and diatoms (*Solenofilomorpha crezeei*).

Normally, two individuals will exchange sperm in one of these two brutal ways.

As if the rigours of copulation were not enough, another quirk of evolution has left the acoelomorphs without channels by which the fertilized eggs may escape into the outside world. Instead, the eggs are released into the outside world via the parent's mouth or a brutal, albeit temporary rupturing of its body wall.

Sexual reproduction is far from being the only option and acoelomorphs display almost the full gamut of asexual reproductive strategies. In some species the parent's body differentiates into a chain of clones that eventually separate from one another to go their separate ways (paratomy). In other species, fragmentation, either spontaneously or as a result of injury, yields a number of bits all of which give rise to complete individuals (architomy), while still more bud off miniature versions of themselves from their lobed posterior (e.g. *Convolutriloba* spp.).

ORIGINS AND AFFINITIES

Acoelomorphs provide an object lesson in how dangerous it is to jump to conclusions in biology. They look and behave like tiny flatworms (see Platyhelminthes pp. 278-293), so when they were originally described in the mid-19th century they were classified as just another, albeit aberrant, form of flatworm. Like their relatives the xenoturbellids, the acoelomorphs have since been pushed from root to branch in the animal tree of life, but we now know that any similarity to some of the flatworms is a result of convergent evolution.

Nowadays the code spelled out by their DNA suggests they are actually a sister group (together with the strange worms) of the echinoderms and hemichordates.

If this is true then the xenoturbellids and acoelomorphs represent lineages of animals that have lost many of the complex features of their ancestors in adapting to specific niches [2]. We must remember that evolution is not simply about ever greater complexity (see Introduction, p. 22), but about adapting to a particular way of life; it is a positive adaptation for a lineage to discard complex structures once they become unnecessary.

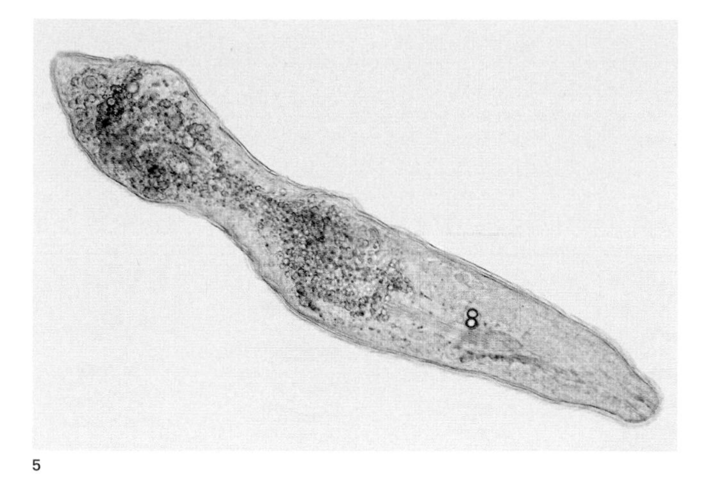

5

6

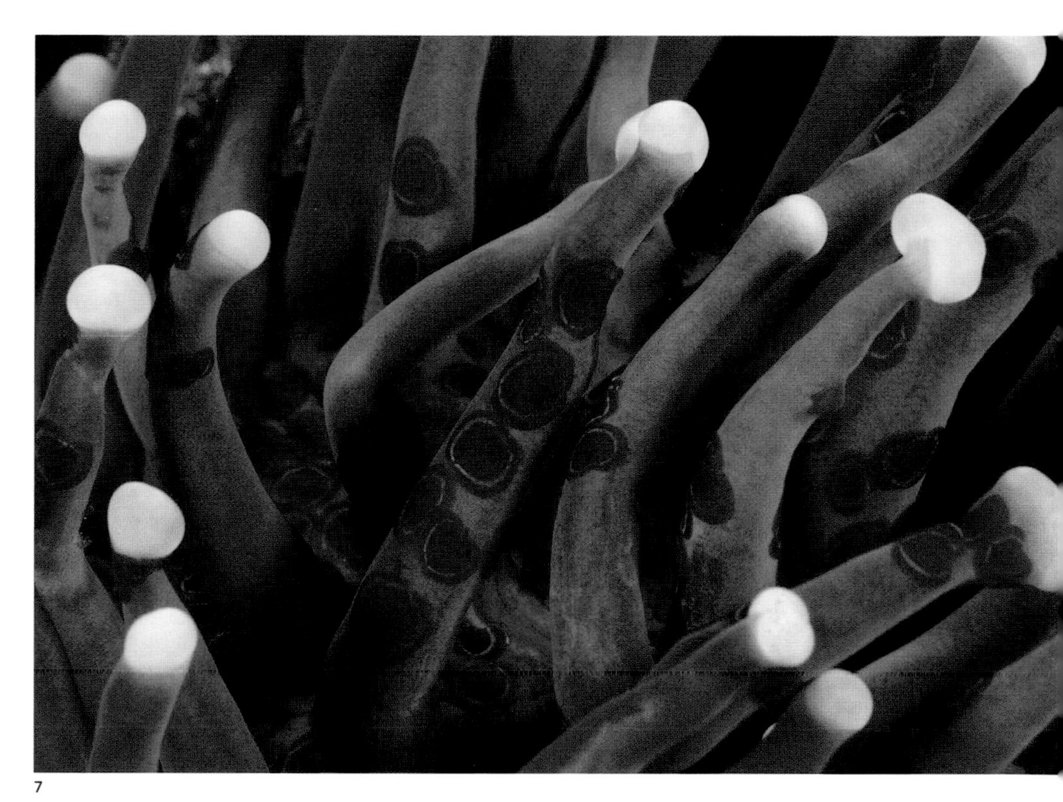

7

8

5 This tiny acoelomorph lives in between sediment grains on the seabed. It moves around using cilia and orientates using the double statocyst (spherical structures) at its head end (*Flagellophora* sp.).

6 An acoelomorph gliding around on the shell of a bivalve mollusc (unidentified acoel).

7 Acoelomorphs are very abundant animals, particularly in marine sediments, although a few species are always found on corals, sometimes in huge numbers. Two types of symbiotic, photosynthesizing dinoflagellates give these *Waminoa* spp. their bronze colour.

8 *Waminoa* spp. on bubble coral.

Chaetognatha

(arrow worms)
(Greek *chaite* =
hair; *gnathos* = jaw)

Diversity
c. 180 species

Size range
~1 to ~12 cm
(0.4 to 4.7 in.)

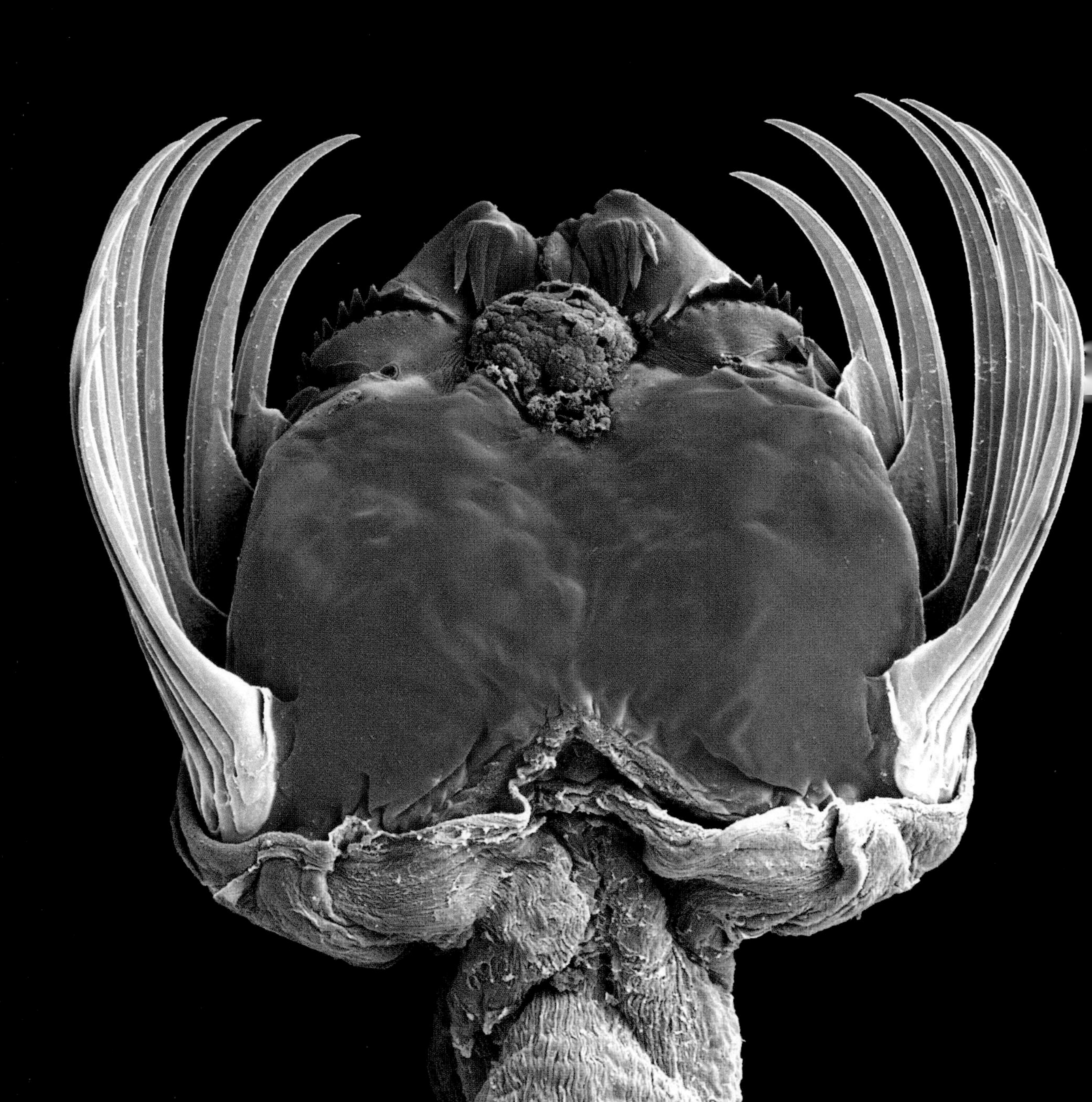

1 This SEM clearly shows the large, chitinous grasping spines of an arrow worm. Two rows of shorter teeth around the mouth can also be seen.

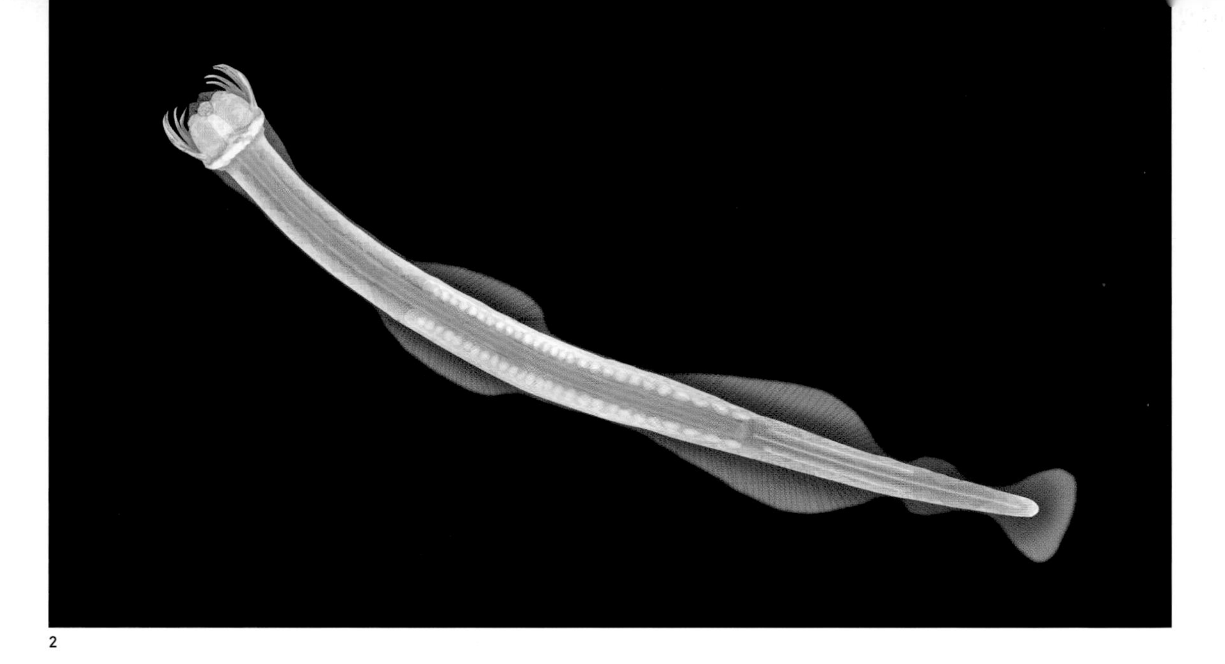

2

Just as the African savannah is a backdrop to an eternal struggle between prey and a host of fearsome ambush predators, so too are the surface waters of the oceans. Among the most captivating predators of the planktonic microcosm are the arrow worms – sleek and transparent beings with a host of unique adaptations for capturing and subduing other animals in this living soup. They are strictly marine and although we rarely see them and few people know of them, they are very abundant. Working out their evolutionary relationships to the other animal lineages has always been something of a headache for zoologists; today, they remain one of the most difficult lineages to place on the tree of life.

FORM AND FUNCTION

What is particularly striking about the appearance of the arrow worms is their astonishing symmetry and transparency [2]. The elongate, glass-clear body with its paired fins looks like something that has been machined, but this body plan is adapted to exploit the bounteous resources of the plankton effectively. At the head end are two rows of chitinous grasping spines, plus one or two rows of short teeth surrounding the mouth [1]. Most of the time the head of the arrow worm, complete with its spines and teeth, is enveloped by a fold of the body wall known as the hood [3].

Arrows worms have a complex arrangement of body cavities and a well-developed gut. They also have a circulatory system, but it seems that its primary role is transporting nutrients around the body rather than gases. They are mobile hunters, and accordingly their nervous system, senses and musculature are sophisticated. The brain, located on the dorsal side of the head, innervates the body via ventral nerves and ganglia. However, the nerves run along the back of the aberrant, deep-sea species, *Bathybelos typhlops*, which is only known from a single specimen. Two eyes, similar to the compound eyes of the arthropods (see pp. 148–77), are located on the upper side of the head; their retinal cells are arranged in such a way that the animal sees through its own transparent tissues. In addition to eyes, arrow worms have numerous short rows of sensory cilia (called ciliary fences) all over the body. These pick up tiny vibrations in the water that betray the presence of prey and enemies. The large muscles drive the bodily motions for swimming, whereas the elegant fins are simply control surfaces.

LIFESTYLE

Almost all arrow worms are ambush predators of plankton. Their prey is often a copepod or a young fish, but they quite commonly dine on other arrow worms too. Using their eyes and sensory cilia they home in on a suitable target before making a darting lunge and unsheathing their grasping spines. The spines and sticky oral secretions grip the prey until the short teeth can pierce it. For particularly tenacious prey the arrow worms have a potent weapon in the form of a toxin derived from bacteria, tetrodotoxin [BOX 1]. This blocks nerve impulses and prevents the prey from struggling. The arrow worm first swallows the subdued prey, then envelopes it in a permeable bag known as the peritrophic membrane before digesting it in the intestine.

2 Illustration showing the typical body form of an arrow worm. Note the paired fins, gut and ovaries (ovoid structures surrounding the gut), and testes (pinkish structures toward the rear).

3

3 The head, including the grasping spines, is normally ensheathed by a fold of the body wall known as the hood. At the very front of this individual you can also make out two rows of short teeth (*Sagitta* sp.).

Some deep-sea arrow worms are known to discharge a cloud of bioluminescent particles from specialized organs on their fins, although it is not clear if this behaviour is intended to confuse predators or startle prey. It is possible that the tiny vibrations produced by the movement of a startled prey animal give the arrow worm a target it to home in on.

All the arrow worms are hermaphrodites and fertilization occurs internally. To exchange sperm, individuals of the same species must come together, which is easier said than done since any animal of similar size is potential prey. To make sure they exchange sperm with another individual of the same species and to avoid being mistaken for food, the arrow worms go through an elaborate courtship dance. The individuals approach each other, swing their bodies from side to side, touch heads and hold themselves erect in the water, face to face. This mating behaviour culminates with a tail-flick from the individual acting as the male, deftly attaching a small packet of sperm (spermatophore) to the body of its partner. The sperm from this packet wiggle their way into the reproductive tract and towards the waiting unfertilized eggs of the individual acting as a female. In most species, once fertilized the eggs drift to the seabed and in as little as one day juvenile arrow worms hatch and start hunting.

Arrow worms have a voracious appetite and some species are known to consume more than two thirds of their own body weight in prey every day. This voracity allied with their ubiquity and abundance in some waters means they are very important components of the marine food web. In various places where people have taken the time to look there can be as many as 70 arrow worms per cubic metre of seawater (or 50 per cubic yard). With such high densities they collectively consume huge quantities of other planktonic organisms and in turn they are consumed by other animals [4].

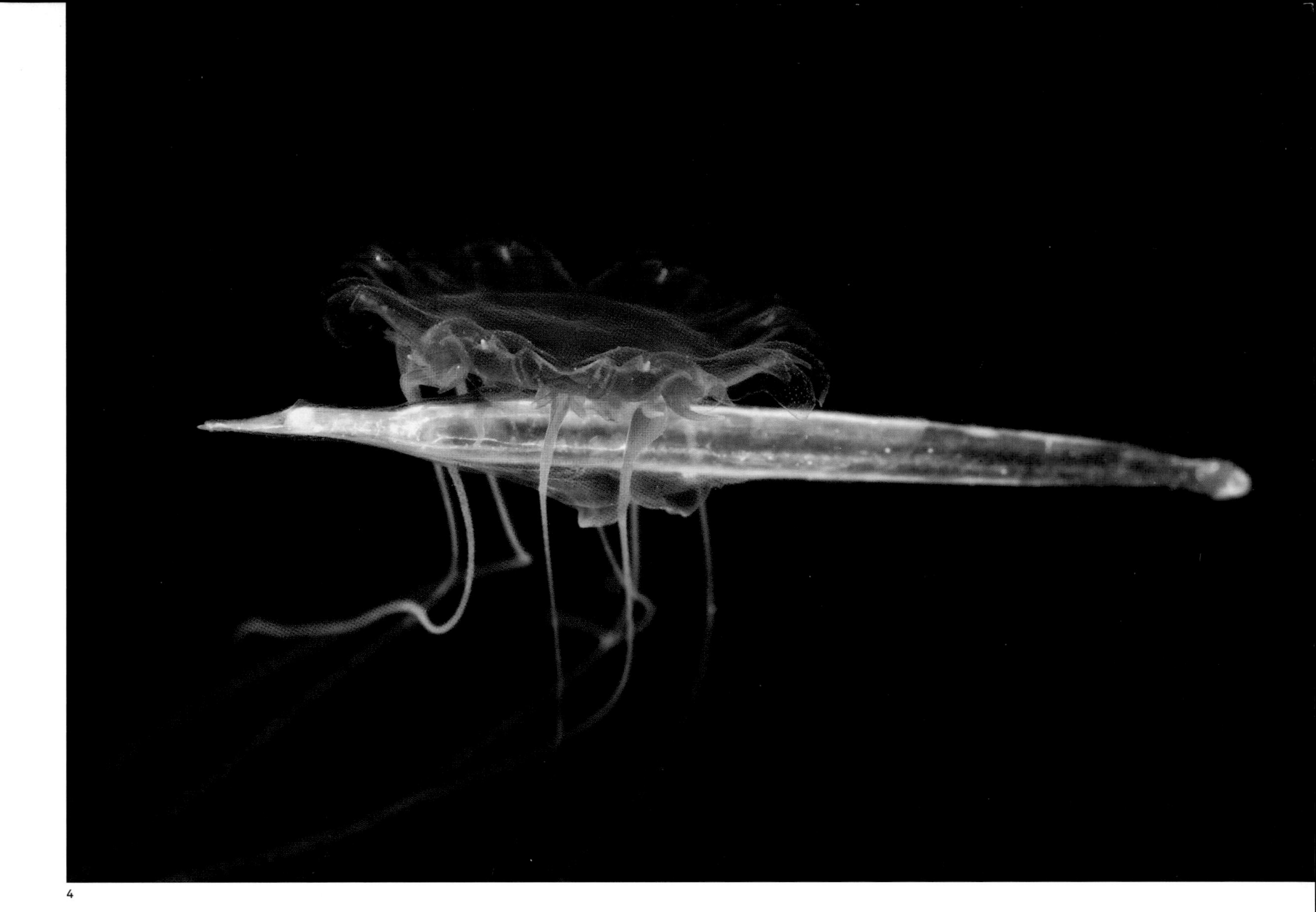

4

4 Arrow worms are voracious, planktonic predators, but they themselves fall prey to countless other animals. Here, the arrow worm *Parasagitta elegans* has been snared by a juvenile jellyfish (*Cyanea capillata*).

ORIGINS AND AFFINITIES

The evolutionary relationships of the arrow worms have mystified zoologists for years and even today, regardless of the deeper understanding afforded by comparing DNA sequences, they are hard to place on the tree of life. Some of their characteristics suggest an affinity with the echinoderms and craniates. On the other hand the presence of chitin in their cuticle, their ability to shed and replace their grasping spines, compound eyes and other features indicate an affinity with the animals that shed their skin in order to grow, such as arthropods, tardigrades and nematodes.

The strange mix of characteristics seen in the arrow worms is living evidence of an ancient split in the evolutionary trajectories of the animal lineages. Working out the relationships of these enigmatic planktonic hunters and their position on the tree of life will help us to piece together some of the key events in the early evolution of the animals.

BOX 1 ***Tetrodotoxin***

Tetrodotoxins are extremely potent neurotoxins produced by several species of bacteria. The bacteria that produce these toxins have been harnessed by a surprising variety of animals – including arrow worms, octopuses, ribbon worms, flatworms, arthropods, fish, amphibians and starfish – as a means of subduing prey and defending themselves. It seems that at some point in the distant past the ancestors of these animals developed genetic mutations that rendered them immune to the effects of tetrodotoxin, which ultimately led to a mutualism between the animals and the bacteria. In some cases this symbiosis evolved to the extent where specialized organs house the bacteria, allowing significant concentrations of the toxin to build up and be dispensed at will.

Nematoda

(roundworms, pinworms, eelworms, threadworms)
(Greek *nema* = thread)

Diversity
c. 24,800 species

Size range
<1 mm to ~9 m
(<0.04 in. to ~29.5 ft)

1 Nematodes are not segmented – what look like segments are in fact just annulations on the body surface. As here, the muscular pharynx is often equipped with a stylet for piercing plant and animal cells (*Mesocriconema* sp.).

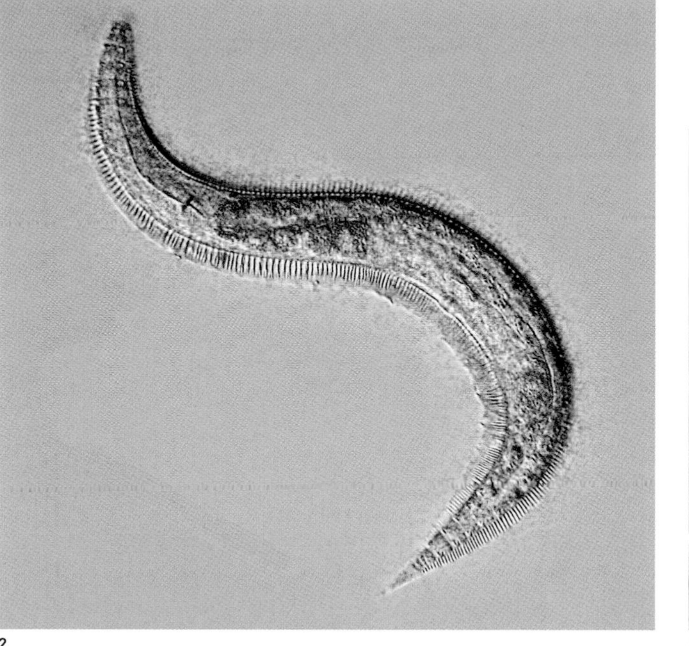

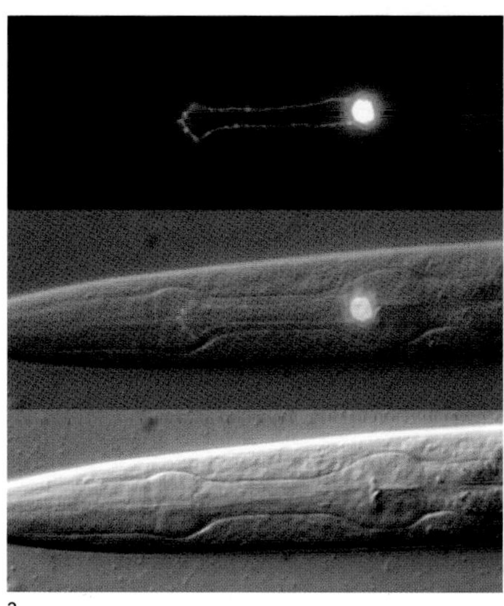

The nematodes are the most widespread and abundant of all the animals, but our understanding of them does not reflect their ecological significance. To date, around 24,800 species have been described, but it has been suggested that there could be as many as 10 million species in total. In discussing their distribution, it is more pertinent to ask where they are *not* found, since they appear to be more or less ubiquitous.

FORM AND FUNCTION

To the naked eye nematodes appear to be rather simple animals and it is true to say that they all share a remarkably restrained general body plan – in essence a cylindrical tube that tapers at both ends [2, 9]. However, taking a closer look through a microscope reveals that they are extremely diverse both in the way they look and the way they live [10–12].

The smallest are much less than 1 mm (0.04 in.) long, but the largest, *Placentonema gigantissima*, from the placenta of the sperm whale, is a 9-m- (30-ft-) long beast. The nematode body is enclosed in a complex, multilayered cuticle, much of which is collagen. They have neither endoskeleton nor exoskeleton; instead they rely on a hydrostatic skeleton.

At the front end of the long thin body there is a mouth surrounded by lips and various sensory structures [13, 14]. On entering the mouth, the next thing food encounters is a muscular pharynx that ends in a tubular gut running almost the entire length of the animal. In some species, especially parasitic and predatory ones, the mouth is furnished with ridges, rods, teeth and even vicious-looking stylets that can be poked out as necessary [1].

Belying their simple shape, nematodes have a well-developed nervous system with a number of nerve cords emanating from a brain [3]. The latter is donut-shaped and encircles the pharynx, an arrangement of central nervous tissue unique to the nematodes and their close relatives. The nervous system receives information from a range of external sensory organs; some aquatic species, both marine and freshwater, even sport simple eye spots.

Nematodes move by contracting four bands of spiralling longitudinal muscles running the length of their bodies [7]. Contraction and relaxation of these bends the body with amazing suppleness and versatility, allowing the nematode to push against its surroundings and propel itself backwards or forwards, as well to swim, lash about and squeeze through tiny gaps. This is one of the advantages of an anatomy based on a hydrostatic skeleton.

Its body being so long and thin, no part of a nematode is ever very far from the outside world; therefore gases and waste can pass in and out of the body without the need for specialized respiratory and circulatory systems.

LIFESTYLE

The nematode form is perfectly adapted to a life spent squirming around in tight spaces. Whether it is the sediment at the bottom of

2 At first glance the nematode body plan seems quite simple, but the nematodes are hugely diverse animals and there could be millions of species waiting to be described (*Bunonema* sp.).

3 Two of the nematode's nerve cells controlling the muscular pharynx have been revealed here by 'labelling' them using proteins that fluoresce green under blue light (*Caenorhabditis elegans*).

BOX 1 ***Extreme nematodes***

Life on earth is limited to the thin biosphere, a living veneer on a giant ball of space rock. However, while this remains true, it has been known for some time that the biosphere actually extends several kilometres beneath the earth's surface.

Until 2011, this extreme habitat was thought to be the preserve of single-celled organisms, but several species of nematode, including a new species (*Halicephalobus mephisto*) were found thriving 3.6 km (2.2 miles) down in water-filled fractures in South African rock. These animals eat bacteria growing on the rocks, reproduce asexually and, almost unbelievably, can tolerate the oppressive heat and pressures at such great depths.

At the other extreme *Cryonema* nematodes thrive at or near freezing point in the small spaces within Arctic sea ice.

4

5

a lake, moist soil or the tissues of an animal or plant, they can move around with ease, poking their way in between grains of sand and even the spaces between individual cells. In fact, their success is largely thanks to their body shape and where it enables them to live – which is nearly everywhere [BOX 1]. With that said, all nematodes are fundamentally aquatic animals and not one species has completely conquered the terrestrial domain. It is true that some of the their life stages, particularly the eggs, are capable of tolerating extreme desiccation, but this is a transient phase, survival rather than conquest, and nematodes' active stages can only survive in water, moisture in soil or around plants and the tissues of living organisms.

For some people, nematodes are the archetypal parasite, but although their exploitation of animals and plants is almost unparalleled, it is worth noting that a large proportion of nematodes are free living. Many are carnivores that capture and consume other small animals, including other nematodes. Some are parasitoids that have a lifestyle bordering on the predatory: when they attack a snail or an insect larva, they multiply explosively, never ceasing to feed until the host is a mass of squirming worms in a skin or shell. In contrast, another group of nematodes survives by grazing on fungi, algae and bacteria. Nematodes often associate with dead and decomposing matter, but in such situations they usually feed on the bacteria and fungi rather than the dead material. Many of the species most harmful to human interests are specialized herbivores, capable of piercing plant cells with their mouth stylets to get at the nutritious contents [1, 4].

The close association of nematodes with other animals has led to the evolution of some incredible lifestyles, not all of them parasitic – some are mutualists. For example, adult female *Zalophora* nematodes live inside the gut of millipedes, but they are not parasitizing the arthropod. If anything, they are the millipede's ally because they prey on other nematodes that *are* parasites [10].

Some nematodes are parasitic for only part of their life, whereas others spend almost their entire life in the body of another animal; some species only parasitize a single host, while others require two hosts to complete their life cycle. Inside the body of their host, nematodes have two main ways of obtaining nourishment. The first, more host-friendly, involves nutrients simply diffusing into the body of the nematode from the fluids of the host [5]. The second, more grisly, entails the nematode nibbling and tearing the tissues of the host to get at its blood.

Being a parasite may seem like an easy option for any animal, but getting into the host in the first place is exceedingly tricky and surviving once there is far from assured. To get around these problems, nematodes have evolved numerous strategies, such as incredible fecundity, host manipulation and a variety of mechanisms for getting around the host's defences [BOXES 2 & 3].

4 A larval root-knot nematode, *Meloidogyne incognita*, penetrating a tomato root. Once inside, the larva establishes a feeding site and causes nutrient-robbing galls to form on cells. As it consumes nutrients, the plant's growth is stunted.

5 Mermithid nematodes develop in the bodies of arthropods before they leave their host for a fleeting free-living existence. This individual (*Hexamermis* sp.) was dissected from a carabid beetle. Relative to its host this parasitic nematode is huge, occupying all of its abdomen.

6 A juvenile *Trichinella spiralis* within its nurse cell, a hijacked skeletal muscle cell.

BOX 2 ***Trichinella spiralis***

One of the commonest and best-known parasitic nematodes also happens to have one of the most remarkable life histories.

At 1.4–1.6 mm (0.05–0.06 in.) long, *Trichinella spiralis* is one of the smallest parasitic nematodes. Its hosts are humans and many other mammals. An initial infection depends on eating meat riddled with the juvenile worms safely ensconced in little nurse cells. The upheaval of being eaten triggers the juveniles to leave their nurse cells and take up residence in the surface of the host's intestine where they grow, mature and reproduce. A gravid female gives birth to hundreds or thousands of live young that end up in the blood of the host and get transported all around its body, ending up in every conceivable tissue.

For juveniles lucky enough to end up in skeletal muscle, this is where the really incredible phase of their life begins. The young nematode penetrates an individual muscle cell and slips inside. Here, much like a virus, it hijacks the cellular machinery of the host to do its own bidding and before long the muscle cell has been converted into a cosy nurse cell complete with an augmented blood supply. Within these capsules the young nematodes have to wait for the host to be eaten, but this can be a long wait, so to sit it out they can enter a dormant state. Exactly how long they can survive is not known – the host's immune system does respond to their presence, slowly calcifying the nurse cell and the nematode within, but living juveniles have been found in muscle tissue almost 40 years after the initial infection.

Exactly how juvenile *T. spiralis* subvert the host's cells to fashion a nurse cell is not known, but it beautifully illustrates the incredible adaptability of the nematodes.

Most nematode species have separate sexes, although there are some hermaphroditic and parthenogenetic species. Males and females of the same species have to find each other, which they achieve by producing and following pheromone trails. Mating itself is quite a tender affair, for the female seeks out the coiled rear end of the male and threads her body through to align her genital pore with that of her mate. Needless to say the male is excited by the female slipping through his lasso-tail, not to mention the alignment of sexual openings, so he protrudes a pair of horny spicules to hold open the female's pore while he sheds his sperm. The sperm of nematodes are unusual among animals in that they lack a flagellum (the distinctive 'tail'); indeed, some of them are reminiscent of amoebae and certain kinds even undergo some metamorphosis once they find themselves in the female. In some species there are two distinct types of sperm: relatively huge ones as much as 0.1 mm (0.004 in.) wide and tiny ones around 0.002 mm (0.00008 in.) wide. It is the job of the latter to fertilize the eggs, but to get there they hitch a ride on the larger, transporter sperm.

Typically, the female will lay eggs after fertilization [8], although some free-living species and many parasitic ones (e.g. *Trichinella spiralis* [BOX 2]) retain the eggs and give birth to live young, a strategy called ovovivipary. As far as we know, the life cycle of every species includes the egg, four juvenile stages and the adult. Each juvenile stage ends in the nematode shedding its cuticle – in much the same way as other cuticle-moulting animals do (e.g., Arthropoda, pp. 148–77).

The ecological significance of the nematodes is overwhelming. They are so diverse and abundant that there can be no interactions in the living world where their influence is not felt. The majority of animal and terrestrial plant species play host to at least one type of parasitic

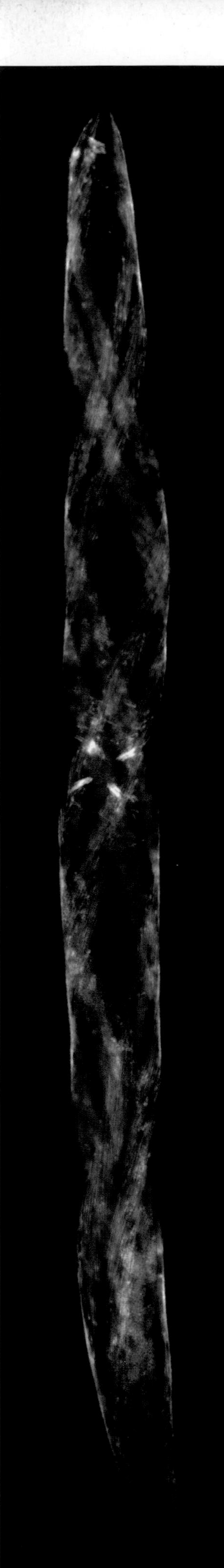

nematode. In this respect alone their influence on nature is immense, since they have a direct effect on the size and health of wild populations of animals and plants.

In feeding on single-celled organisms and smaller animals, nematodes are key in connecting decomposers and primary consumers to the more lofty parts of the food chain. In almost every habitat nematodes make up a considerable proportion of the total animal biomass. In a mere teaspoon of marine mud there may be more than 1000 nematodes represented by at least 36 species, while just the top few centimetres of a single acre of farm soil can support anywhere between 3 and 9 billion living, active nematodes. These big numbers illustrate just how important nematodes are, yet when we take a walk through a woodland or stroll along the beach, few of us would take a moment to reflect on the inconspicuous little worms beneath our feet that keep the living world ticking along.

ORIGINS AND AFFINITIES

Carboniferous rocks 300 to 360 million years old from Scotland and Illinois bear the fossilized remains of ancient nematodes. Although these fossils do not reveal a great deal about the evolution of the lineage, since they look so similar to living species, they do give us an insight into the lives of these animals. The Scottish specimens were found between the cuticle layers of an ancient sea scorpion. Rather than parasitizing a living animal it seems the nematodes were simply exploiting the single celled organisms that were decomposing a dead body.

Regarding their affinities, it is now clear that the closest relatives of the nematodes are the Nematomorpha (see pp. 134–37). Together, these animals are part of the branch that also includes the Arthropoda, Tardigrada and

7 A nematode has four muscle bands that spiral the length of its body (revealed here using proteins that fluoresce green under blue light). The 'cross' of muscles in the centre of the image surrounds the vulva. In orange, snaking its way through the animal, is the gut (*Caenorhabditis elegans*).

8 The sculpted eggs of a nematode. The eggs of many species are incredibly resistant to extreme conditions and can remain viable for years, much like the seeds of a plant (*Ascaris lumbricoides*).

›9 *Foleyella* sp., a frog parasite.

Onychophora. As a whole, the nematodes are very poorly known, with countless secrets still to offer up. One species, *Caenorhabditis elegans*, is, however, one of the best studied animals of all. It was the first multicellular organism to have its genome sequenced, and we know what every one of its embryonic cells will end up doing in the adult animal. The very fact that each cell in the life history of any multicellular organism could have such a precise future mapped out for it is a mind-boggling discovery.

BOX 3 ***The giant roundworm***

The giant roundworm, *Ascaris lumbricoides*, is one of the monsters of the parasitic nematode world. Adult females are up to 50 cm (20 in.) long and 6 mm (0.25 in.) wide. As adults they reside in the intestinal tract of perhaps 1.4 billion humans.

This species has an elaborate life history, but even more impressive is its fecundity and the durability of its eggs. The ovaries of the female contain as many as 27 million eggs at a time, 200,000 of which can be laid every day, finding their way into the outside world via the faeces of their host. The eggs have thick, sculpted shells that are formidably resistant to all manner of chemicals and extreme conditions [8]. Even after ten years languishing in soil, for example, more than half the deposited eggs are still viable.

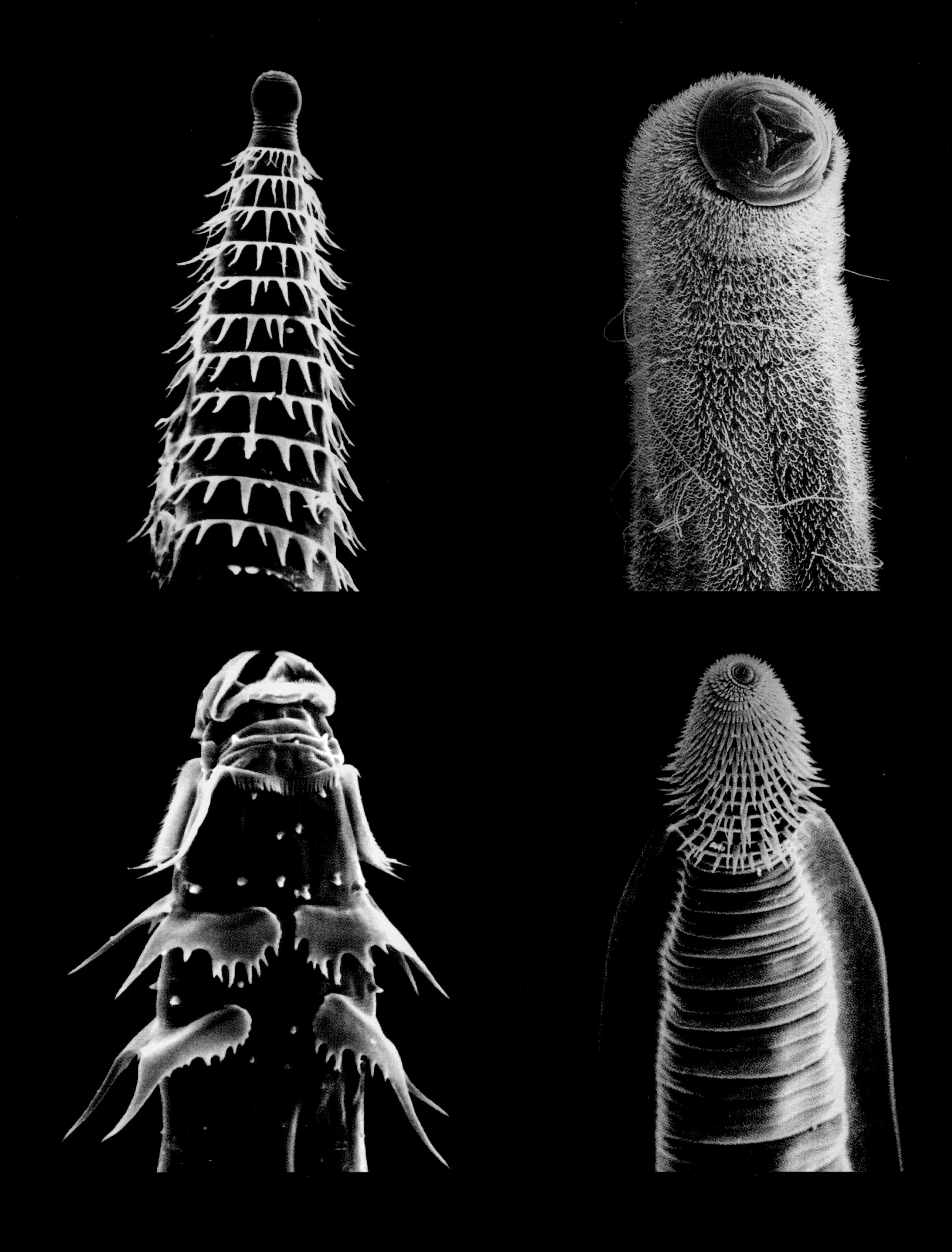

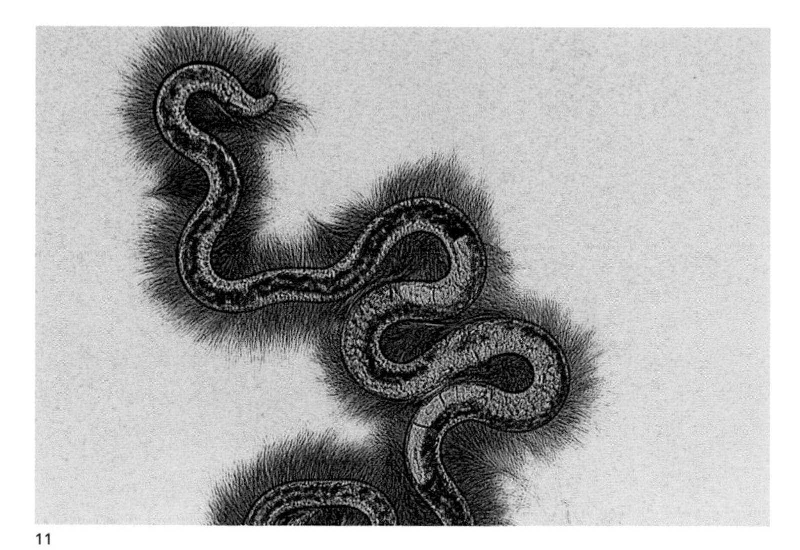

11

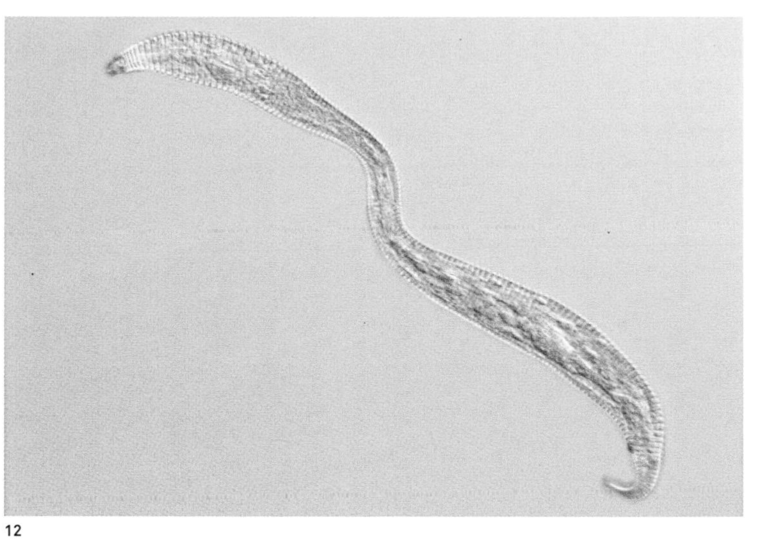

12

‹ 10 These SEMs reveal just how diverse the nematodes can be in terms of appearance. These parasitic species inhabit the intestines of large, tropical millipedes, where they also fall prey to other nematodes (i.e., *Zalophora* spp.). Clockwise from top left: *Carnoya* sp. female, *Rhigonema tomentosum*, *Carnoya fimbriata* and *Heth* sp.

11 A marine nematode (unidentified stilbonematid). The 'hairs' are actually filamentous, symbiotic cyanobacteria growing on the nematode, which it consumes. To supply their symbionts with the reduced sulphur compounds and oxygen they need, the nematodes migrate through marine sediments.

12 A marine nematode, *Epsilonema* sp.

13 The head of the parasitic nematode, *Toxocara canis*, showing the three lips. Humans, often children, can become infected with the eggs of this nematode via the faeces of an infested dog or contaminated soil. The swallowed eggs hatch in the intestines and the larvae then migrate to organs such as the liver, lungs, brain and eyes.

14 Close up of the head of *Toxocara canis*; note the sensory structures on the lips.

13

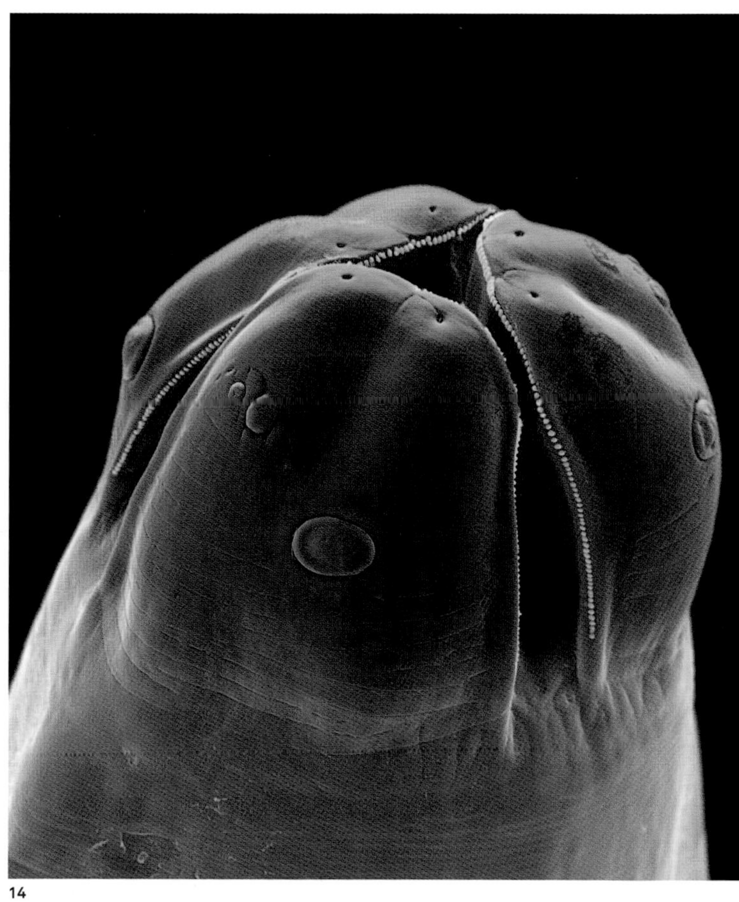

14

Nematomorpha

(Gordian worms, hairworms, horsehair worms)
(Greek *nema* = thread; *morph* = shape, form)

Diversity
c. 350 species

Size range
~5 cm to ~1 m
(~2 in. to ~3.3 ft)

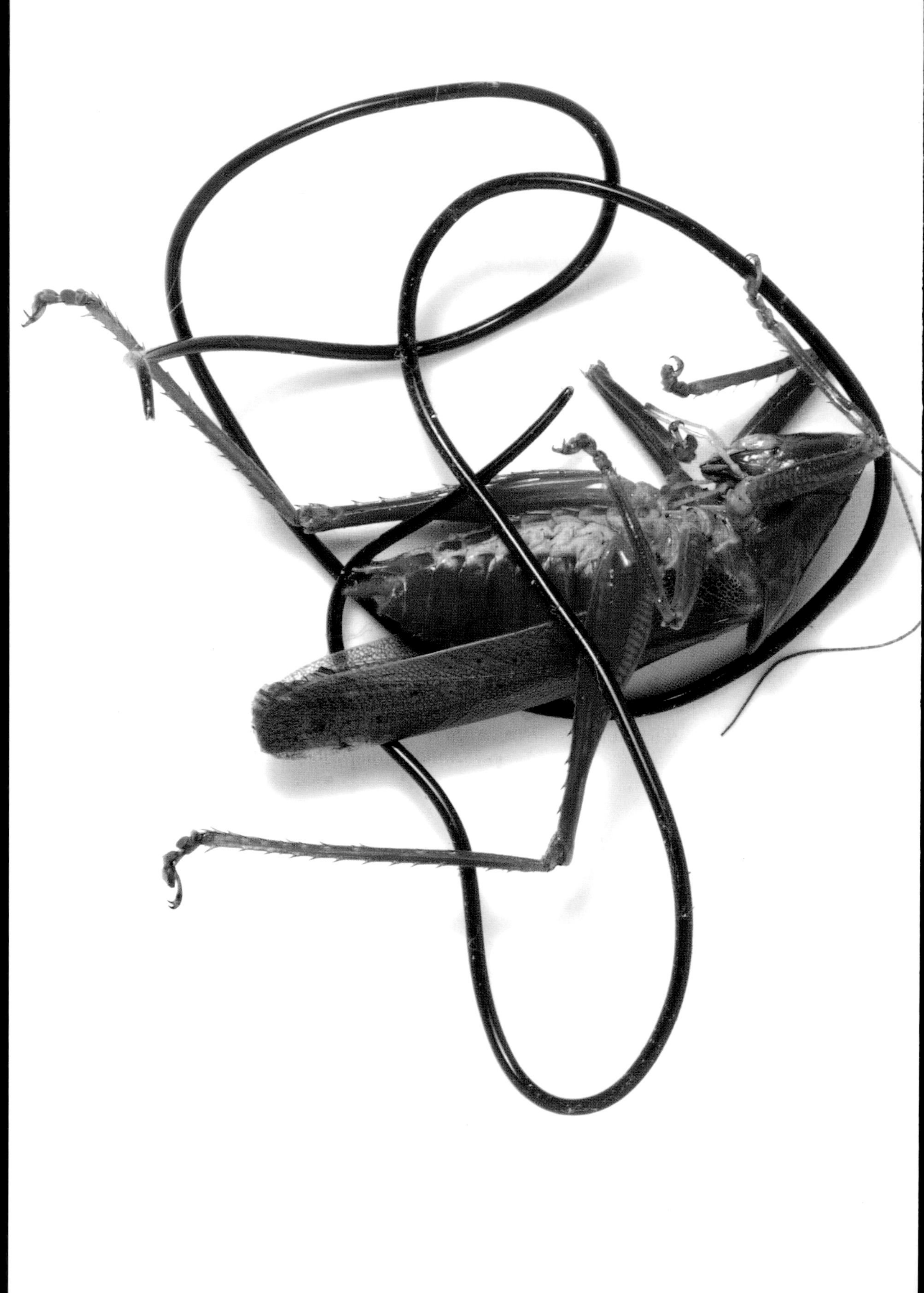

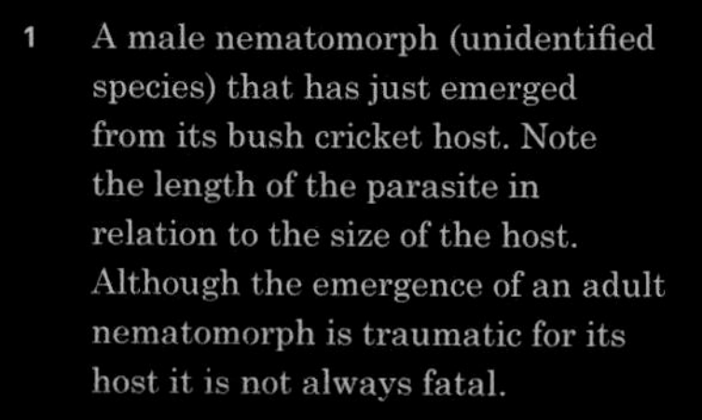

1 A male nematomorph (unidentified species) that has just emerged from its bush cricket host. Note the length of the parasite in relation to the size of the host. Although the emergence of an adult nematomorph is traumatic for its host it is not always fatal.

The short-lived adult nematomorphs are decidedly hair-like, longer and thinner than even the nematodes (see pp. 126–33), and have a propensity for getting tangled. They are also one of the few animal lineages where *all* the representatives are parasitic, spending most of their lives inside the bodies of a variety of hapless arthropods. Not a lot is known about the intricacies of their life history and even though they are represented by relatively few species their impact on other animals may be as far-reaching as their long, thin bodies.

FORM AND FUNCTION

To the naked eye, the adult nematomorphs are essentially featureless; this is why in the past they were often confused with hairs [1]. There is no distinct head and the extremely slender body is enclosed in a thick cuticle that under magnification is actually finely sculptured. Male nematomorphs can be identified by their bifurcated posterior end, which helps to keep a firm grip on the female during copulation [4]. The rarely seen marine *Nectonema* species have well developed bristles along the upper and lower margins of their body that help them swim [3].

Like the nematodes, nematomorphs have muscle bands running the length of their bodies and their nervous system is centred on a doughnut-shaped brain encircling the pharynx. In adapting to a parasitic way of life, adult nematomorphs have become little more than long, sinuous gamete factories. They do not ingest anything so the gut is reduced and usually there is no mouth or indeed an anus. There are no signs of any excretory or respiratory organs because their body shape permits the diffusion of gases and waste across the body wall. The gonads, on the other hand, are impressive; a pair of cylindrical sacs extending the length of the animal.

The nematomorph larva is so very different to its parents and is such a crucial stage in the life of these animals that it deserves a separate description. In some respects it resembles a chubby fly maggot, but at its head end there is a proboscis armed with stylets [2]. The proboscis protrudes under hydrostatic pressure, very much like when a water balloon is squeezed. At the base of this proboscis are two or, more commonly, three whorls of recurved spines. Inside the body of the larva there is a big gland, probably important in penetrating the body of the host, and a so-called pseudointestine. It is not known if this pseudointestine gives rise to the vestigial intestine that is seen in the adults.

LIFESTYLE

All nematomorphs are parasites. The adults have a fleeting existence as free-living animals, but most of their life is spent inside the body of variety of arthropods, including beetles, caddis flies, dragonflies, cockroaches, crickets, crabs and millipedes. In temperate habitats the commonest hosts are grasshoppers, crickets and beetles, while mantids and cockroaches are favoured in the tropics. The adults of almost all nematomorph species are found in fresh water or moist soil. The aquatic species can swim by whip-like thrashing of their body, but most of the time they just coil around bits of aquatic vegetation. With the exception of a newly discovered, parthenogenetic species, there are separate sexes and when a male finds a female he coils tightly around her posterior end in preparation for transferring a packet of sperm.

The sperm are unusual in that they are rod-shaped and they lack a flagellum. After the sperm have reached the eggs and fertilized them, the female unloads her reproductive cargo as a long string, usually either straight into the water or by sticking them to underwater vegetation. Like many parasites, especially those that deposit their eggs in such a way that they encounter their hosts largely by chance, nematomorphs are extremely fecund. In many species the females can produce more than six million eggs, and the egg strings may be several times the length of the female worm. In the wild, numerous adults, desperate to mate, will coil around each other to form an impenetrable tangle of bodies. This is where the common name 'Gordian worm' comes from, in reference to Alexander the Great and the legendary Gordian knot.

Eventually the larvae hatch and set about seeking a host, but this is where the biology of these animals becomes a little hazy. Some species appear to require only a single host, others may need two or perhaps even more hosts to complete their development, but for the most part we do not really understand the intricacies of their life cycles.

To complete their development it seems that nematomorph larvae typically rely on a very passive approach, in others words getting inadvertently swallowed by a suitable host. The larvae of some species are able to see out unfavourable conditions by encysting

2

2 Illustration of a nematomorph larva. This is the infective stage of the life cycle. Note the stylet-tipped proboscis, anterior gland and pseudointestine.

won vegetation or in the soil. The formation of cysts allows the larvae of some species to survive in a state of dormancy when the water in an ephemeral aquatic habitat has dried up, and until they are inadvertently swallowed by an arthropod, perhaps a millipede. There are even some species that are known to encyst in other parasites, such as flukes (see Platyhelminthes, pp. 278–93), making them hyperparasitic. Larvae of nematomorph species that depend on being swallowed must then bore a hole in the host's gut and wriggle into the body cavity where they will continue their development in earnest.

In the tropics, the life cycles of a few species have been scrutinized and it seems the larvae find their way into an aquatic insect such as a mosquito larva, where they promptly encyst to sit out their ride to the next host. They remain in the body of the larval mosquito all the way through metamorphosis and are eventually carried aloft as the fly takes to the air. With any luck and perhaps with a little coercion from the parasite, the mosquito will blunder into a waiting praying mantis and get eaten. This suits the larval nematomorph perfectly as the mantid will host its development through to adulthood.

Inside the body cavity the larvae metamorphose into juvenile worms and absorb all the nutrients they need from the haemolymph of the host. Then they take on the exceedingly slender form of the adult. There are few host–parasite relationships where the latter is so massive relative to the host. The nematomorph grows and grows until it occupies all of its host's body with the exception of the head and the limbs. Naturally, this exacts a heavy toll on the host and many of its abdominal organs are simply crowded out and pushed to the sides of the body cavity [1]. Coiled up tightly, bathed in the haemolymph of its host, the adult parasite begins the last and most macabre phase of its life cycle. By secreting different chemicals the nematomorph effectively hijacks the brain of the host, altering its behaviour so that it seeks out and takes the plunge into whatever body of water it can find, natural or otherwise – exactly where the parasite needs to be, but not the sort of behaviour you would expect from any self-respecting cricket or millipede.

The nematomorph, sensing the final straight of its tortuous life cycle, heaves and writhes to tear the thin membrane between the abdominal plates around the host's anus. The adult parasite wriggles free out into the water, leaving its now useless vehicle to flounder, but not always to perish. It seems that some hosts are made of stern stuff and even after several weeks of unwillingly nurturing the development of a massive parasite and being subjected to its grisly exit they somehow survive, albeit with withered gonads of little use for reproduction.

Nematomorphs are difficult animals to study. Even the large adults are very short-lived and easily overlooked, so it is no surprise that their impact on the populations of other organisms is hard to gauge. Studies have shown that they may infect anywhere between 1.6 and 58 per cent of a host population, but this very much depends on the location and the time of year. Their ability to influence the behaviour of their hosts – a subject zoologists have only scratched the surface of – has implications for the ways in which lots of smaller animals interact with one another and other organisms, perhaps changing the dynamics of whole ecosystems.

ORIGINS AND AFFINITIES

Unlike the nematodes, nematomorphs have a very limited fossil record. Various specimens from German and Italian deposits 34 to 56 million years old have been reported, but their identification as nematomorphs has been disputed. Protruding from the anus of a 25-million-year-old cockroach encased in Dominican amber is what is thought to be the oldest known nematomorph, so it seems they have been making the most of their particular niche for some time.

Nematomorphs are very similar to mermithid nematodes. They resemble each other both in form and in lifestyle, although they have their own suite of defining characteristics. For example, the structure of their cuticles is very distinct and so are their larvae. Currently, the nematodes and nematomorphs are considered to be very closely related, but did the latter evolve from a mermithid-like ancestor or are the nematodes and nematomorphs derived from a common ancestor? There is some evidence to suggest that the second scenario may be correct.

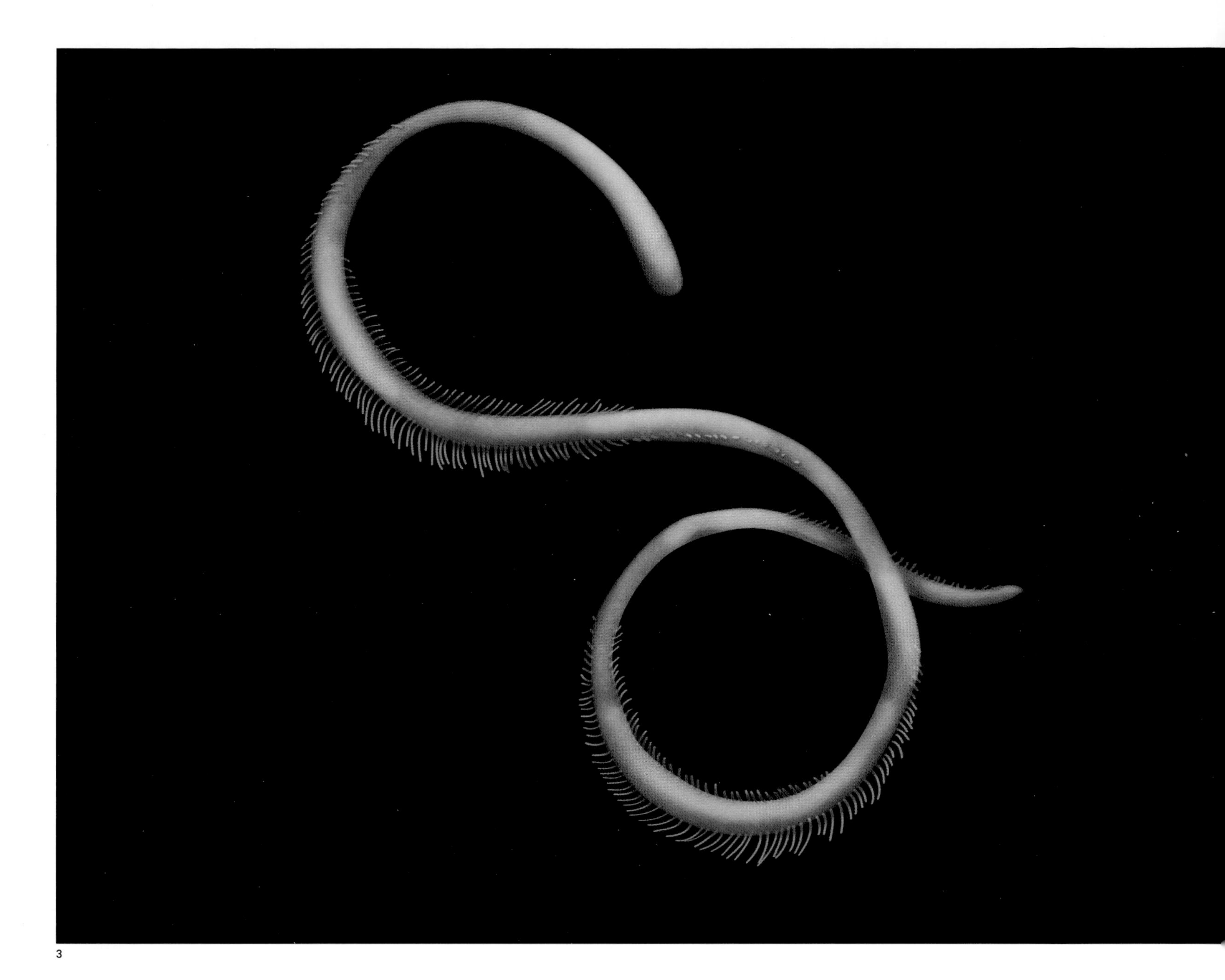

3

4

3 One group of nematomorphs are marine animals. The short-lived adults have well-developed bristles along their upper and lower margins that help them swim (*Nectonema* sp.).

4 Male nematomorphs can be identified by their bifurcated rear end. During copulation the males and females coil around one another, often forming tangled masses.

Tardigrada

(water bears)
(Latin *tardus* = slow;
gradus = step)

Diversity
c. 1160 species

Size range
~0.08 to ~2 mm
(~0.003 to ~0.08 in.)

1 Water bears have a tough, sculpted cuticle that gives them a formidable appearance. The spines visible on this individual are sensory structures known as cirri (*Echiniscus granulatus*).

2

2 The characteristic features of water bears include four pairs of stubby, clawed legs and a mouth cone (*Echiniscus granulatus*).

With their plump appearance and ponderous gait, it is no wonder these animals are commonly known as water bears. Exotic as they look, these tiny, captivating creatures are surprisingly common. They are aquatic, being found both in marine and freshwater habitats, but their small size and resilience also allows them to thrive in unusual and extreme places; some occur, for example, in the thin films of water enveloping the mosses growing on a wall, and others in the scalding water of hot springs.

FORM AND FUNCTION

Behind an indistinct and rather short head, the water bear body has four trunk segments each bearing a pair of stubby legs – their most distinctive feature [1, 2, 5, 6]. These little legs are equipped with claws or tiny adhesive discs. The animal moves them through a combination of muscular force and internal fluid pressure. The epidermis of some species secretes a tough body armour of articulating, chitinous plates, and often these plates have evolved into all sorts of elaborate shapes. As the animal grows, it periodically sheds this tough outer covering.

Just inside the mouth there are a pair of very sharp stylets that the animal can protrude at will. Indeed, it can extend the whole mouth a little to form a cone-shaped turret, thereby giving the stylets a bit more range. Once the stylets have done their work, in other words piercing the cells of other organisms, the muscular, bulbous pharynx gets busy sucking and draining the fluids of those cells [7, 9].

Most water bears are so tiny that they have no need for specialized respiratory and excretory organs – oxygen and waste products simply diffuse across the body wall. However, in those species that have adapted to moist terrestrial habitats, getting rid of waste products is more of a problem since there isn't sufficient water to wash these noxious chemicals away. To get around this problem, specialized excretory organs have evolved from the gut tissue of these species. They concentrate the wastes into a solid form that can be eliminated without lots of water.

A ladder-like arrangement of nerve fibres connects four pairs of ventral ganglia to a brain that has several distinct lobes. The fibres relay signals to the muscles and they collect information from a variety of sensory structures throughout the animal's body, including long, hair-like structures known as cirri [1, 6]. In many species those sense organs also include simple, cup-shaped eye spots.

LIFESTYLE

Water bears are animals of the water. Even those species found in terrestrial habitats are restricted to the moisture surrounding other objects. Marine species commonly are found living in the spaces between sediment particles on the seabed. Terrestrial ones are found in micro-habitats as diverse as the weed in a pond, the miniature pools that accumulate in epiphytic plants growing high in the canopies of tropical forests, and the lichen encrusting a headstone.

Some are herbivorous, piercing plant and algal cells with their sharp stylets, while others are predatory, hunting out animals such as nematodes and rotifers, again using their stylets to puncture the bodies of the victims before sucking them dry.

What really sets the water bears apart is how tough they are. Wherever you are reading this, you are probably no more than a few metres away from what are among the most resilient of all animals, with a tenacity for life that makes most other species look like weaklings [BOX 1].

In almost all marine water bear species the sexes are separate, giving occasion for the evolution of some interesting reproductive strategies. For example, in some cases the male seeks out a female that is in the process of shedding her skin, taking the opportunity to deposit his sperm into the space between the soft female and her partially shed skin [8]. When the female is ready to wriggle free of her old skin she deposits her eggs into the ready-made egg case, where they are fertilized by the male's sperm. With the exception of some hermaphroditic species, many if not most freshwater species reproduce via parthenogenesis, so males are unnecessary. The eggs of freshwater species, often beautifully ornate, can be thin- or thick-shelled depending on the conditions. The former are produced when the prevailing environmental situation is good, while the latter are tough and able to cope with extreme cold or desiccation [10]. When circumstances are favourable the young hatch into miniature versions of the adults, and continue with their water bear lives.

The small size of the tardigrades and their ability to survive periods of unfavourable conditions as dried out husks or durable eggs means that they are very easily dispersed by running water, the wind and by other animals. As a result they are numerous and more-or-less ubiquitous wherever films of water can persist, even fleetingly. Our understanding of their ecology in the wild is very limited, but as

BOX 1 ***Cryptobiosis***

Dormancy, in its many forms, is a common phenomenon in the animal world. Many creatures have evolved reproductive strategies, such as resistant eggs, to survive extreme conditions, but the water bears, tiny and overlooked though they are, must be the toughest animals on the planet, because when their habitat dries up they can enter a state of death-like, suspended animation, known as cryptobiosis (Greek for 'hidden life'). In this state they can tolerate temperatures ranging from close to absolute zero (much colder than liquid nitrogen) up to 120°C (250°F), huge doses of radiation, and pressures ranging from hard vacuum to 6000 atmospheres, which is about ten times the pressure in the oceanic abyss.

Entering the state of cryptobiosis often takes less than an hour, during which time the water bear becomes an almost featureless speck known as a tun. Dehydration is the key to this process, with glycerol and a simple disaccharide – trehalose – gradually replacing water molecules. These substances take over the role of water in maintaining the structure of large molecules such as DNA and proteins, and cellular features such as organelles and cell membranes.

A water bear in a state of cryptobiosis is essentially a dried out husk in which the spark of life is vanishingly faint – the basic processes of metabolism fall to almost immeasurable levels (0.01 per cent of normal). Although the animal is on the cusp of death, it quickly bounces back when water returns, apparently none the worse for wear. This ability not only makes the tardigrades incredibly tough, it also makes them very long lived. It is not known exactly how long these animals can remain dormant, but it could be centuries, possibly even millennia.

Many species of rotifer and some nematodes that have adapted to the same micro-habitats as water bears also use cryptobiosis, in effect travelling through time from one period of suitable conditions to the next.

3

4

3 A water bear (*Macrobiotus richtersii*) entering the dormant state known as cryptobiosis.

4 When cryptobiosis is complete, the water bear is little more than a withered husk (or tun) (*Milnesium tardigradum*).

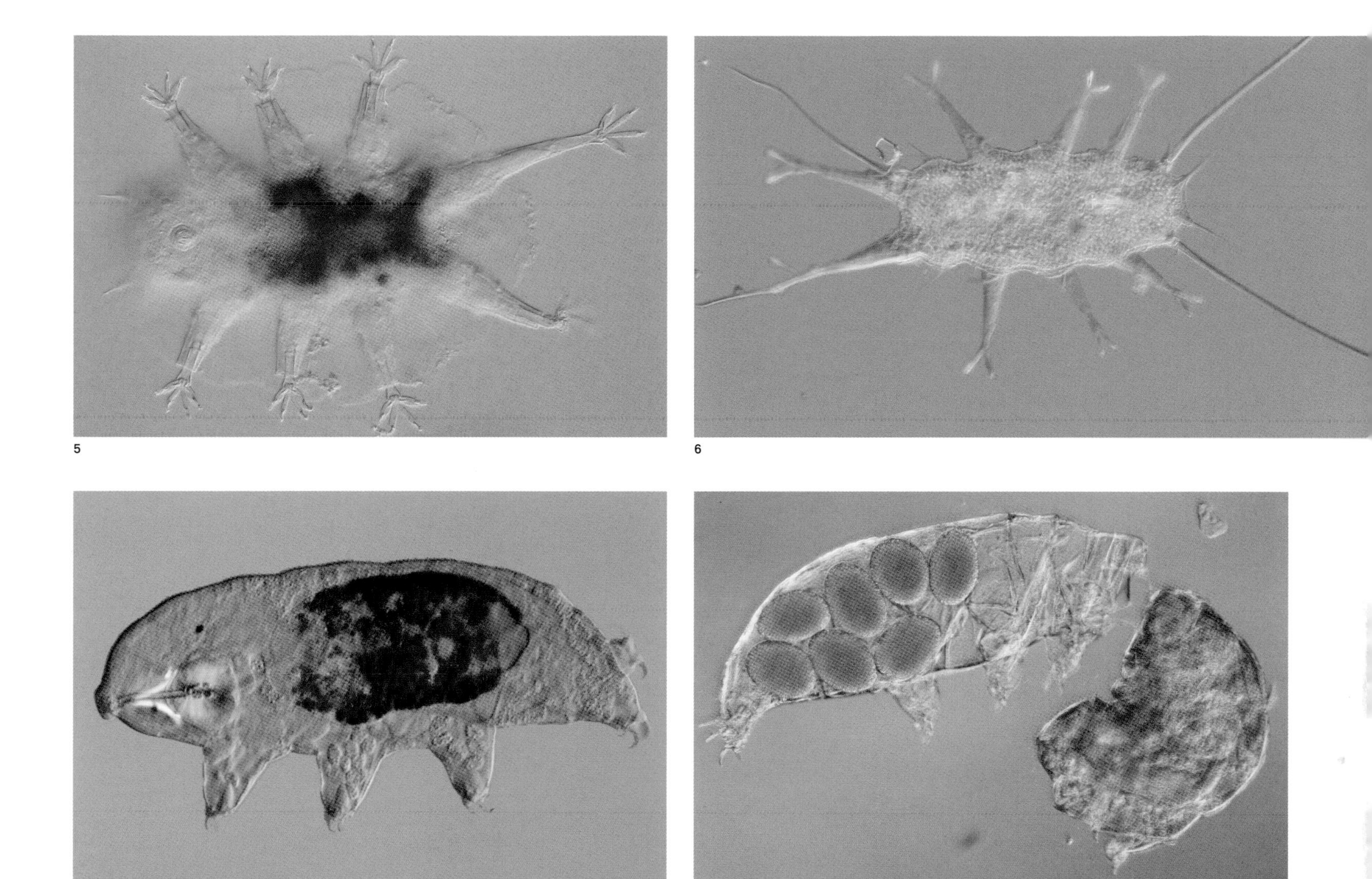

5 Some of the marine water bears have extremely ornate cuticular projections, including protective plates on their backs (*Halechiniscus* sp.).

6 This marine water bear, *Tanarctus* sp., has very long cirri.

7 At the anterior end (left) of this water bear you can see the distinctive muscular pharynx (larger, circular structure) and the piercing stylets (curving, whitish structures) (unidentified species).

8 Some water bear species use their shed cuticle as an impromptu cocoon for their eggs. Here, *Milnesium tardigradum* has deposited seven eggs into its shed skin. Note the size of the eggs in relation to the adult.

they are so abundant they are an important, albeit overlooked component at the base of the ecosystems in which they are found.

ORIGINS AND AFFINITIES

Precious little fossil evidence exists to show how the water bears evolved. Specimens entombed in amber are known from the Cretaceous period, but the fact that they resemble modern forms shows us that we would need to go much further back in time if we wanted to reveal the early evolution of these creatures. The arrangement of their mouthparts as well as those of their relatives, the velvet worms (see pp. 144–47), has prompted some scientists to suggest these two lineages may have stemmed from bizarre animals of the Burgess Shale known as anamalocarids.

To muddy the waters even further, water bears have a raft of interesting characteristics that have excited and confused zoologists for decades. Before the understanding and techniques to compare DNA sequences had been developed, some scientists considered water bears to be something of a 'missing link', along with the velvet worms, between the annelids and the arthropods. Others grouped them with the rotifers and a grab bag of other hard-to-place animals in a now defunct group known as the aschelminthes.

Today, evidence from DNA sequences supports the theory that water bears are the very close relatives of the arthropods and the velvet worms. The more we find out about the affinities of the water bears, the better placed we will be to understand the origins of the arthropods (see pp. 148–77) – the most diverse and arguably the most successful animals on the planet.

9

9 The small size of water bears allows them to live in the incredibly thin water film that surrounds plants, lichens and soil particles on land (*Paramacrobiotus craterlaki*).

› 10 An SEM reveals the hidden beauty of a thick-shelled, dormant water bear egg (*Paramacrobiotus kenianus*).

Onychophora

(velvet worms)
(Greek *onyx* = claw;
pherein = to bear)

Diversity
c. 180 species

Size range
~5 mm to ~15 cm
(~0.2 to ~6 in.)

1 The velvet worm's long, fleshy antennae are its primary sensory structures, extremely sensitive to touch and odour/taste (unidentified species).

2

3

2 Velvet worms are segmented animals, discernible here from the regular spacing of the stubby legs (*Ooperipatellus insignis*).

3 The claw-like jaws are just visible poking out of the mouth of this velvet worm as it feeds on a mosquito (unidentified species).

With features reminiscent of various other groups of living animals, the velvet worms have intrigued generations of zoologists, some of whom thought they represented living, breathing evidence of an important transitional step in animal evolution. Today, we know this is not quite so, but that is not to say they are any less interesting, or less engaging, for not being a 'missing link'. Living fossils they certainly are, that much is beyond contention, and their biology, chock-full of peculiarities, is nothing short of fascinating.

FORM AND FUNCTION

Superficially, a velvet worm is reminiscent of a caterpillar. The soft, velvety body, topped off with a rather indistinct head, is segmented, although this is not easily discernible on the outside apart from the regular spacing of the 13 to 43 pairs stumpy legs, each of which terminates in a pair of claws (hence the name Onychophora, 'bearing claws') [1, 2, 4, 5, 8].

Inside the animal, the segmentation is much more obvious because the circulatory, excretory and nervous systems are organized into a series of repeating units corresponding to the positions of the appendages. The mouth, on the underside of the head, conceals a pair of sharp jaws [3]; the animal sucks up its food by the action of its muscular pharynx, and from the pharynx the food proceeds into a long, straight gut that is divided into three parts: the fore-, mid- and hind-gut (thus resembling the typical gut of the arthropods).

The body is protected by a cuticle composed largely of chitin. In common with the arthropods, this chitinous covering extends deep in to the mouth and anus of the velvet worm, serving as a tough, abrasion-resistant lining for the fore-gut and the hind-gut. The mid-gut is not lined in such a way, so to protect the delicate membrane from being damaged the food is enveloped in a special chitinous tube, known as the peritrophic membrane, that is continuously secreted as food passes through, and is excreted with the waste. The cuticle also lines the incredibly thin channels known as tracheae that pipe air from the atmosphere into the velvet worm's tissues. In contrast to the arthropods, the tracheae of velvet worms are more numerous, unbranched or little branched, and serve only the nearest tissues, functioning far more locally than typical insect tracheae, for example.

The cuticle has one important flaw – its unyielding nature; only limited growth is possible between moults, and major changes in size must take place while the new cuticle is soft and pliable. However, unlike insects, which only shed their cuticle until they reach adulthood, the velvet worms go through this process every two weeks or so for their whole life. During shedding, and until the new cuticle hardens, the animal is extremely vulnerable and there is also the added complication of shedding the lining of the gut and the tiny air channels.

Like the arthropods, velvet worms have what is known as an open circulatory system. Rather than blood coursing through a network of ever smaller vessels, the tissues are immersed in colourless blood circulated by the rhythmic contractions of a long, tubular heart running the length of the animal. In addition, their blood also moves through a system of channels, which is located between the cuticle and the muscles of the body wall. Respiratory pigments based on copper rather than iron (e.g., haemocyanins) enhance the oxygen carrying capacity of the blood.

At the base of each stubby leg there is a pore connected to a specialized excretory organ, a structure that elsewhere in the body evolution also has modified into secretory glands (in particular the salivary glands).

Curious and unique are a pair of slime jets, the ends of which are highly modified limbs just visible as a pair of small, fleshy turrets beneath the velvet worm's beady eyes. The sticky slime, proteinaceous in nature, is produced by a pair of very sizeable glands running almost the whole length of the animal. An onychophoran can eject its slime with some vigour in the general direction of prey or enemies, immobilizing them. When bloated with slime these glands account for around 11 per cent of the weight of the animal.

4

5

Velvet worms have a brain and it connects to the rest of the body via a pair of long nerve cords, which in turn link together in a ladder-like arrangement that innervates all parts of the animal, including a battery of sense organs. There are touch- and vibration-sensitive bristles on the body surface and small, albeit structurally complex eyes.

LIFESTYLE

All extant velvet worms dwell on land, an accolade this lineage shares with no other. However, their mastery of the terrestrial way of life is imperfect, and most of them are rainforest animals restricted to the dark, moist confines of leaf litter or decaying logs. Even those species that commonly venture out from shelter only do so under the cover of darkness when it is cooler and more humid. Caught in a dry, hot atmosphere they quickly desiccate and die.

All known species are highly effective predators with a raft of adaptations for detecting, catching and dispatching just about any animal that crosses their path, as long as it is of a similar size. Velvet worms are hardly what you would call speedy, but their slime jets enable them to snag their quarry from a distance of as much as 15 cm (6 in.). Once the slime hits home it loses water via evaporation and quickly begins to solidify as the protein chains link together forming a very sticky glue. This is an effective way of tackling prey and enemies from a distance, but the sheer quantity of protein involved means that it is very expensive in terms of energy and material; therefore, the velvet worm cannot afford to miss and to make its efforts worthwhile it must go for relatively large prey.

Once the prey is trapped in these hardening threads, the velvet worm can amble over and start feeding, which it does in a rather grisly way. Two claw-like jaws are poked out of the mouth to slice the prey's body permitting the entry of copious amounts of saliva, the digestive enzymes in which quickly begin to digest the still living prey, turning its innards into a soup that the velvet worm ingests over several hours [3, 7].

In all cases courtship ends with the male producing a little packet of sperm – a spermatophore. In some species the male probably uses his penis to insert this directly into the female's vagina ('probably' is appropriate because no one has actually seen this happen). In other species the males have a specialized structure on their head that can be squeezed out to grasp the spermatophore. The male manoeuvres into position and places his gamete-loaded head adornment against the female's vagina, whereby she completes the embrace by clinging onto his head with her rearmost legs. Even more bizarre are the species that have dispensed with the intimacies of inserting the sperm into the female's reproductive tract. Instead, the males simply attach the spermatophore to the female's body; cells in the female's blood detect its presence and start releasing enzymes to create an opening in her skin and the underside of the sperm packet. The sperm wriggle from the spermatophore into the female's blood and make a bee-line for the ripe eggs in her ovaries. The head structures

4 It is easy to see why these animals were once wrongly assumed to represent an evolutionary intermediate between the annelids and arthropods. They are soft-bodied with skin that is dry and velvety, hence their common name (unidentified species).

5 The stubby legs of a velvet worm terminate in a pair of grasping hooks (unidentified species).

6

7

of some male Australian velvet worms may even be used to make a hole in the female's body to provide a route of entry for the sperm.

Some velvet worms lay eggs, but most give birth to live young – miniature replicas of the adults – which are gestated for a year or more. The developing animals are nurtured inside their mother either by secretions from her uterus or the yolk in their egg. Among the former, there are even species with a placental structure that delivers the uterine secretions directly to the developing embryo.

Alongside some of their fascinating reproductive behaviours it seems that some velvet worm species live in small, female dominated societies [6]. Up to 15 individuals live together in these little groups and although they hunt together the dominant female gets to feed first, alone, before the others get a chance. Not only do they work together to bring down relatively large prey, but they also react aggressively to individuals from another group, suggesting that these mini-societies are composed of closely related animals.

ORIGINS AND AFFINITIES

The earliest fossils thought to be the remains of a velvet worm are 300 million years old, although some of the Cambrian Burgess Shale animals, notably *Aysheaia*, bear a striking resemblance to the velvet worms and have therefore been interpreted as a probable early ancestor. The unusual distribution of this lineage points to an ancient heritage. One group occurs in Australasia, Southern Africa and South America – landmasses that were once linked together in the enormous continent known as Gondwanaland. The ancestors of these velvet worms were living in the ancient forests of this huge landmass long before the forces of plate tectonics tore it apart. The other group of velvet worms are found in the Neotropics, the Antilles, West Africa and parts of Southeast Asia, also as a result of continental drift.

For some time many zoologists believed velvet worms were a 'missing link' between the annelid worms and the arthropods and going on their physical characteristics alone you can see why they came to this conclusion. However, in recent years DNA analysis has shown us that the true evolutionary relationships of the animals are more complex than morphology suggests. It seems the velvet worms and the arthropods are in fact closely related, but the annelid worms are on a distant branch of the tree.

8

6 Some velvet worm species exhibit a degree of social behaviour. In these female-dominated groups, small numbers of related velvet worms live and hunt together, although very little is known about how they interact since it is so hard to observe their normal behaviour in the wild (unidentified species).

7 A velvet worm making short work of an unfortunate mosquito. Note the predator's small eyes (unidentified species).

8 The velvet worms are all terrestrial, but are typically only encountered in moist microhabitats, such as the inside of rotting logs (*Macroperipatus* sp.).

Arthropoda

(millipedes, centipedes, insects, crustaceans, arachnids)
(Greek *arthron* = joint; *podos* = foot)

Diversity
c. 1.2 million species

Size range
~0.1 mm to ~3.5 m
(~0.004 in. to ~11.5 ft)

1 SEM of an adult cat flea (*Ctenocephalides felis*). These parasitic animals live on the bodies of mammals and birds, piercing their host's skin with their mouthparts to feed on its blood. In adapting to this niche, they have secondarily lost their wings, but have evolved incredible jumping abilities. Like all arthropods, they have an elaborate exoskeleton – one of the keys to the success of this group of animals.

2

3

2 Crustaceans, such as this tadpole shrimp, typically have a very tough exoskeleton that affords a good deal of protection from predation and water loss (*Triops longicaudatus*).

3 Spiders are supremely adapted terrestrial predators. Many species use silk in various ways to catch other animals, but jumping spiders like this one are active hunters (*Saitis barbipes* feeding on fly).

The arthropods form the largest lineage of animals, accounting for around 80 per cent of all known animal species. One could make a case for describing the animals as 'arthropods and others'. Certainly the most diverse animals on the planet and arguably the most successful, they range in size from minute wasps, small enough to parasitize the eggs of other insects, and microscopic crustaceans and mites, scarcely visible to the naked eye, to giant spiders whose legs would span a dinner plate and deep-sea crabs with a body as big as a football and legs spanning more than 3 m (10 ft) [12–21]. Being so varied and numerous, the arthropods are fundamentally important in all the world's ecosystems. In browsing, predating, scavenging, parasitizing and getting eaten by other organisms they profoundly affect the movement of nutrients and energy through these systems. In myriad seemingly insignificant ways they keep life on earth ticking over, living out their lives in often strange and sometimes even mind-boggling ways.

To attempt to summarize the arthropods in just a few paragraphs is a futile exercise, an insult to their incredible diversity – instead we will take a whirlwind tour of some of their extraordinary lives and the characteristics that have made them so successful.

FORM AND FUNCTION

Arthropods are visibly segmented animals. In the millipedes and centipedes this segmentation is at its most apparent – with long bodies composed of a series of repeating units [10, 22]. Segmentation in several animal lineages has proved to be one of the most important structural principles behind the body plans of complex or large organisms, and segmentation in all the arthropods has been adapted to special functions. For example, multiple segments may fuse to form larger compartments, such as the head, thorax and abdomen of an insect. Sprouting from these segments are appendages that have evolved into a bewildering array of mouthparts, legs, wings and gills [9, 11, 23–30].

A tough outer shell – the exoskeleton – forms a capsule that provides rigidity, solid points of attachment for the muscles, protection for the delicate internal organs and a good degree of waterproofing to prevent fluid loss [1, 2, 11, 16, 18, 31–34]. Comprised of chitin for toughness and sclerotin for rigidity, and often further reinforced with inorganic compounds, for example in the crustaceans, this suit of armour is one of the keys to the success of the arthropods; similarly to the endoskeleton in the case of the vertebrates, the exoskeleton not only enabled the arthropods to grow large and significant in aquatic environments, but also it was instrumental in allowing them to colonize the land. Aquatic animals can often afford to be flimsy, even jelly-like, because water mitigates the relentless tug of gravity, but for a life on land any pioneer has to be able to support its own weight and retain moisture. It was the special characteristics of the arthropod exoskeleton that first made life and movement on land possible for them.

BOX 1 ***Mantis shrimp compound eye***

The eyes of mantis shrimps, each one separated into three different sections that form clear images and perceive depth, can be swivelled in every direction. They are equipped with at least 16 different types of light sensitive cells, compared to four in humans. These cells enable these large, predatory crustaceans to see over 100,000 colours (about ten times more than us), as well as infrared, polarized light and four types of ultraviolet light. We can only speculate how these crustaceans perceive the world, but they may see things in high-definition, psychedelic detail.

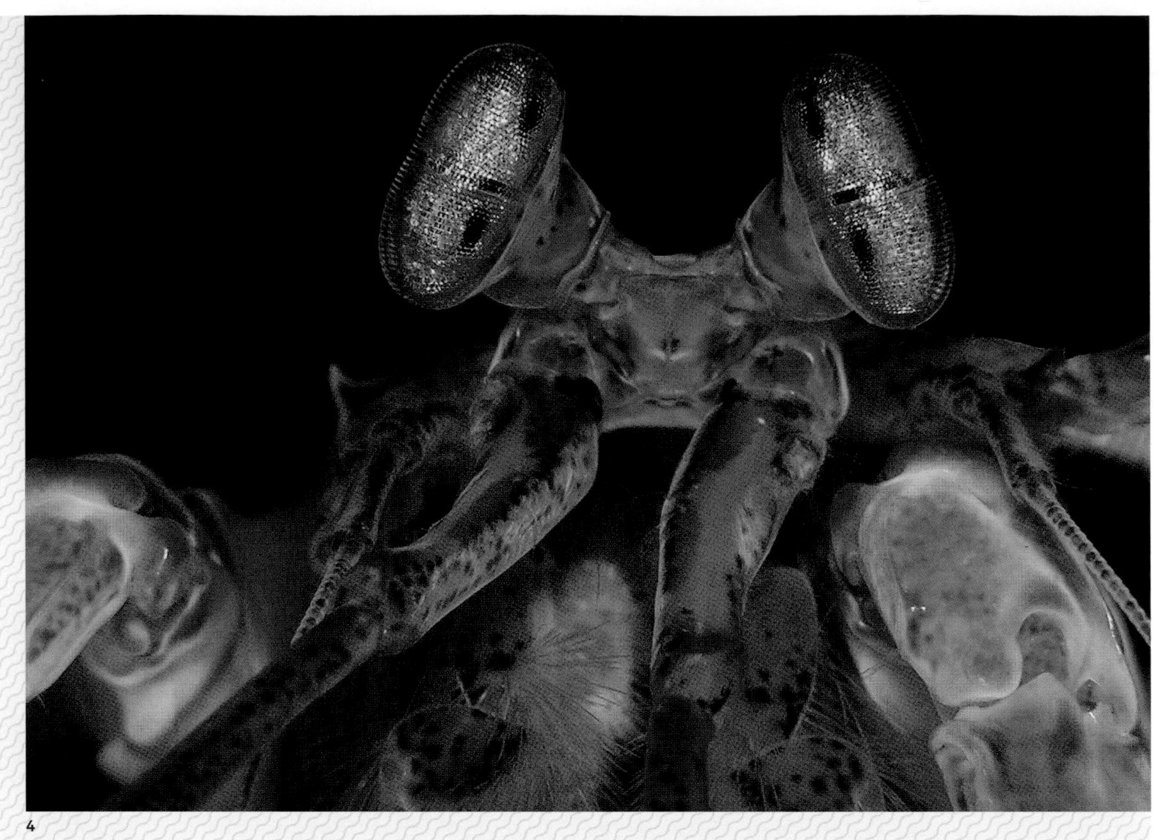

4 Mantis shrimps have among the most sophisticated eyes of any animal (*Squilla aculeata*).

In many respects the general anatomy of the arthropods is similar to that of the velvet worms (see Onychophora, pp. 144–47). For example, they must shed their cuticle to grow, gas exchange is accomplished through the tiny, branching tubes known as tracheae, specialized organs rid the body of metabolic waste, and with the exception of a few vessels the circulatory system is open (the tissues are bathed in haemolymph – the arthropod equivalent of blood). However, evolution has furnished arthropods with some incredible adaptations that are all their own.

Perhaps the most obvious of these is the compound eye, a complex arrangement of lots of tiny facets, each equipped with a lens and light sensing cells [4, 35–38]. These compound eyes give a unique view of the world as every single unit sends a signal to the brain that is processed into an image. The picture formed by an average arthropod eye must be like some form of mosaic as each unit only captures a limited amount of detail. But as an object passes from one unit to the next, the ability of the whole compound eye to detect movement is second to none. Arguably, the compound eye is at its most spectacular in the mantis shrimp [BOX 1]. However, the eyes of mantids, dragonflies, hawk moths, flies and myriad others are no less marvellously sophisticated.

Insect flight depends on unique wings – gossamer outgrowths of their exoskeleton [BOX 2]. The winged insects are the most accomplished small flying animals on the planet, and much as the flight of large animals uses principles differing from some of those that larger insects rely on, tiny insects in turn use principles differing from those underlying the flight of their larger cousins. If you watch hoverflies during the summer you can't help but be amazed by their aerial abilities [5]. As they dart this way and that and hover with incredible precision over a flower, there can be no doubt that you are a privileged observer of one of the most supremely adapted flying animals.

Flying, big eyes, and a host of other activities and senses found among various arthropods necessitate excellent control and coordination, so most arthropods have very sophisticated nervous systems and some have surprisingly well-developed brains. Complex behaviours are common and even problem solving is not beyond the abilities of some arthropods, notably jumping spiders [3] – those of the genus *Portia* appear

BOX 2 ***Insect wings***

Insects evolved the ability to fly at least 350 million years ago – more than 100 million years before any other animals, including the pterosaurs. That is a staggering period of time, and one that is hard to fully comprehend; the dinosaurs, for example, have 'only' been extinct for about 65.5 million years.

The wings of insects are unique because they are not modified legs, like those of a bird or a bat. Instead, they are outgrowths of the exoskeleton. Exactly how they evolved is a bone of contention: one theory is that they may have evolved from gills that adorned the thorax of an ancient aquatic insect; alternatively, small outgrowths of cuticle that helped stabilize primitive insects when they were hopping or falling may have been the precursor to the insect wings that we know today.

Whatever their origin, insect wings are not just hinged flaps; they are extremely complex structures. Their surfaces have wrinkles and pleats that reinforce the wing, channel air efficiently, and permit certain kinds of wings to fold when not in flight. In flight, these seemingly delicate structures twist and deform to provide very efficient lift and thrust that is far beyond the capabilities of current human engineering.

5

6

5 The ability of insects to use their wings to take to the air is perhaps most refined in the hoverflies, of which there are something like 6000 species (this one is *Episyrphus balteatus*). Complementing their advanced aerobatic skills, many insects have developed unexpectedly sophisticated ways of navigating using external cues, and they at least rival and probably exceed the capacity of birds to migrate huge distances (and over similar time periods too).

6 The toughened forewings (known as elytra) of a beetle protect the delicate flying wings and the soft abdomen, but they are too heavy and inflexible to be of any use in propelling the animal through the air. This beetle (*Photinus pyralis*) is one of the many types of 'firefly', which produce light from their abdominal organs (the pale segments visible in this photo).

to be capable of calculated decision making, especially when they are hunting other spiders, their favourite prey.

Another characteristic unique to the arthropods is their ability to produce silk [BOX 3]. Arachnids and certain insects are the only living things that can make this wonder substance, which is as strong as Kevlar. It has been said that a strand of spider silk the same thickness as a pencil could bring a jumbo jet in flight to a halt. Of the arthropods that make and use silk it is the spiders that have become the true masters. Their ability to produce and manipulate this substance has become so refined that they use their threads effectively as an extension of their bodies. They weave it into complex shelters, webs, traps, snares, protective sacs and more [7, 42, 44].

LIFESTYLE

The diversity of arthropod lifestyles is nothing short of staggering, and there is no easy way of generalizing the ways in which they live. They occupy just about every ecological niche there is, from tiny detritus munchers on the seabed, sessile filter-feeders on the rocky shore and a dazzling array of terrestrial plant eaters to aerial predators, parasites, parasitoids and hyperparasitoids [8, 39–41, 43–50, 53].

Not only are arthropods extremely diverse, they are also true pioneers: they were the first animals to colonize the land. As we have seen, their exoskeletons were crucial in allowing them to take advantage of this (at the time) alien world, and of the animals today it is only the insects, arachnids and some of the craniates that have fully exploited the terrestrial domain. The various worms and molluscs that have any sort of independent presence on the land are far more constrained in their ability to fully exploit their environment because of the limits of their body plan.

Insects and arachnids have also adapted to some of the driest places on earth. In some desert environments, scorpions can make up more than 85 per cent of the total predatory arthropod biomass, with a density of 1000–5000 individuals per hectare (or roughly 2500–12,500 per acre) [18]. Just how scorpions can thrive in such marginal habitats is something of a mystery, but it is known that they convert food into arachnid biomass very efficiently. The metabolism of these desert dwellers is very slow and many species can survive for a year a more

BOX 3 ***Silk Structure***

Silk is a protein; that fact determines its properties and structure. All arthropods that produce silk synthesize it in glands as a liquid. As the animal draws this fluid through the increasingly narrow tubes of its spinnerets, the process aligns the amino acid chains in the silk, producing distinctive threads. When scientists first produced silk protein artificially, it proved to be useless: the spinning process is vital to the structure of the finished product. The difference between spun silk and silk protein is as great as that between a house and a pile of bricks. Each fibre of silk has crystalline and stretchy regions and the combination of these yields a material that is both flexible and extremely strong.

7 Highly complex structures can be constructed from silk fibres, including webs and other prey capture devices.

BOX 4 ***Marine insects?***

Although there are some 1.2 million species of insect, only around 1400 occur in brackish or marine habitats – and these are very much restricted to the peripheries of the ocean realm. A few species of the so-called sea skaters (*Halobates* spp.) are oceanic, but they occupy the junction between water and air, skimming around on the sea's surface, so they can hardly be described as aquatic.

Why should there be so few marine insects when they've managed to colonize just about every other habitat so successfully? No one really knows, but it could be that there was simply no ecological 'space' for any potential colonists. The ocean was already brimming with crustaceans and other creatures occupying and exploiting all the available niches in the marine environment.

8

8 Barnacles are highly specialized and successful filter-feeding crustaceans. Protected by its exoskeleton, the animal lies on its back using modified legs (or cirri) to collect edible matter suspended in the water (*Balanus nubilus*).

without food, but when they do eat they gain as much as one third of their body weight from one meal thanks, in part, to external digestion. Their combination of frugal metabolism, low-energy hunting technique (ambush) and their willingness to take a range of prey, permits scorpions to live in habitats that simply are too harsh for other animals.

Interestingly, the insects and many of the arachnids, perhaps the most well adapted of all terrestrial animals have never managed to colonise the oceans in the same way as the mammals did [BOX 4].

The crustaceans, in contrast, are at their most diverse in aquatic habitats, and in particular in the ocean [8]. Some marine crustaceans, particularly the parasitic copepods, have gone on evolutionary tangents, in their adult forms exhibiting few outward signs that betray their arthropod identity. A juvenile, female Sarcotaces is unmistakably a copepod, but after it has tunnelled into the flesh of its fish host it undergoes a radical transformation that sees it grow from a larva perhaps 1 mm (0.04 in.) in length into a 45-mm (1.75-in.) segmented blob entirely encapsulated in a fluid-filled cyst derived from the host's tissues. If this weren't bizarre enough, pressed against the side of this cyst by the bulk of the relatively gigantic female are many minuscule males, all of whom vie to fertilize the eggs of their huge, sac-like mate [51].

Similarly, *Sacculina*, a relative of barnacles, has to be seen to be believed. Like *Sarcotaces*, the immature female *Sacculina* larva looks normal enough, but after it latches on to its host, an extremely unfortunate crab, any outward semblance to a crustacean rapidly disappears. She becomes something of a living syringe, injecting her essence – a creeping mass of cells – into the host [52]. These cells make straight for the central nervous system of the crab and once there they set about ramifying, tumour-like, through its entire body, nourished by its haemolymph and destroying its reproductive system. Body-snatched, castrated and doomed, the crab spends the rest of its miserable existence as a surrogate mother to the parasite's brood [55, 56]. Almost as poetic justice, *Sacculina* is in turn parasitized and castrated by yet another peculiar crustacean, *Danalia*, the adult females of which are little more than small, pendulous sacs.

On land, the arthropods have taken organization to another level with the evolution of complex societies, which are at their most sophisticated in the wasps, bees, ants and termites [60, 66]. Some wasp colonies may be made up of a few dozen individuals, whereas a honeybee hive may contain several thousand bees and some ant colonies may be 15-million strong. A colony of eusocial insects functions as a single super-organism in which each and every worker and soldier is completely dispensable. If a worker ant dies while out foraging or in defence of the nest it is no more a problem for the colony than a cat losing a single hair. Eusocial marine species are very rare [57]; it is only recently that shrimp colonies formed of a single female and several hundred males have been discovered living inside particular species of sponges (see Porifera, p. 43).

As the majority of arthropods are rather small animals there's no shortage of other creatures that relish eating them, a selective pressure that has driven the evolution of

9 Diplurans, such as this *Parajapyx* sp., are extremely common, but typically remain unnoticed as they are tiny animals (usually in the range of 2–5 mm (0.08 – 0.2 in.) long) living in leaf litter and soil. This and related species use their pincer-like cerci to catch their prey.

9

innumerable defences [54, 58, 59, 62]. In the sea there are crustaceans that use bioluminescence to startle and confuse predators, while on land insects evade and deter their enemies at every turn. Witness the bombardier beetles, capable of squirting boiling, noxious chemicals in the face of would-be attackers, the rove beetles that exude a dispersant enabling them to skim across the calm surface of a pond at breakneck speed to escape danger [61], and the huge variety of grasshoppers, butterflies, moths, beetles and bugs that gorge themselves on plants laden with toxins – these chemicals, intended to keep herbivores at bay, are perversely sequestered by the insects to ward off their own enemies.

Reproduction, like every other aspect of arthropod life, is bewilderingly diverse, although here there are a few generalizations we can make. Normally there are separate sexes, although some of the most fecund species reproduce without the need for males. The humble aphid is the Russian doll of the animal world as the founding female of a colony carries a daughter embryo and within this embryo another embryo develops; three generations in the body of one tiny animal all produced via the process of parthenogenesis.

Any high degree of parental care among the arthropods is rare, major exceptions being certain arachnids and eusocial insects, whose females put a lot of time and effort into looking after their offspring [63]. Gravid females typically lay lots of eggs and development often involves a larval stage [64]. The most successful insects spend most of their life as larvae, the sole purpose of which is to eat, grow and lay down the reserves that will make the fleeting adult, reproductive phase as successful as possible [66]. Between comes the pupa, a motionless and vulnerable stage where the tissues of the larvae are liquefied and reordered to build the adult in a process known as metamorphosis. Looking at a maggot and the fly it will eventually become, it is impossible not to marvel at this extraordinary transformation.

Some insects have such elaborate life cycles that the female needs to lay huge numbers of eggs to ensure any of her offspring make it through to adulthood. A female trigonalid wasp lays thousands of eggs into the margins of leaves in an attempt to have at least one of her offspring swallowed by a caterpillar. However, to make things even more complicated not any old caterpillar will do – it has to have itself been parasitized by another type of wasp or fly larvae because these are the prey of the trigonalid larva.

Being extremely diverse and numerous the arthropods are of unparalleled ecological importance, although much of what they do goes on unseen. For example, hordes of crustaceans make short work of whale carcasses on the seabed and wave after wave of specialist insects strip a dead animal on land of its flesh until nothing is left but bare bones. Grisly this scavenging may be, but without it the cycle of life would quickly grind to a halt.

In almost every ecosystem arthropods exist in huge numbers, eating and getting eaten by other organisms [65, 67, 68]. In the tropics, the leaf-cutter ants, colonies of which can comprise millions of individuals, collectively harvest huge amounts of plant tissue, stripping large areas of forest to cultivate large underground fungus gardens. In boreal forests, millipedes are fundamentally important in the recycling of rotting vegetation, consuming up to 36 per cent of the conifer leaf litter. In the sea, the largest animal that has ever lived, the blue whale, grows to its enormous size on a diet of small shrimps called krill. These little crustaceans can support the huge appetites of enormous mammals because they exist in such vast numbers, aggregating in enormous swarms hundreds of metres long. These colossal crustacean aggregations, however, are nothing compared to a swarm of Rocky Mountain locusts sighted in the United States in 1874. This swarm was estimated to be about 2900 km (1800 miles) long and 180 km (110 miles) wide, and it probably contained in the region of 12 trillion insects with a total weight of 27 million tonnes. Sadly this particular species of insect was extinct by 1902, its vast populations powerless to withstand the spread across the continent of intensive farming practices.

The land-living arthropods, particularly the insects, have become intimately entwined with terrestrial plants, so much so that neither can exist without the other. In pollinating plants, insects perform a vital role and most of the flowers, so desirable to us, are actually elaborate insect-attracting beacons. The relationship is at its most intimate between some kinds of figs and their fig wasps. Each species of fig has its own species of fig wasp responsible for transporting pollen between plants, and these peculiar little insects see out most of their complex lives within the confines of a fig. As a further example of the inextricably linked lives of terrestrial arthropods and terrestrial plants, some orchids lure their

pollinators – male bees and wasps – with false promises of sex. The orchid produces flowers that are astounding mimics of the female insect both in terms of appearance and smell. In attempting to mate with the decoy the amorous insect inadvertently picks up the plant's pollen sacs, which it carries to the next orchid it visits.

In other ways too, both for better and for worse, the insects, terrestrial plants and fungi affect one another. Tiny beetles carrying fungal spores tunnel into trees and infect them with their cargoes. The now-diseased tree begins its long demise and in doing so it attracts countless other insects. Trees, sometimes huge numbers of them, affected in this way will eventually fall, opening up areas of forest to regrowth and thus changing the dynamics of an ecosystem.

Arthropods also have the most direct negative impact on humans of any animals. They feed on us, our livestock and our plants and in doing so they transmit a huge variety of pathogens. Mosquitoes transmit the parasites (*Plasmodium* spp.) that cause malaria, a disease that killed at least 1.2 million people in 2010 alone, mainly in Sub-Saharan Africa.

Still, whatever negative impacts arthropods have on our human lives, we have to remember that these are massively outweighed by their positive impacts, mostly unseen, that make the living world what it is today.

ORIGINS AND AFFINITIES

The tough arthropod exoskeleton has left a detailed fossil record that extends back hundreds of millions of years. Some of the enigmatic creatures from the Burgess Shale have been interpreted as early arthropods, and the trilobites were clearly members of this group. The latter were hugely successful animals of the ancient seas, present from the early Cambrian to the late Permian, a duration of some 270 million years. The very first land animals, known from fossils more than 400 million years old, are arthropods, highlighting just how long ago these animals pioneered a terrestrial way of life. The vertebrates seem to have followed their example only some 30 to 60 million years later.

Until fairly recently the long-held belief was that the arthropods and the annelids were very closely related lineages. This idea was based on their similar segmentation, and the presence of chitin in the cuticle of both groups. The ability to sequence DNA turned this theory on its head, and the arthropods and the annelids are now thought to sit on different branches of the animal family tree. Today, the arthropods are considered to be closely allied with all the other animals that shed their skin in order to grow, whereas chitin occurs in a number of creatures outside that branch.

In addition, until just 20 years ago, some zoologists were of the opinion that the Arthropoda actually comprised three distinct animal lineages: the Chelicerata (horseshoe crabs [47, 53], arachnids and sea spiders), Atelocerata or Tracheata (insects, millipedes, centipedes, springtails, etc.) and Crustacea. However, DNA comparisons strongly suggest that that all these animals have evolved from a common ancestor. Indeed, it seems the insects may actually be crustaceans that adapted to a life on land long ago.

10 Of all the arthropods, segmentation is at its most obvious in the millipedes and centipedes (SEM of an unidentified millipede).

11 With ~400,000 described species, the beetles are far and away the most diverse group of animals. Why this should be so is not clear, but two important factors are their generally robust exoskeletons and their heavily reinforced forewings, which form a carapace over a beetle's relatively soft abdomen (*Hololepta* sp.).

12

13

14

15

12 Ants, bees and wasps are among the more well-known arthropods. Note the numerous pollen grains that have been trapped by this bee's dense covering of branched chitinous hairs (setae), underlining the intimate relationship between insects and flowering plants (unidentified solitary bee).

13 The segmented arthropod body is clear to see in these very elongated crustaceans (unidentified caprellids).

14 Mites range in size from 0.1 to 30 mm (0.004 – 1.2 in.). They are extremely diverse and are of huge ecological importance. Around 54,600 species have been described, but there could be at least another 1 million species out there. As they are so small it is hard to appreciate their beauty, but a false colour SEM reveals this peacock mite (*Tuckerella* sp.) in all its glory.

15 In the arachnids the segments of the body have fused into two distinct regions (cephalothorax and opisthosoma). Pedipalps (pincers) and chelicerae (mouthparts – between the pincers) are clearly visible here (unidentified pseudoscorpion).

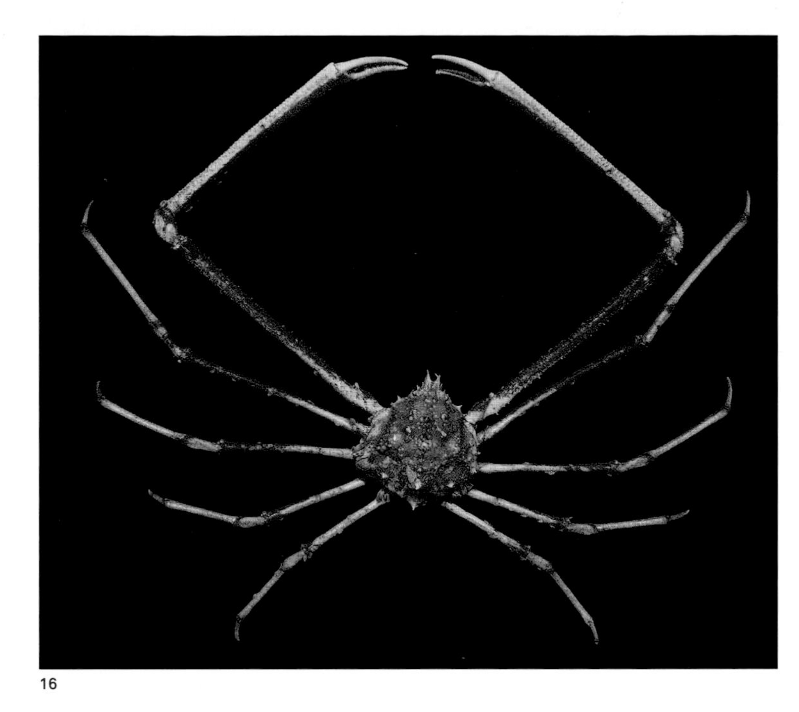

16

17

16 With a body as big as a football and a leg span of more than 3 m, some species of spider crab are at the limit of how large arthropods can grow under current conditions (*Rochinia crassa*).

17 Springtails and their relatives are very common terrestrial animals that fulfil an important ecological role in the decomposition of organic matter. A tail-like appendage, the furcula, is held under tension beneath the body and can be released to propel the animal very rapidly over relatively large distances (unidentified sminthurid).

18 Scorpions are ancient and very successful arthropods, particularly in arid habitats. The long abdomen terminates in a sting, which is used to subdue prey along with the massively enlarged pedipalps that form a pair of formidable pincers (*Pandinus imperator*).

18

19

20

21

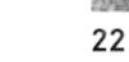

22

23

19 Sea spiders (pycnogonids) are abundant, but rarely seen marine animals closely related to arachnids and horseshoe crabs. The gut of a sea spider has numerous outgrowths that extend into its very long limbs

20 Marine arthropods are every bit as diverse and beautiful as their terrestrial counterparts, as highlighted by this harlequin shrimp (*Hymenocera elegans*).

21 In terms of diversity and the breadth of niches they occupy, crustaceans are to the sea what insects are to the land. The elaborate, branching appendages of this small shrimp, *Microprosthema plumicorne*, are antennae.

22 Segmentation of the body and appendages is very obvious in this centipede (*Scolopendra cingulata*). Curving beneath the head are the 'fangs', which are actually modified legs.

23 Mantis shrimps have raptorial or club-like appendages that are extended at terrific speed to impale or bludgeon their prey (*Squilla aculeata*).

›24 The forelimbs of some crustaceans have evolved into elaborate pincers used in feeding and communication (*Coralliocaris graminea*).

›25 Snapping, or pistol shrimps have asymmetrical pincers, the larger of which works like a pistol hammer to generate a very powerful pressure wave that is used to stun prey, such as fish, and for communication (*Alpheus armillatus*).

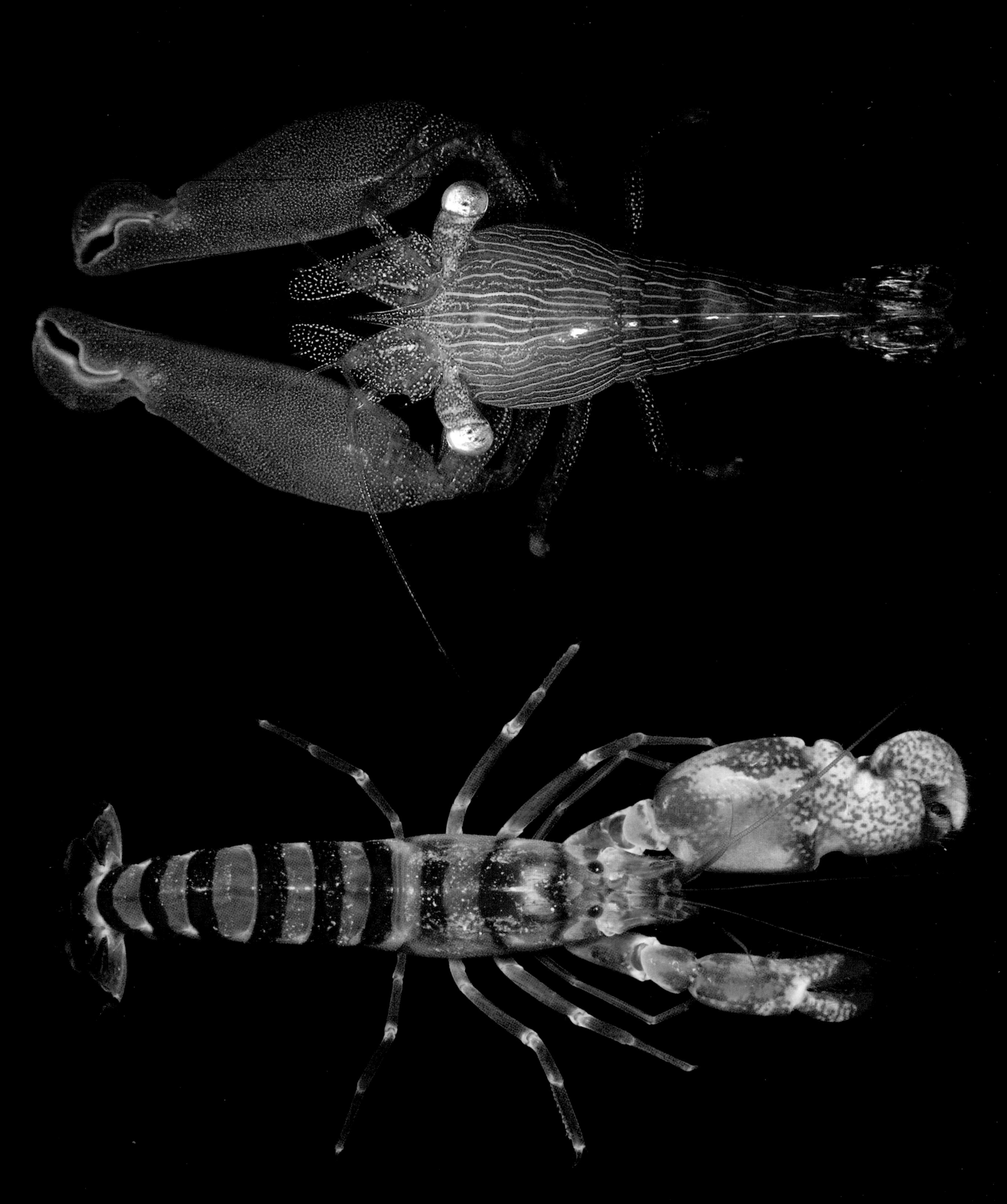

27

‹ 26 Segmented appendages, like the limbs and antennae of this marine amphipod (*Acanthostepheia malmgreni*), are a defining feature of the arthropods.

27 This mantis fly beautifully demonstrates the variety of structures that outgrowths of the arthropod body have evolved into, such as mouthparts, sensory appendages, walking limbs, grasping limbs and wings (*Climaciella brunnea*).

28

29

28 The leaf-like appendages sprouting from the abdominal segments of this mayfly nymph (unidentified heptageniid) are gills – one theory is that insect wings evolved from thoracic gills. The 'buds' that will give rise to the adult's wings are clearly visible covering the first part of the abdomen. Many insect species have an aquatic life stage with numerous adaptations for maintaining a grip on slippery rocks in fast-flowing water. This one has a flattened, hydrodynamic shape and very strong limbs.

29 Chelicerae are the characteristic mouthparts of the arachnids, horseshoe crabs and sea spiders. This arachnid (unidentified species) is a solifugid – these have the largest chelicerae, relatively, of any arachnids. They are typically desert specialists.

30 SEM of the chelicerae of an opilionid (*Phalangium opilio*). The eyes of opilionids, commonly known as harvestmen, are normally mounted on a small turret (purple structure, top).

30

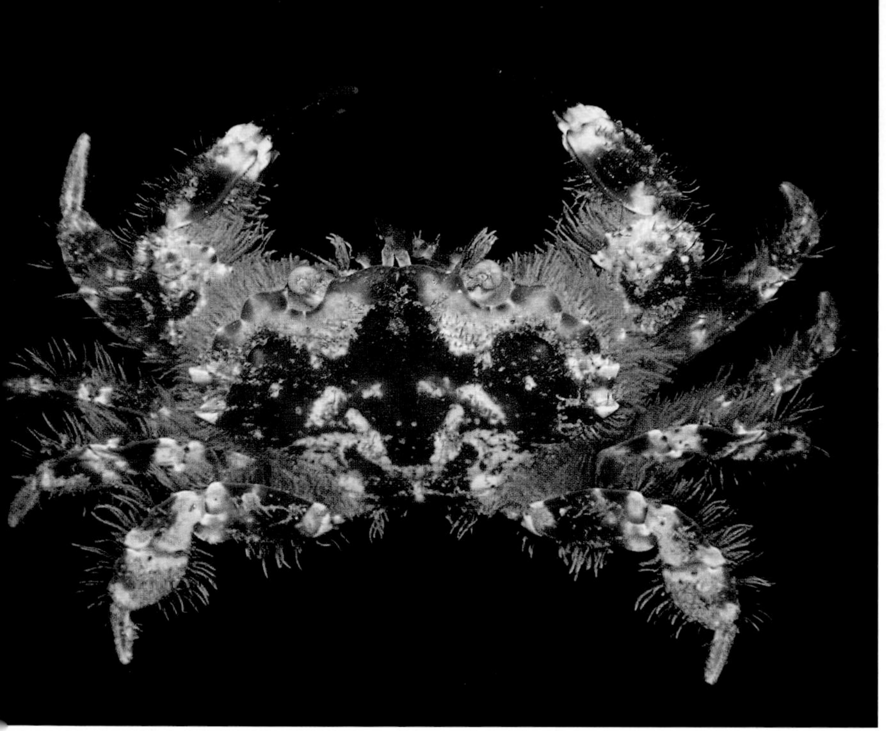

31

32

31 In amongst the other organisms of the coral reef this brightly coloured crab (*Lophozozymus incisus*) is very well camouflaged.

32 In many crustaceans the arthropod exoskeleton is heavily reinforced with calcium carbonate making a formidable suit of armour (*Petrolisthes laevigatus*).

33 A huge variety of extensions and outgrowths can develop from the arthropod exoskeleton, as exemplified by this crab (*Pilumnus vespertilio*).

34 Some crabs have flattened limbs, an adaptation for swimming or burrowing (*Ashtoret lunaris*).

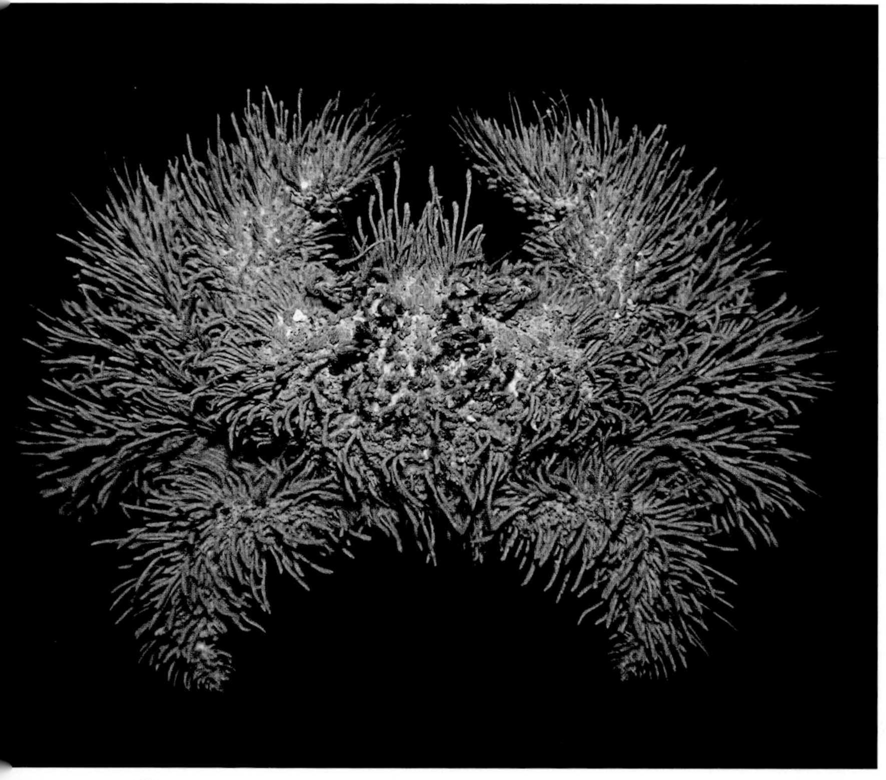

33

34

35

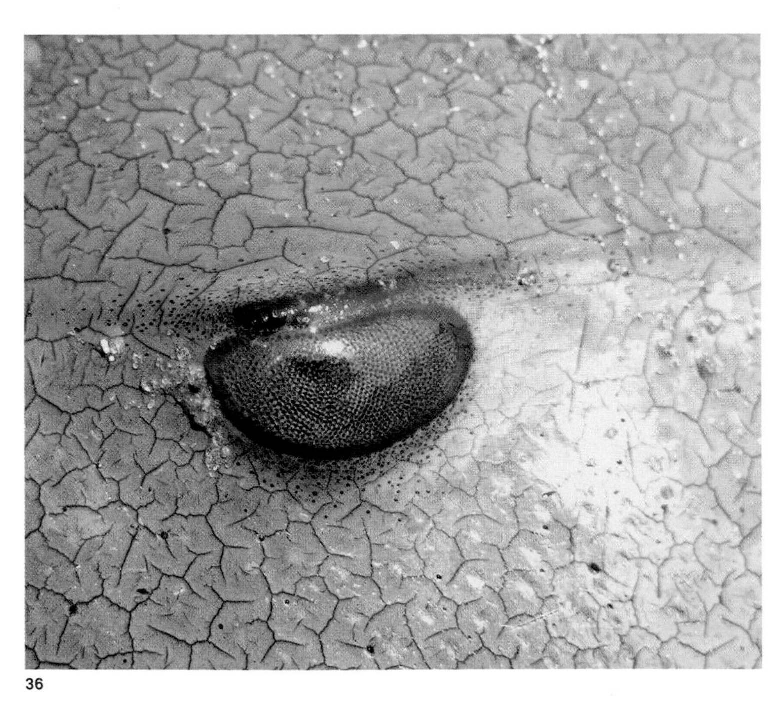

36

35 The compound eyes of a cynipid wasp (unidentified species). Some insects have simple eyes in addition to compound eyes, three of which can be seen on the top of this wasp's head.

36 The compound eye of a horseshoe crab (*Limulus* sp.) is made up of approximately 1000 units, each of which consists of a lens overlying 8–14 photoreceptors.

37

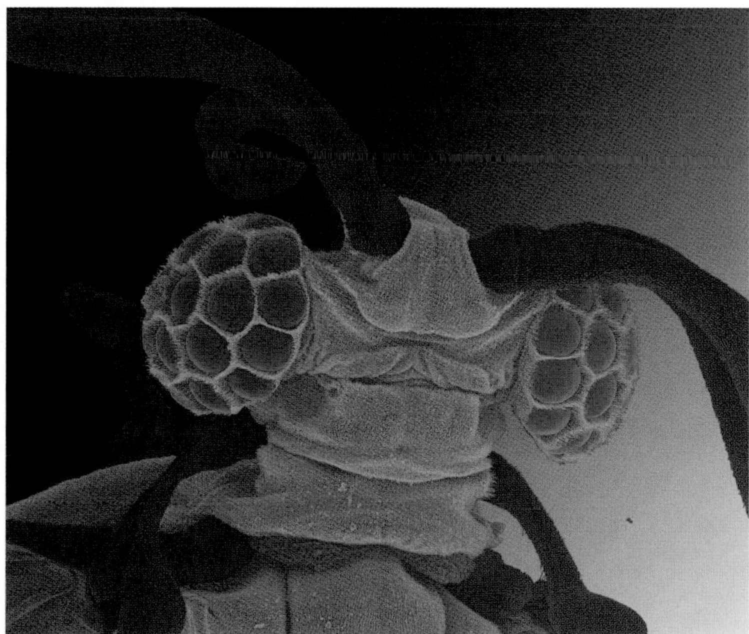

38

37 The individual hexagonal units making up the compound eye of a parasitoid wasp. Compound eyes made up of many discrete units are characteristic features of many arthropods (unidentified chalcid).

38 The compound eyes of an adult male strepsipteran. These structures are unique among the living arthropods in that they combine principles of both compound and simple eyes. In addition to their strange morphology, the strepsipterans have among the most outlandish life histories of all the terrestrial arthropods. They are parasites that develop in the abdomen of other insects, often sterilizing the host and changing its behaviour. The grub-like adult females never leave the host, so the very short-lived males must find a partner and then mate with her via a tiny pore at the base of her head.

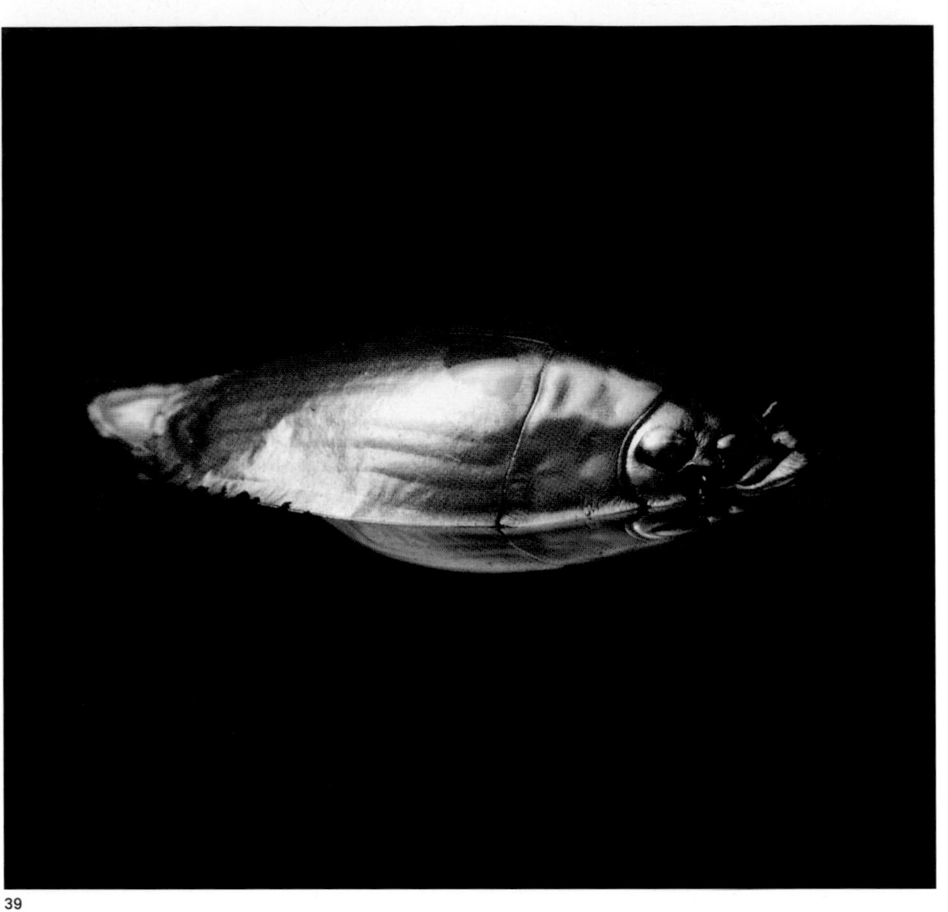

39

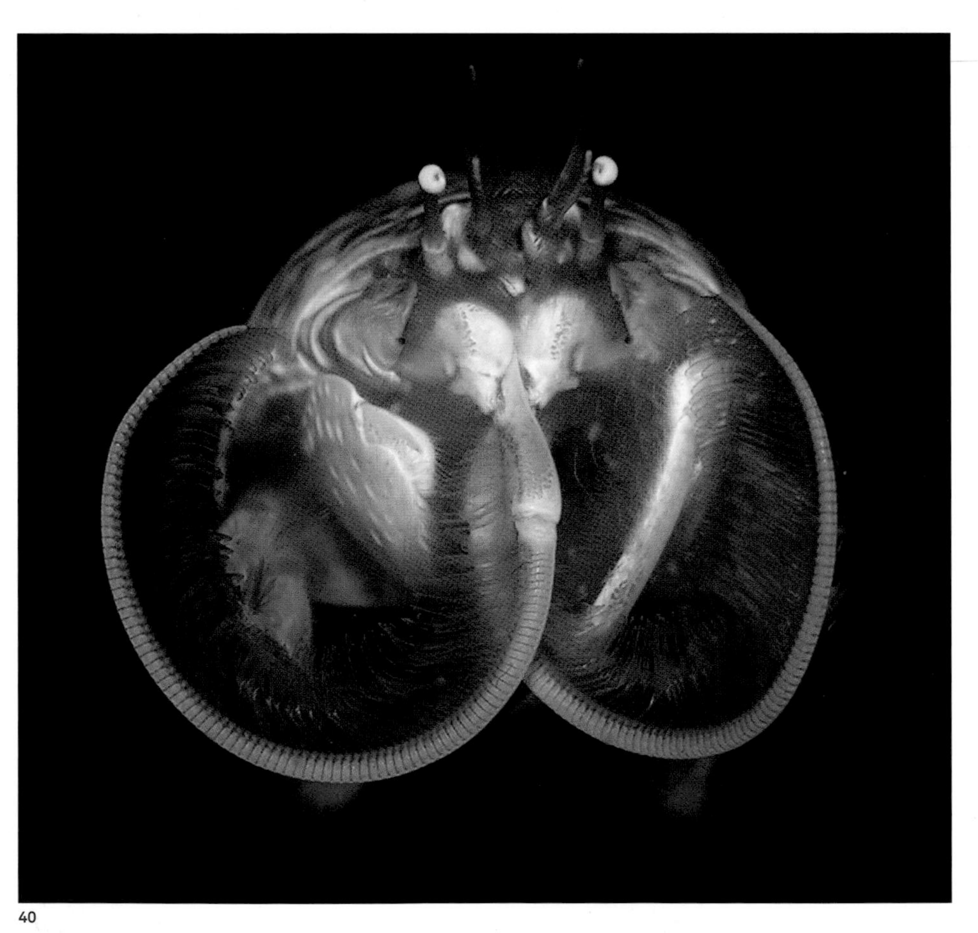

40

39 Insects, such as this whirligig beetle (*Dineutus sublineatus*) are much more successful in freshwater habitats than they are in marine ones. The eyes of whirligig beetles are divided into two, enabling them to see above and below the water simultaneously.

40 *Emerita* spp., commonly known as mole crabs, burrow in the shoreline sediments and use their elaborate antennae to filter food.

›41 Ectoparasites have evolved some impressive adaptations for latching onto their hosts. This ectoparasitic crustacean has elaborate suckers (modified mouthparts; top) and robust spines at the base of its forelimbs. To feed, the parasite repeatedly thrusts a stylet (the thin, tapering structure between the suckers) into the host to break up the tissues and release fluids (*Argulus foliaceus*).

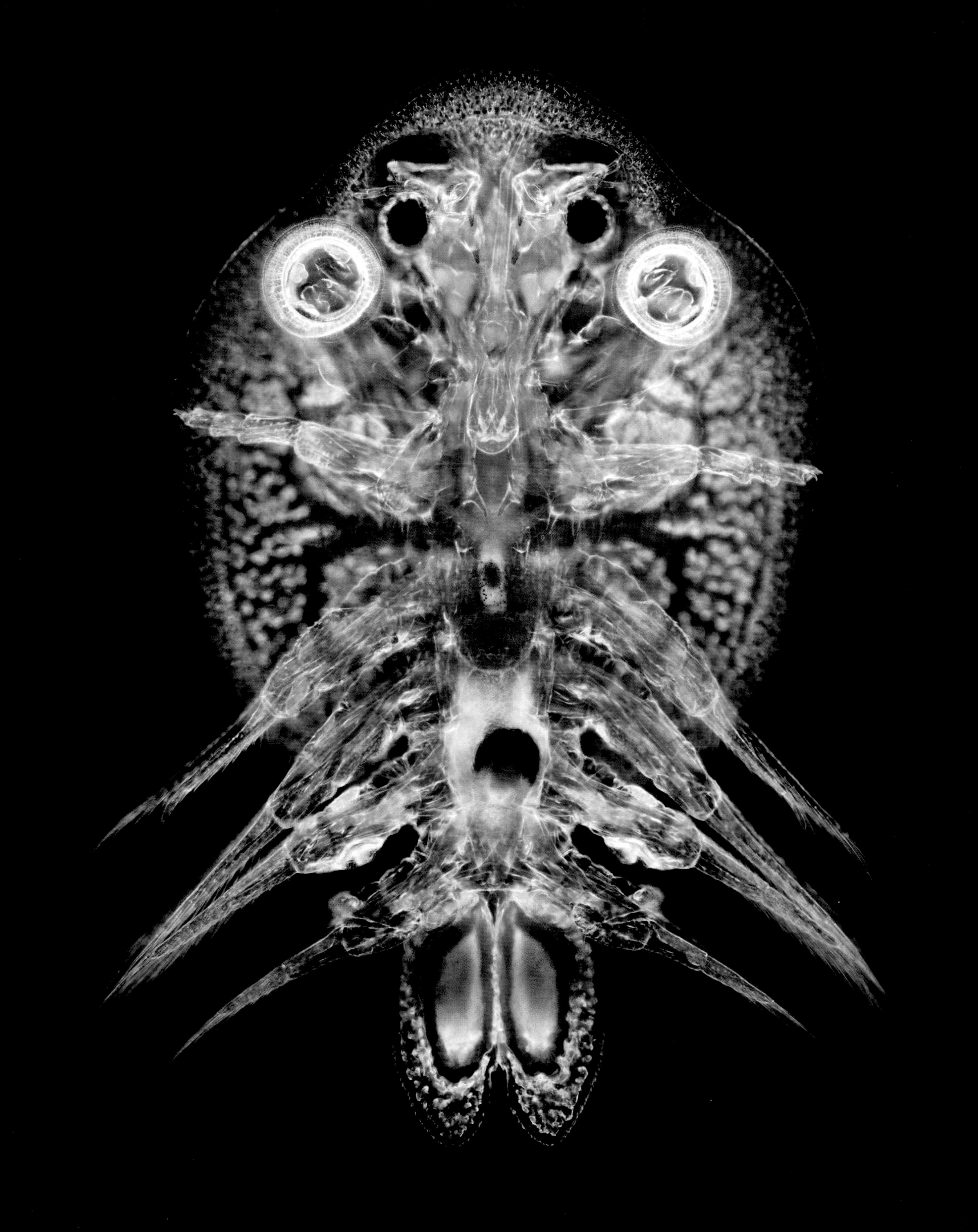

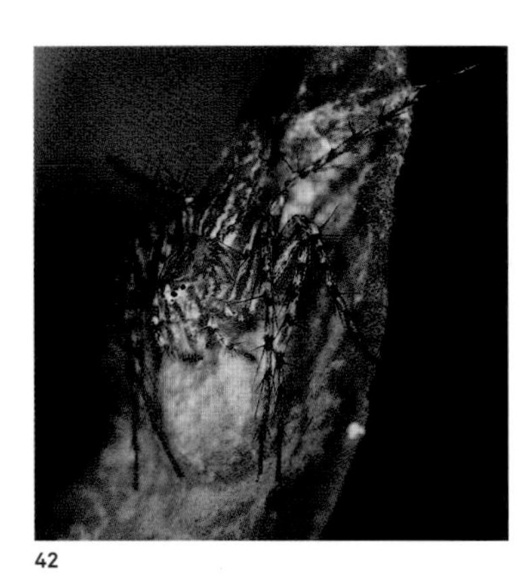

42

43

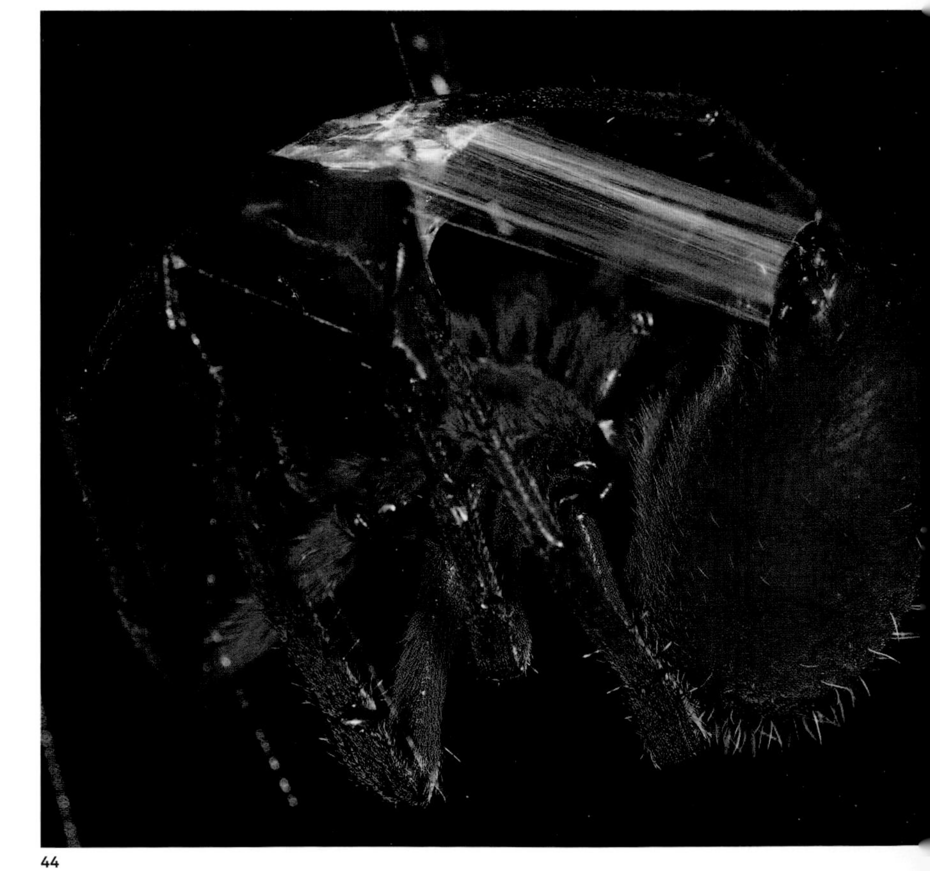

44

42 Many arthropods use silk to construct a protective cocoon for their eggs. This spider has constructed its cocoon on a leaf, which itself is suspended from a taught line of silk (unidentified oxyopid).
43 Phoretic mites clinging to the body of an opilionid (both species unidentified).
44 An orb-web spider wrapping its prey (unidentified species).
45 This parasitoid (*Aphidius ervi*) is laying an egg inside an unfortunate pea aphid (*Acyrthosiphon pisum*). Parasitoids, unlike parasites, kill, sterilize or consume their host. This way of life is very common among the insects.

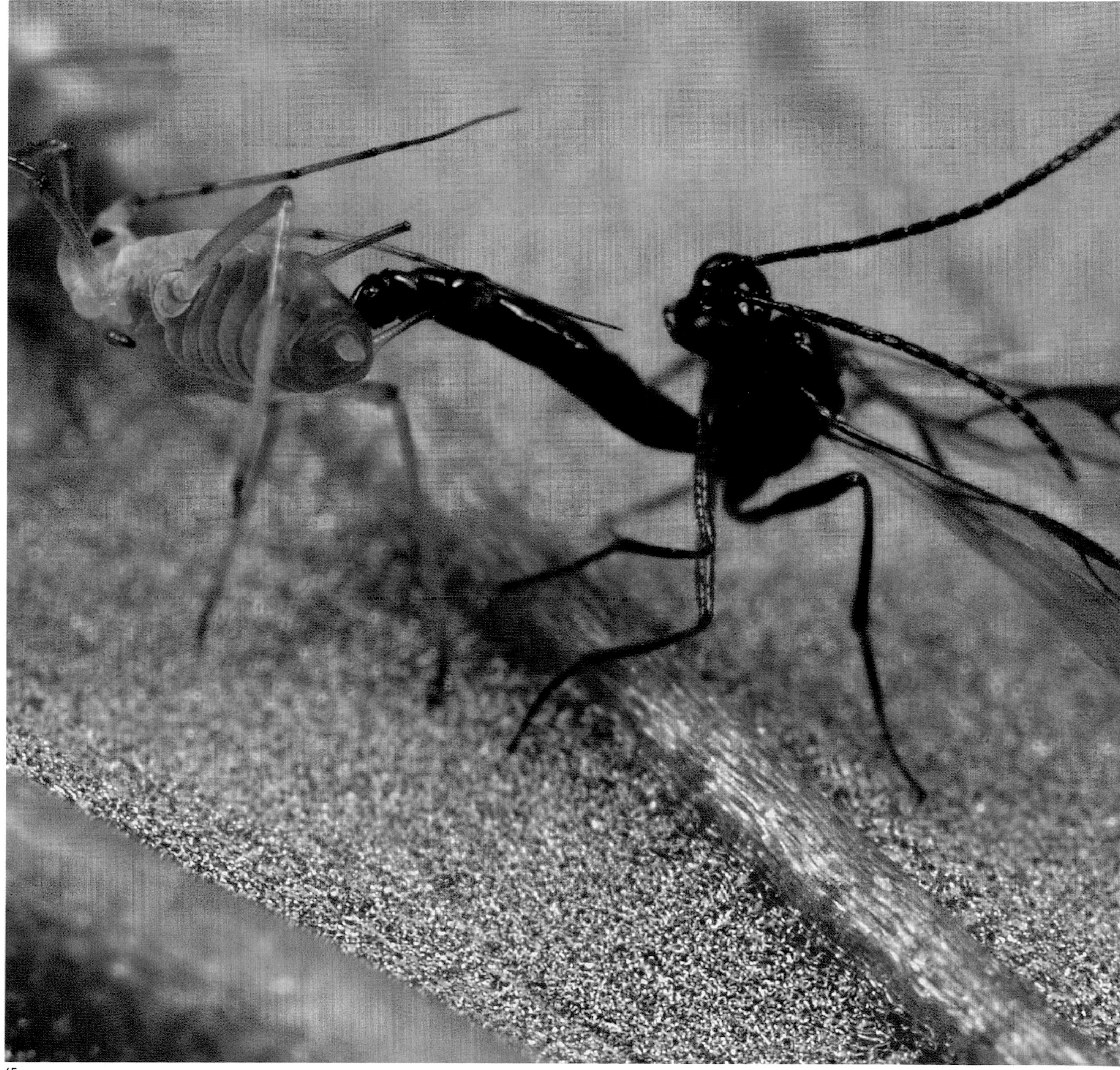

45

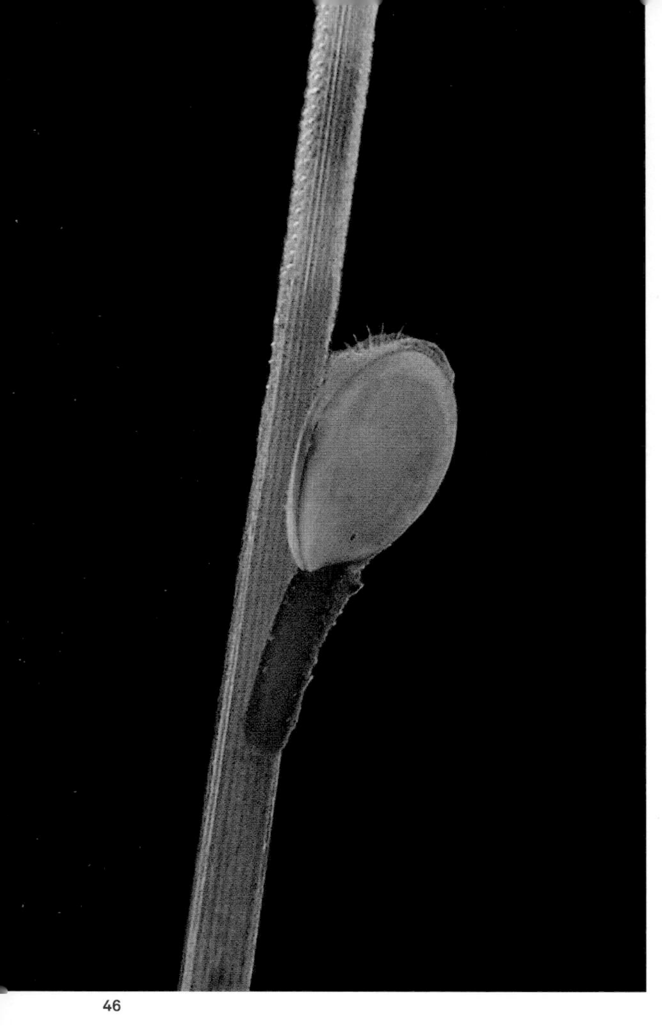

46

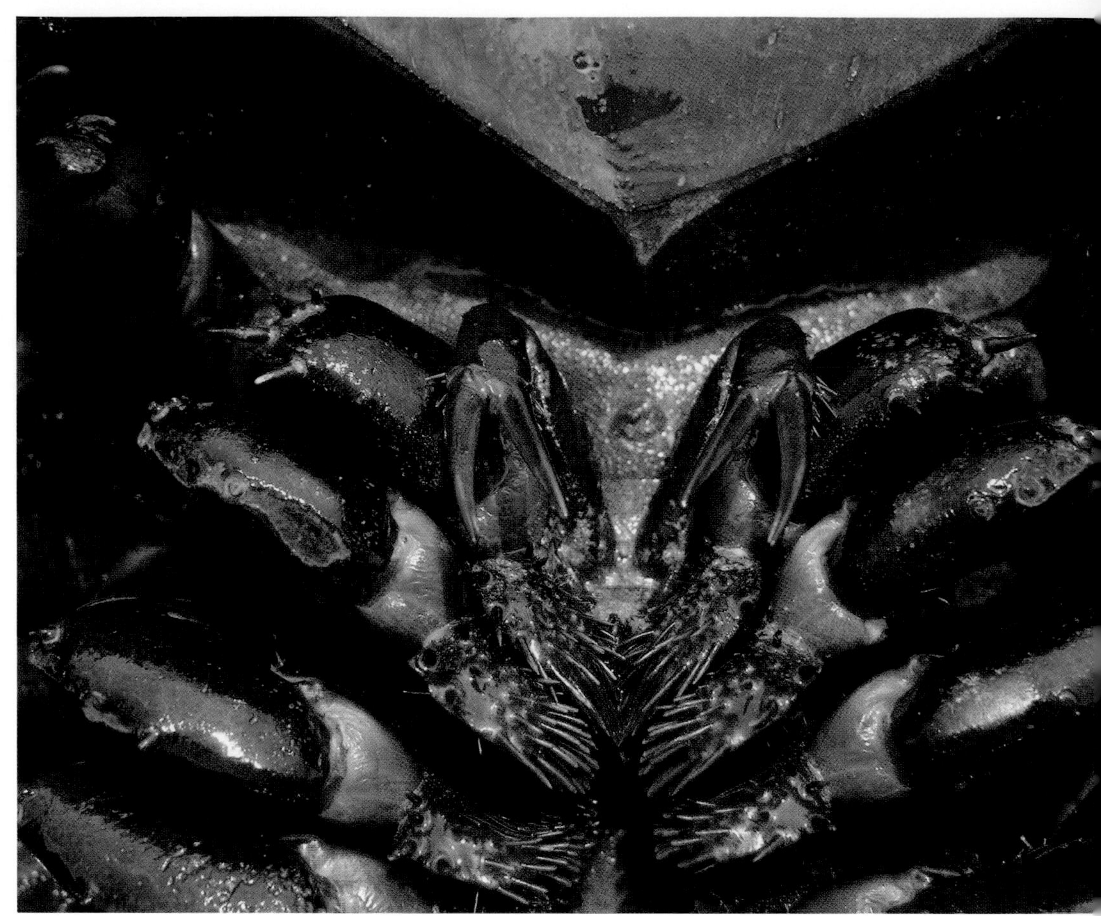

47

48

46 Many arthropods live on the bodies of other animals without causing any harm. This tiny goose barnacle (*Lepadomorpha* sp.) is attached to the spine of a sea urchin.

47 The underside of a horseshoe crab showing the pincered chelicerae between the walking legs (*Limulus* sp.).

48 Instead of simply eating the leaves of plants, many herbivorous insects have become specialist sap-suckers that use sharp, straw-like mouthparts to get at the sugary fluid (*Oncometopia orbona*).

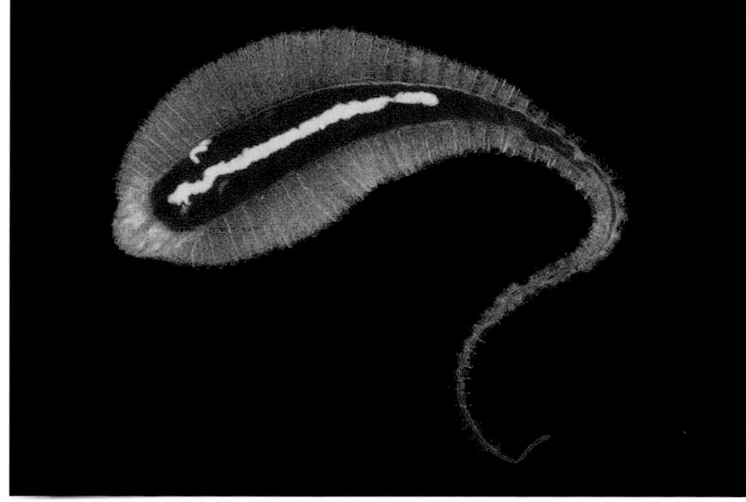

49

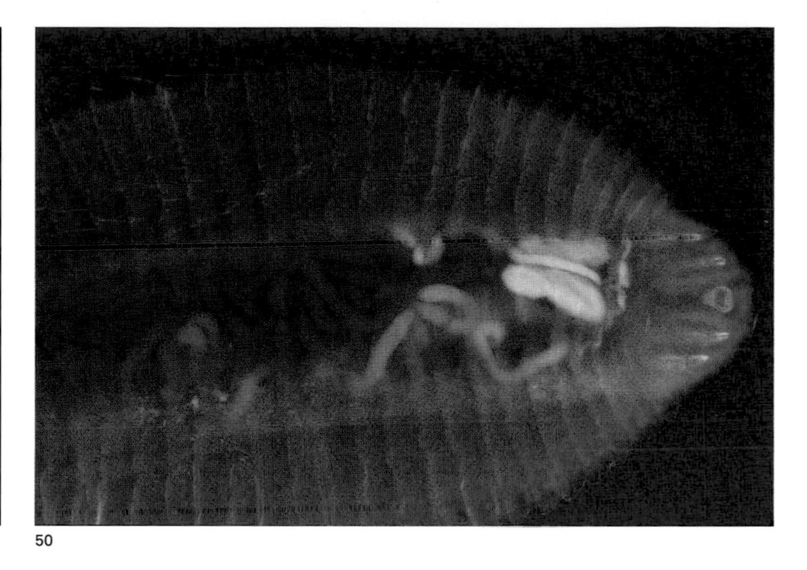

50

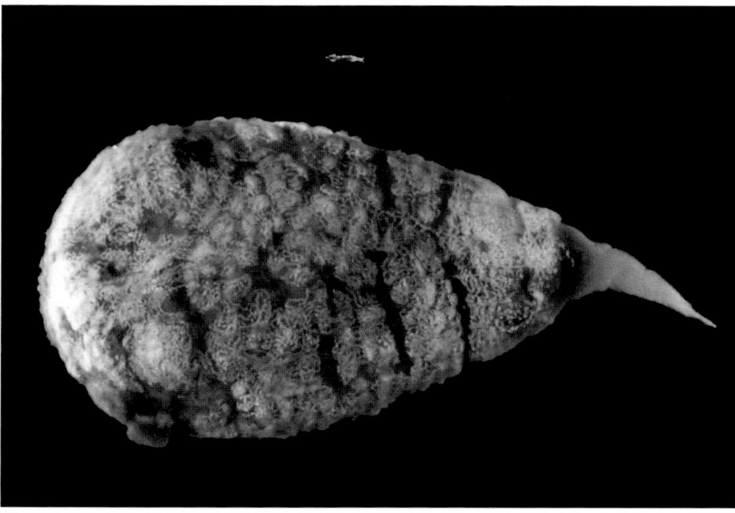

51

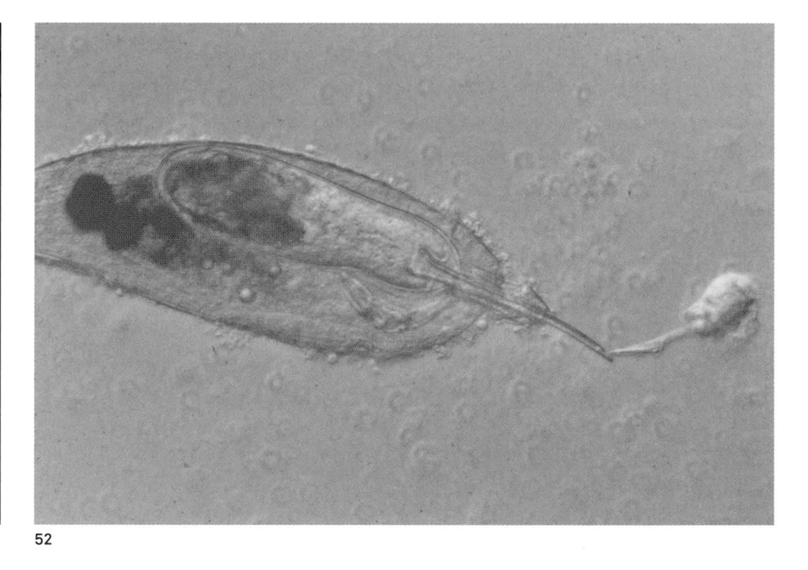

52

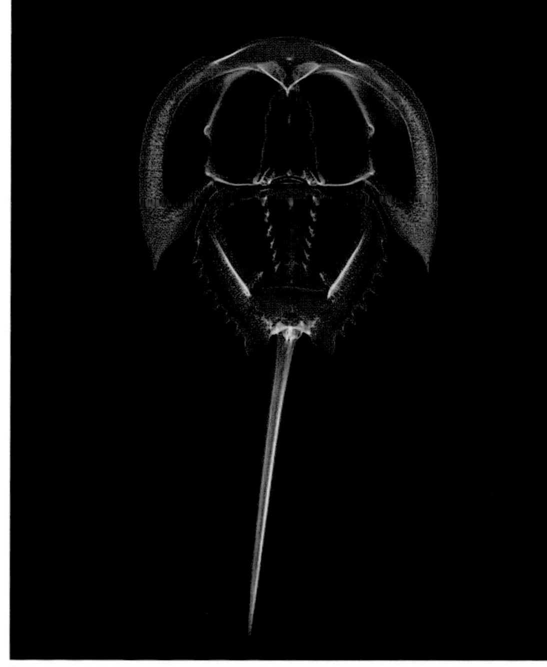

53

49 Pentastomids were once considered to be a distinct lineage of animals, but we now know they are actually very specialized, endoparasitic crustaceans. They live in the nasal passages, sinuses and lungs of snakes, crocodiles, lizards, birds and mammals, where they feed on blood, epidermal cells and mucus. This species (*Linguatula arctica*) lives in the nasal cavity of reindeers.

50 Close-up of the reindeer pentastomid *Linguatula arctica*. The two pairs of anchoring hooks surrounding the mouth (right) are the vestiges of typical, arthropod limbs.

51 *Sarcotaces* spp. are bizarre parasitic copepod crustaceans that, as adults, have lost almost all the typical outward arthropod traits of their ancestors. The male (top, tiny and dart-like) is dwarfed by his relatively gigantic mate.

52 One stage in the life cycle of *Sacculina* spp. is a larva that is little more than a living syringe. It injects a mass of cells (right) into the host, which will go on to develop into the root mass-like adult.

53 Confusingly, horseshoe crabs are more closely related to arachnids and sea spiders than to crustaceans. They have existed, superficially unchanged, for hundreds of millions of years (*Limulus polyphemus*).

54

55

56

57

58

54 Instead of pretending to be dangerous, some arthropods mimic objects that are simply unpalatable. This katydid (*Pycnopalpa* sp.) is pretending to be a diseased leaf.

55 *Peltogastrella gracilis* is a close relative of barnacles that parasitizes other crustaceans, in this case a hermit crab (*Pagurus edwardsii*). The yellow, sausage-shaped structures are the parasite's brood sacs.

56 This unfortunate shore crab has been parasitized by *Sacculina carcini*, another close relative of barnacles. The parasite grows through the victim as a root-like mass, belying its arthropod affinities. Here, the brown brood sac of one of these parasites pokes out from beneath the abdomen of its host.

57 As aquatic animals can be so much more difficult to study than their terrestrial counterparts, we still know very little about how these animals live in the wild. For example, these tube-building, amphipod crustaceans may live cooperatively in their little colonies (*Ericthonius difformis*).

58 Many insects can secrete substances to prevent desiccation, to camouflage themselves and to keep parasitoids and predators at bay. Some sap-sucking insects secrete a powdery, wax-like substance to this end (unidentified flatid hopper).

59

59 Mimicry is an effective form of defence and one that is rampant among arthropods, particularly the insects. From above, many tree hoppers are reminiscent of ants, which are normally avoided by many predators (*Heteronotus* sp.).

60 Ants have evolved extremely complex societies. Here, two small workers of a leaf-cutter ant (*Atta* spp.) protect one of their larger sisters from parasitic flies while her jaws are occupied carrying a piece of a leaf back to the nest. A fungus is cultivated on the plant matter and it is that the ants feed on.

60

62

63

‹ 61 Arthropod mouthparts have evolved into a huge variety of forms. The telescoping labium of this small rove beetle (*Stenus clavicornis*) can be shot out under fluid pressure to catch prey, such as springtails.

62 Plants produce a variety of noxious chemicals to deter herbivorous animals. Not only can many insects withstand these toxins, but they can also assimilate them to keep their own enemies at bay, often broadcasting their defences with bold colours and patterns (unidentified chrysomelid).

63 Some arthropods are dedicated parents. In sea spiders it is the male who is responsible for looking after the eggs. He clings onto the egg mass with specialized legs and the young may remain with him for a while after hatching (unidentified species).

64

64 This cyprid larva will eventually settle and develop into a barnacle. A planktonic larval stage allows aquatic arthropods to disperse over huge distances.

65 Sea spiders are typically predators of other benthic, marine animals. This individual (*Nymphon grossipes*) is feeding on a hydroid polyp. Note the numerous amphipods hitching a ride.

65

66

66 Stages in the life cycle of the ant, *Pseudomyrmex gracilis*. Adult (left); larval stages (second to fourth from left); pupal stage (first and second from right). The resting stage (pupa) is unique to the insects and it involves an incredible reorganization of the tissues – transforming the larva into the adult.

67 Complex interactions, mostly unseen, take place between different arthropod species. Here, a predatory mite feeds on a bark louse (unidentified species).

› 68 This small robber fly (*Holcocephala fusca*) is a specialist predator of other flying insects. To catch such prey requires acute vision, a sophisticated nervous system and incredibly fine muscle control.

67

Priapulida

(penis worms)
(Latin *priapulus* = little penis)

Diversity
c. 20 species

Size range
~0.5 mm to ~40 cm
(~0.02 to ~16 in.)

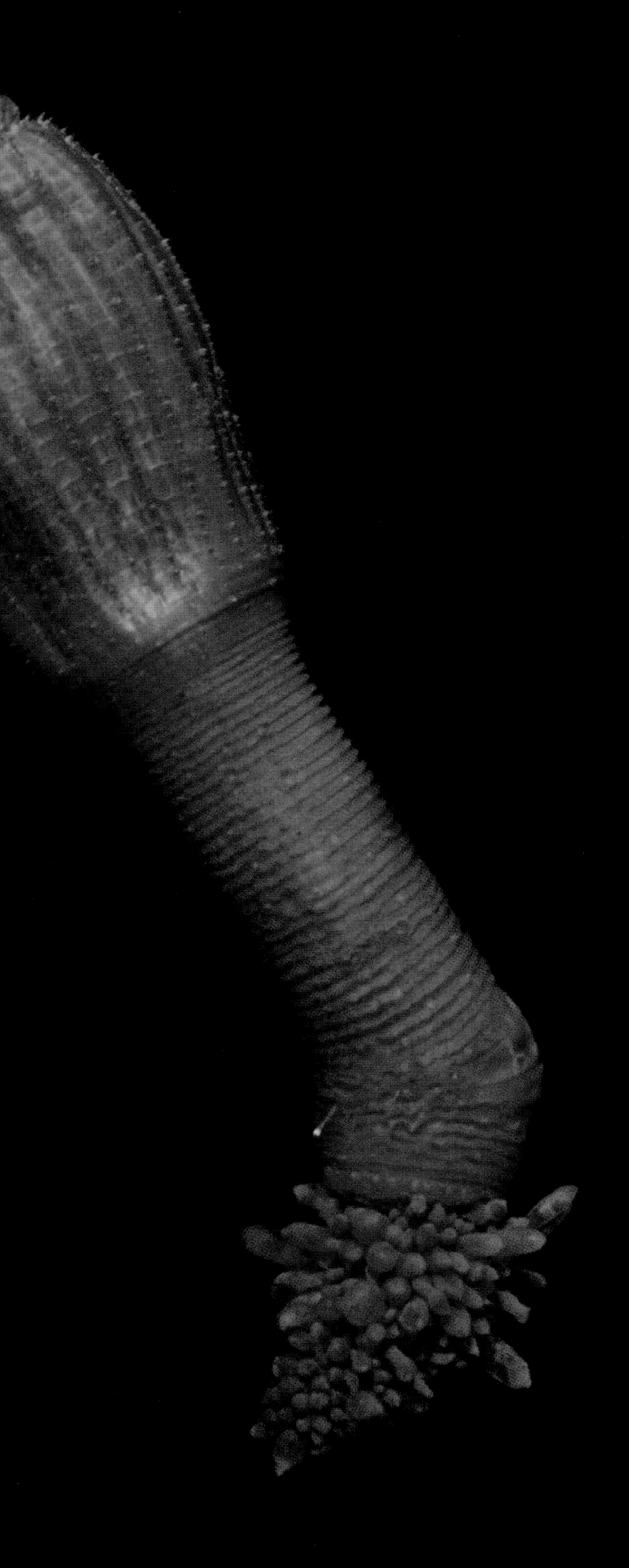

1 *Priapulus caudatus*, one of the more commonly seen priapulids. The bulbous proboscis is clearly visible as are the branching, posterior appendages, which may function in respiration.

2

2 The well-defined proboscis, trunk and branching caudal appendages of a priapulid (*Priapulus* sp.).

Priapulids, or penis worms, have not attracted much attention from the zoology community as a whole, so we know very little about how they live in the wild. Exclusively marine, they are found throughout the oceans, but always on or in the seabed.

FORM AND FUNCTION

The chunky worm-like body of a priapulid has no distinct head, but it is divided into a large trunk, an anterior proboscis and sometimes one or two tail-like appendages dangling from the posterior end [1, 2, 3, 5]. The scientists who named the first described species likened them to a human penis, hence the common and scientific names. In many species the body, particularly the proboscis, bristles with various outgrowths including small spines known as scalids that are unique to the priapulids and their close relatives, the loriciferans (see pp. 182–85) and kinorhynchans (see pp. 186–89).

Inside the body there are well-developed muscles, a straight gut, kidneys and a large cavity filled with pinkish blood. The blood contains two types of cell, one of which is loaded with a respiratory pigment. Exactly how gas exchange is accomplished is a mystery, but in many species the tail-like appendages may have a respiratory function.

A rudimentary brain, nothing more than a nerve ring encircling the mouth, is connected to a single nerve cord that extends the length of the body. Nerves ramify from this cord. They reach out to numerous, tiny sensory structures on the body surface, including the scalids. There are no eyes or any other obvious sensory organs.

LIFESTYLE

Exclusively benthic, priapulids have been found all over the world and at all depths. Some species move through mud and sand by using their muscles to constrict their body and force blood into proboscis whereby by it extends out into the sediment. Forcing yet more blood into the proboscis makes it swell, serving as an anchor, while muscles running the length of the animal pull the rest of the body forward. Other species simply spend most of their time in a loosely assembled tube with the mouth more or less flush with the surface of the seabed. Unsurprisingly, not a great deal is known about what they eat or how they go about catching it. Some species are thought to be carnivorous, using their eversible proboscis to engulf other benthic creatures such as polychaetes, which are snared and swiftly despatched with substantial pharyngeal teeth. *Maccabeus* spp. have spiny tentacles, which are thought to close over prey [4], while other priapulids just gorge themselves on delicious sand and mud, digesting any microorganisms that may be adhering to the sediment particles.

There are separate sexes and in some species the males and females are noticeably different. Larger species simply shed their sperm and eggs into the water for external fertilization, while internal fertilization is thought to be the norm amongst the smaller

3 Some of the priapulids inhabit the tiny spaces between sediment grains on the seabed. *Tubiluchus* spp. have a very long and thin caudal appendage and a proboscis wreathed by scalids.

3

4 A tube-dwelling, predatory priapulid (*Maccabeus* sp.). This species lies hidden in the sediment with just its anterior, spiny tentacles poking above the surface. The tentacles close over small prey animals to capture them.

5 SEM of the mouth of the meiofaunal priapulid, *Meiopriapulus fijiensis*, surrounded by short tentacles.

species. The larvae that hatch from the eggs are very similar to the adults, but they have no posterior appendage and the trunk is enclosed in a tough little suit of armour known as a lorica. The lorica also can accommodate the retracted front part of the body if needs be. The larvae have to shed their cuticle as they grow and the lorica eventually disappears as they metamorphose into adults.

As we know so little about these animals we can only speculate on their contribution to the ecosystems in which we find them. In some places they can be very common; some 85 adults and 58,000 larvae per sq. m (or 70 adults and 48,000 larvae per sq. yd) of seabed have been recorded in some places. Many species also thrive in anoxic sediments suffused with eggy, highly toxic hydrogen sulphide; few other animals can tolerate such a harsh environment. Where priapulids are abundant they must have an impact on the populations of other benthic organisms, not only by consuming and being consumed, but also because large numbers of them churn up the sediment by squirming through it. In these ways they are likely to make a significant contribution to the movement of energy, oxygen and nutrients through these seabed habitats and accordingly do their bit for the ecology of the world's oceans as a whole.

ORIGINS AND AFFINITIES

For such a small lineage of animals the fossil record of the priapulids is surprisingly substantial. Animals interpreted as their ancient ancestors are known from the Burgess Shale. Numerous trace fossils, tracks very similar to the spoor of living priapulids, are known from Cambrian rocks; if the interpretations are correct, they indicate that this body plan was well established 500 million years ago. Indeed, not only does it appear that the forerunners of living species abounded in ancient seas, but it seems that the considerable variety in form of the few species that we see today is no more than an echo of their once diverse past. Evidence suggests that as animal diversification gained pace long ago in the Cambrian seas, the priapulids were outcompeted by other creatures, such as polychaete worms, that exploited the opportunities of the seabed more effectively.

Loricifera

(brush-heads)
(Latin *loricus* = corset, girdle; *fero* = to bear)

Diversity
c. 30 species

Size range
0.25 to 0.85 mm
(0.01 to 0.03 in.)

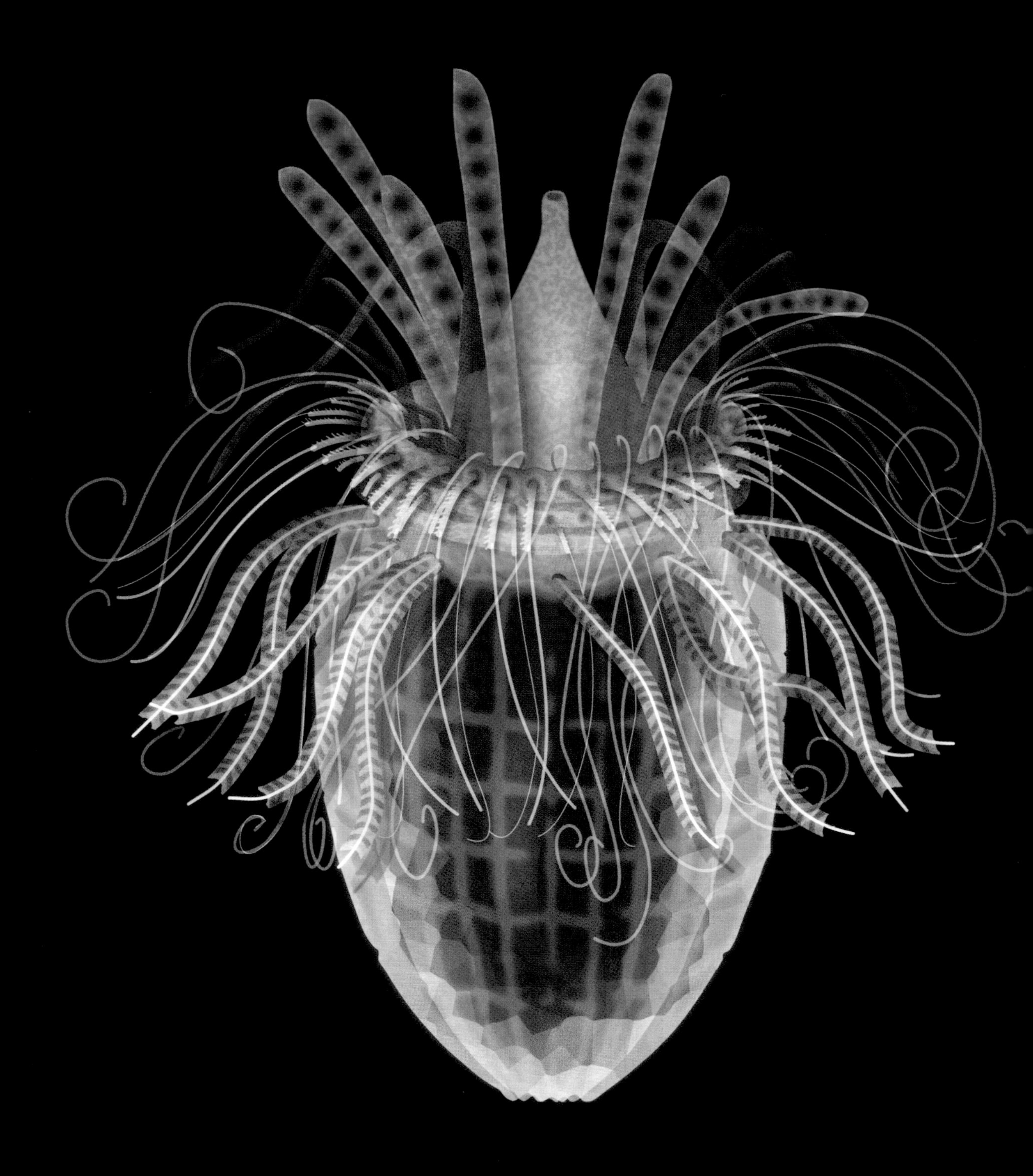

1 Much of the body of a brush-head is shielded by chitinous plates. The head bears a mouth cone equipped with stylets and numerous scalids (*Pliciloricus* sp.).

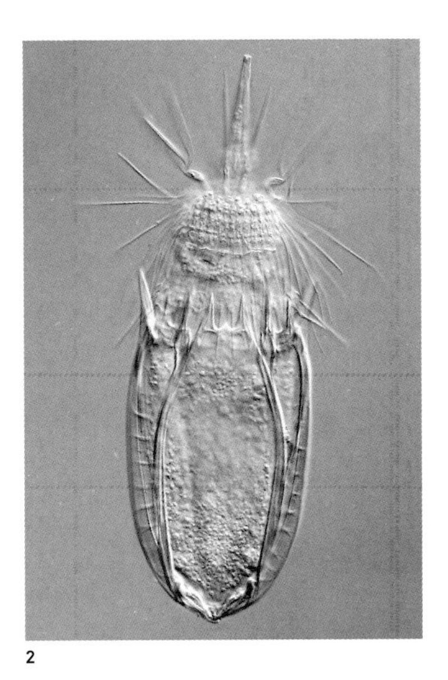

The brush-heads are yet another group of recently described and enigmatic marine animals. Even the giants among them are minute animals that spend their life in amongst the sediment grains of the seabed. They were first seen in 1974 and formally described by the zoologist Reinhardt Kristensen in 1983. This scientist and his colleagues have discovered all sorts of incredible species in their investigations of the meiofauna, the myriad tiny animals inhabiting the spaces between aquatic sediment particles.

FORM AND FUNCTION

A fraction of millimetre long, the brush-heads are surprisingly complex animals with a body that is composed of around 10,000 tiny cells. The body is divided into a complex anterior structure (known as the introvert), a thorax and a trunk, which is enclosed in chitinous plates forming a suit of armour – the lorica, from which the scientific name of these animals derives [1]. Sprouting from the introvert is a telescoping mouth cone equipped with several stylets and as many as 300 large and elaborate spines, known as scalids, arranged in nine rings [2, 4, 5]. At the base of the introvert is a complex, bulbous pharynx, above which, filling much of the introvert, is a disproportionately large brain reaching out to the rest of the body via ten nerve cords. The numerous scalids are thought to be sensory.

Brush-heads are sufficiently small for gases to diffuse freely in and out of their body across the body wall; however, waste products are excreted via specialized cells that in fact are part of the animal's gonads. Muscles can pull the introvert into the safety of the lorica, squeeze the trunk and move the scalids around. The gut is well developed and is divided into distinct sections: oesophagus, mid-gut and hind-gut.

LIFESTYLE

One barrier to increasing our understanding of these animals is the fact that very few live specimens have been observed. The adults adhere tenaciously to sediment particles using adhesive glands at their rear end of the body and the only way to get them to relinquish their grip is to dunk them in freshwater. Needless to say, this does not do them the world of good – the osmotic shock kills them. Dead specimens are fine if we just want to look at morphology, but in terms of understanding how an animal lives they are not good enough. But we can speculate a little based on what we know about their morphology. The mouth cone, its stylets and the large sucking pharynx suggest that brush-heads may be carnivores that drain their prey of fluids. The elaborate scalids may be used to taste and smell their environment, to move around and to capture food.

Like everything else about these animals, what we know so far about the ways in which they reproduce is surely just the tip of the iceberg. Separate sexes are common and the males and females are often quite different. The eggs of some species are fertilized internally, but mating has never been observed. The life cycles are often elaborate, and can vary within the same species, seemingly dependent on whether the individuals in question are well fed or not. Asexual life cycles are known where the larva outstrips the adult in size and precociously develops an ovary itself before transforming into a cyst-like state. That ovary gives rise to several eggs, all parthenogenetic offspring of the neotenous larva that is now a cyst. These eggs develop into larvae and the cyst that spawned them disintegrates, liberating the numerous young, which go on to moult, grow and repeat this odd asexual routine again until the juvenile stages are extremely common, much more so than the adults. This reproductive strategy is reminiscent of the parasitic Platyhelminthes (see pp. 278–93), where a single egg can spawn huge numbers of juveniles.

The so-called Higgins larvae of some brush-head species are equipped with long spines and paddle-like structures known as 'toes' that they use to propel themselves through the water during their fleeting time as pelagic animals [3, 6]. The toes of other species that never leave the benthic realm are long, slender and equipped with adhesive glands for maintaining a good grip on the substrate.

Although around 30 brush-head species have been identified to date, hundreds or very probably thousands more await discovery and identification. They have been discovered living at all depths from locations all over the world, and now zoologists know how to look for them it turns out they can be rather common animals of the seabed. Some species have even been found 3000 m (10,000 ft) down in the Mediterranean Sea, living out their entire lives in sediment completely lacking in oxygen. Mitochondria, the power-plants of most cells, are useless in the absence of oxygen, so these brush-heads are equipped with alternative, tiny power-plants, commonly seen in single celled

2 Loriciferans are very difficult to study since they adhere tightly to sediment grains. Almost all of what we know about them comes from dead specimens (*Nanaloricus* sp.).

4

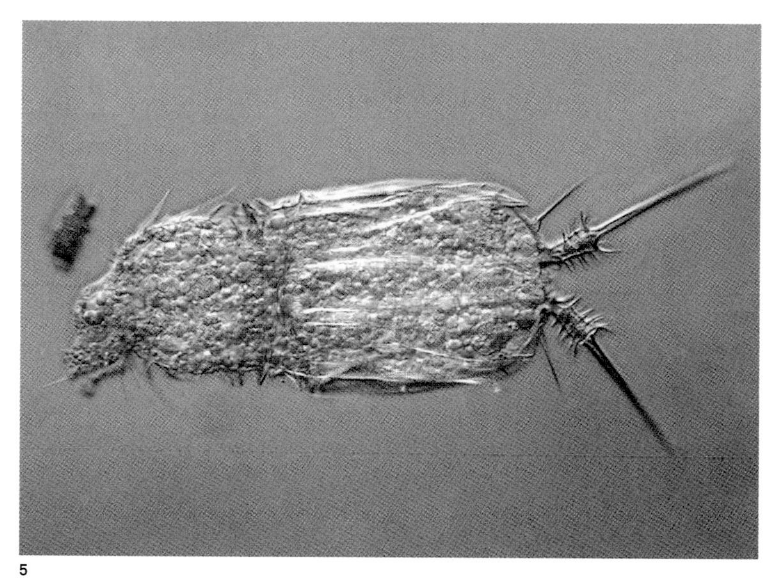

5

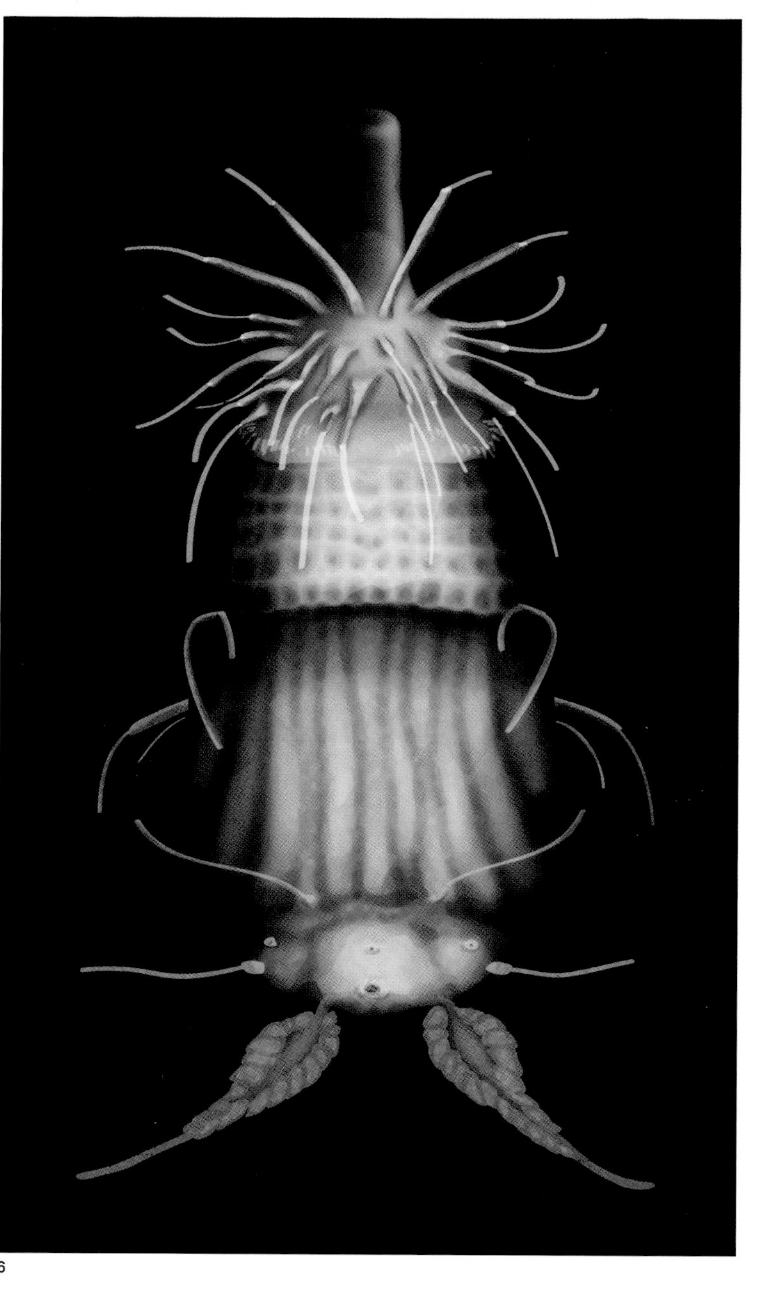

6

‹3 The immature stage in the life cycle of a loriciferan is known as a Higgins larva. In some species these are free swimming (*Armorloricus elegans*).

4 The bristle-like structures known as scalids, surrounding the mouth cone, are a characteristic feature of these animals (*Rugiloricus* sp.).

5 Along with a huge range of other tiny animals, loriciferans occupy the labyrinthine channels and spaces between sediment grains on the seabed (*Armorloricus elegans*).

6 Brush-head larvae have distinctive paddle-like 'toes'. Depending on the species these are used for anchoring the animal to the substrate or swimming (*Nanaloricus* sp.).

organisms, which can generate cell fuel in anoxic conditions.

As the lives of brush-heads in their hidden world are essentially unknown, we can only speculate on how they might contribute to the dynamics of the oceanic ecosystem. For all we know they may be numerous enough to be both important predators and prey.

ORIGINS AND AFFINITIES

Being such recent additions to the roll call of animal life, it will come as no surprise the heritage of the brush-heads is completely unknown. They are such small animals that fossils have never been found, but their morphology provides us with many valuable clues as to their affinities. Many of their characteristics are shared with the priapulids (see pp. 178–81) and the kinorhynchs (see pp. 186–89), another lineage of very small, benthic animals.

When they first came to light it was suggested that brush-heads were nothing more than larval priapulids capable of reproduction. It is true that brush-heads are very similar to the priapulid larvae, but this resemblance is superficial. If this proposed relationship was real, the loriciferans would not have larval stages and the complex life cycles we see. Rather, these two lineages, as well as the kinorhynchs, share a common ancestor – an ancient animal that may have resembled a brush-head larva.

Kinorhyncha

(mud-dragons)
(Latin *kinein* = to move;
rhynchos = snout)

Diversity
c. 180 species

Size range
~0.1 to ~1 mm
(~0.004 to ~0.04 in.)

1 Mud-dragons are tiny animals of marine sediments. Their heads bear seven concentric rings of characteristic spines known as scalids (*Echinoderes* sp.).

2

2 In this SEM the scalids are clearly visible (*Echinoderes spinifurca*). Along with the other animals that inhabit the tiny spaces and channels between sediment grains on the seabed, the mud-dragons are very poorly known and much of how they live in the wild is a mystery.

Living out their lives burrowing in mud or negotiating the endless channels and cavities between grains of sand, the mud-dragons are another lineage of small, enigmatic marine animals. Although they are often very numerous and easier to find than the brush-heads, there is still a great deal we do not know about these little creatures.

FORM AND FUNCTION

Rather worm-like, the head of a mud-dragon bristles with seven concentric rings of scalids [1, 2, 6]. Nestled in the centre of these scalids is a mobile mouth cone topped off with a wreath of nine stylets [7]. To extend this mouth cone, the animal applies fluid pressure from within, and retracts it using muscles. The trunk comprises 11 segments, each of which is clad in two or three tough plates [3, 4]. Collectively they form an inelastic cuticle that the animal must shed intermittently as it grows. Along the length of the trunk there is a variety of sensory structures, spines and adhesive tubes.

Issuing from a well-developed brain, which forms a ring around the pharynx, are two long nerve cords serving the rest of the body, plus a number of nerves serving the head. Along the length of the long nerve cords, corresponding to the animal's outwardly visible segmentation, there are swellings known as ganglia – amounting to a mini-brain for each segment. The mud-dragon's sensory organs feed information from the outside world to this central nervous system; they include the scalids and numerous receptors studding the body surface. Some species even have simple eyes equipped with lenses.

The body cavity is full of fluid, abounding with mobile cells. Like in other small animals, muscles constrict this cavity to shunt the fluid around the body for circulation, and also for purposes of extending the mouth cone in feeding and locomotion. In contrast to the situation in their close relatives the brush-heads (see pp. 182–85) and the priapulids (see pp. 178–81), the mud-dragon's excretory organs are separate from its substantial gonads.

LIFESTYLE

The mud-dragons use their scalid-adorned heads to edge their way through sediment. Muscles in the trunk constrict and move blood into the animal's front end, forcing it to protrude rapidly and thereby sweeping the scalids backwards and pulling the animal forwards. Muscles then retract the head back into the trunk so that

3

4

the locomotory cycle can start again [5]. The sediment in which they nose around is alive with all sorts of edible matter, including single-celled organisms and detritus. The stylets around the mud-dragon's mouth seize morsels of food before the suction of the muscular pharynx draws them into the body for digestion. Alternatively, food is trapped between the scalids and is sucked up from there.

There are separate sexes, but exactly how they go about making more mud-dragons is something of a puzzle. The females have discrete structures for receiving and storing sperm; therefore, we assume that the eggs are fertilized internally. The males of some species are known to produce tiny packets of sperm (spermatophores), but exactly how they transfer them to the female is not well understood, because copulation has only been observed in one species. Caught in the act, these mating mud-dragons were seen facing different directions, linked by their posterior ends. The actual mechanics of copulation were obscured by a brownish mass of mucus thought to be some kind of spermatophore. Spines towards the rear of the male may lock the sexes together during mating, as well as holding open the pore that leads to the female's ovaries.

Mud-dragons are ubiquitous, but are rarely seen. As well as living free on and in the sediment of the seabed, they have been encountered on seaweed, inside sponges and even on the bodies of other animals closely associated with marine sediments. They have been found in sediment samples from the seashore all the way down to 5 km (3 miles) below the waves. They normally occur in the first few centimetres of sediment, but the species of the shore are sometimes found up to 60 cm (24 in.) beneath the surface. In some places it is not unusual to find 45,000 individuals per sq. m (or 37,000 per sq. yd) of sediment, so they are by no means rare. New species are coming to light all the time and it is estimated there could be as many as 10,000 species of mud-dragon in total. The seabed is an enormous habitat, the largest on earth, so there must be many, many billions of mud-dragons all munching their way through tonnes of other organisms and edible detritus. We can only speculate on their contribution to the natural economy of the oceans, but it must be considerable.

ORIGINS AND AFFINITIES

The fossil record of the mud-dragons is non-existent. They are so small that the odds of finding fossil remains and being able to identify them as belonging to this lineage are very long indeed. The shared morphological characteristics of the priapulids, the brush-heads and the mud-dragons sees them grouped together as a three-pronged twig on the branch that includes all the other animals that moult their cuticle in order to grow. Only when we have managed to sequence the DNA of the mud-dragons and brush-heads will we be afforded a clearer understanding of their evolutionary relationships and their position on the animal family tree.

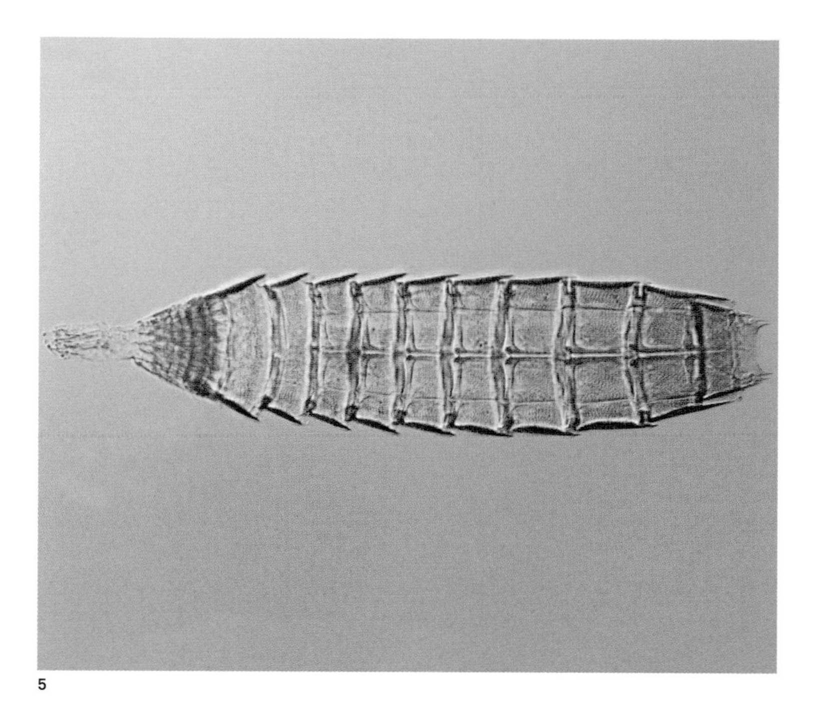

5

6

7

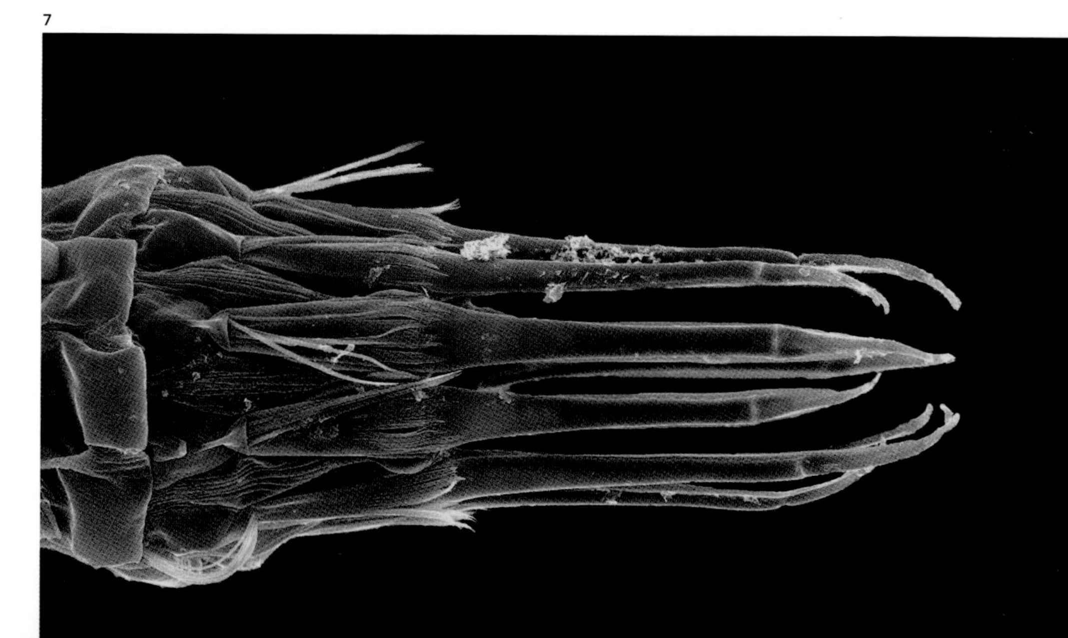

3 Illustration showing the generalized form of a mud-dragon.
4 The body a mud-dragon is clad in a series of cuticular plates (*Pycnophyes* sp).
5 In this image the head with its numerous scalids is retracted back into the trunk (*Echinoderes rex*).
6 The head of a mud-dragon showing the numerous scalids and the nine long stylets of the mouth cone (*Wollunquaderes majkenae*).
7 Close-up of a mud-dragon's mouth cone stylets (*Wollunquaderes majkenae*).

Ectoprocta

(bryozoans, moss animals)
(Greek *ecto* = outside;
proct = anus)

Diversity
c. 5500 species

Size range
~0.5 mm to ~1 m (colonies)
(~0.02 in. to ~3.3 ft)

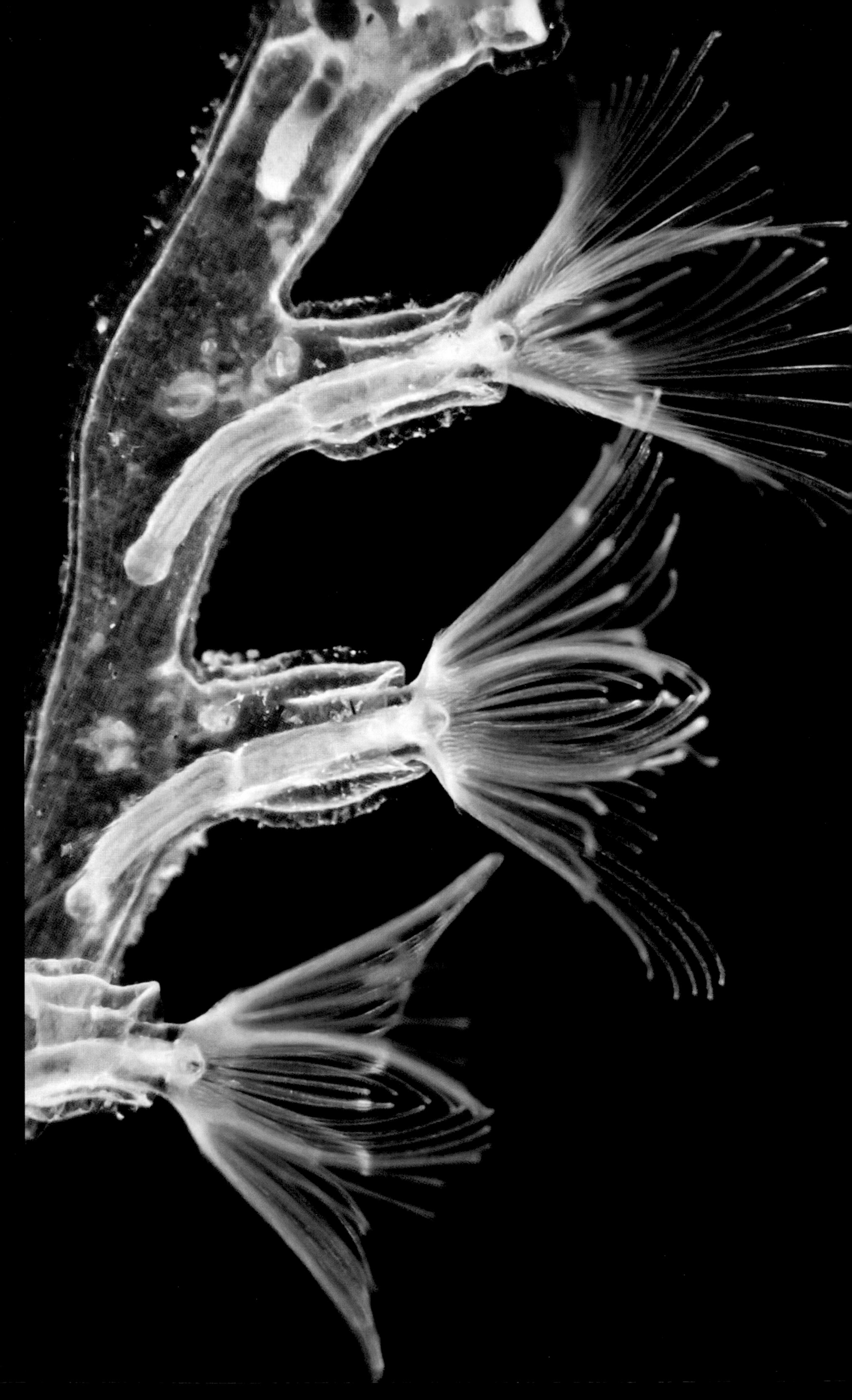

1 Ectoprocts are typically colonial beings made up of functional units (zooids). Each unit is equipped with a crown of feeding tentacles (lophophore) and a well-developed gut. In this image you can see the tubular exoskeleton secreted by the zooids and the strands of tissue linking all the zooids in the colony. The zooid at the very top of the page has been resorbed leaving a brown mass of waste material (*Plumatella repens*).

2 Most ectoprocts are colonial animals. Depending on the species these can be branching, flattened or convoluted structures. They are common, albeit easily overlooked animals of marine and freshwater environments (*Bugula* sp.).

Ectoprocts are successful, remarkable and very common animals, although few people will have knowingly seen them. Their colonies, which can sometimes be surprisingly large, are very inconspicuous and easily overlooked – and even when they are noticed, they are often mistaken for peculiar, aquatic plants (hence their common name, moss animals); but any confusion quickly disappears when you see the builders of these colonies gingerly extending their feeding tentacles into the water. These are unmistakably animals.

Ectoprocts are typically sessile, filter-feeding animals that build their colonies on just about any underwater substrate, natural or otherwise, including rocks, the fronds of algae, the bodies of other animals and even the hulls of ships. There are even some species that form free-living colonies. Although predominantly marine, there are around 90 species that live in fresh water.

FORM AND FUNCTION

Apart from a few aberrant, sand-dwelling species that live a solitary existence [4], all the ectoprocts are colonial animals and for all intents and purposes the colony *is* the animal [6]. The zooids, the units of the colony, are intimately linked and dependent on each other, since they are the clones of an original, founding zooid – the charmingly named ancestrula. Each zooid is a functional creature in that it has an elaborate crown of tentacles for collecting food, a substantial gut for processing this food and gonads for reproducing. But linking each and every one of these zooids is a complex web of living tissue relaying nutrients, waste and information around the colony.

The business end of the zooid is its elaborate crown of ciliated tentacles that funnels edible particles suspended in the water down to the central mouth and large U-shaped digestive tract beyond, much of which consists of a capacious, three-part stomach [1]. The zooids have no heart, respiratory organs or excretory organs. Their small size means that gases can freely diffuse across their body wall, while the problem of waste is dealt with in a very interesting way. Some waste products of metabolism simply diffuse across the body wall and into the surrounding seawater, but a novel solution to the remaining waste is just to let it build up in the tissues of older zooids that are due for replacement, until the body cavity of each is brimming with dying cells loaded with waste. The zooid may wither and have much of its matter resorbed for use elsewhere in the colony, but it is far from dead. In a burst of regeneration, a new zooid forms around the waste-laden remnants of its old body and either voids the brown mass of waste into the sea or retains it in its new body cavity [1, 3].

Each of the vulnerable zooids is encased in an exoskeleton, a protective capsule that can be brittle, gelatinous, rubbery or even rocky depending on the species and the relative proportions of its constituents (chitin, other polysaccharides, proteins and calcium carbonate). Pores in the exoskeleton are the conduits through which the zooids of the colony connect to each other. The colonies themselves can be anything from small, motile assemblages to bushy, branching structures composed of up to 2 million zooids, although the most common are the encrusting forms, one zooid thick, that adhere to grains of sand, rocks, seaweeds and the shells and exoskeletons of other animals [2, 7–9].

In many types of ectoproct there is a division of labour amongst the zooids. The standard issue zooids have the task of feeding the colony, while others are employed for protection. Two types of these defensive zooids are known. The first is a very odd little thing since it looks like a tiny bird's head, complete with a 'beak' and a 'mandible', both of which are actually an elaboration of the zooid's exoskeleton. This bird's head zooid uses its beak to good effect to grab any organisms or bits of debris that settle on the colony and to pinch any predators who think the colony is an easy meal. The second, highly modified defensive zooid bears a long bristle, known as a vibraculum, that is moved with muscles. Not only can these bristles be used to sweep settling organisms and debris from the colony, but some species forming small colonies use them to move around in and on the sand [10].

Each zooid has a nervous system emanating from a simple brain and nerve ring at the base of the crown of tentacles. Via the pores of the exoskeleton the nervous systems of neighbouring zooids are able to exchange electrical impulses with one another, allowing the whole colony to feed and respond to stimuli in synchrony.

LIFESTYLE

Ectoprocts feed on microscopic organisms and other bits of edible matter suspended in the

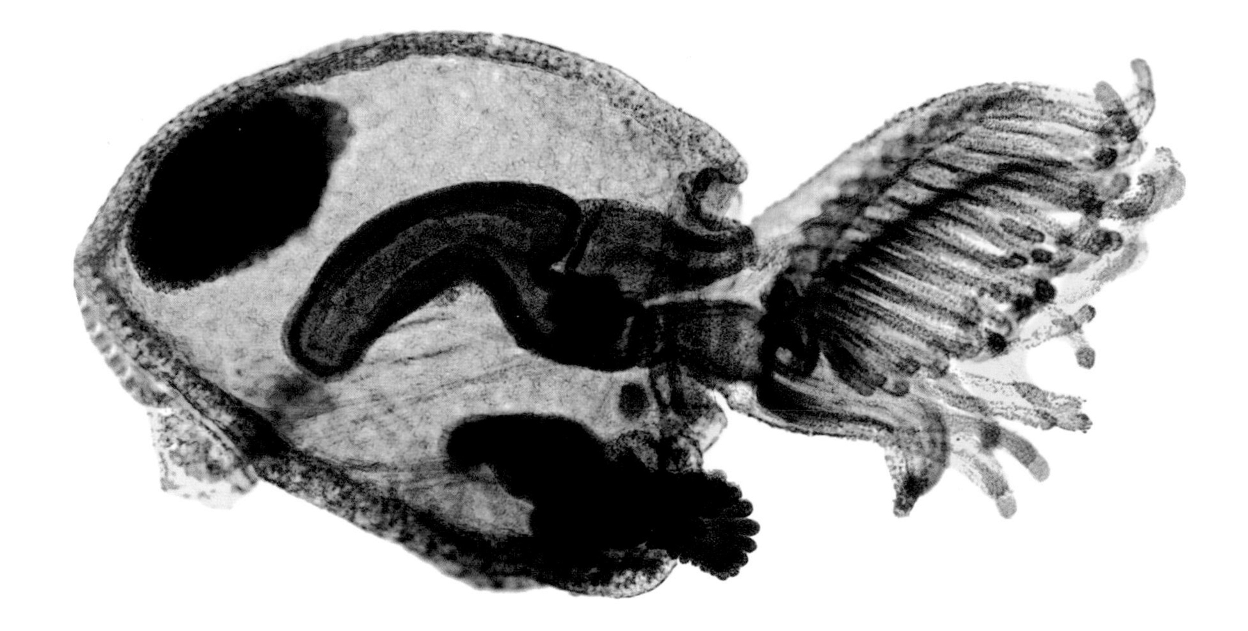

3

water. Cilia on the surface of the tentacles draw water and food toward the mouth, an action enhanced by rapid, inward flicks of the tentacles. Some species are occasional carnivores, ensnaring small animals floating in the water by closing the tips of their tentacles around the quarry to form a cage.

Reproduction in these colonial animals is complex. Most species are hermaphroditic, but there is a tendency for zooids to function as either males or females depending on their age. With zooids in a colony all freely producing sperm and eggs there is risk of inbreeding, since sperm produced by one zooid may fertilize the eggs of a neighbouring zooid in the same colony. To minimize this risk, male and female zooids may be separated from one another by either time or space, and sperm can be shed in a way that is directed away from the colony.

Following fertilization, the large, yolky eggs are normally retained in the body of the zooid as the embryo develops, sometimes even nourished by a placenta-like arrangement. Some species exhibit an intriguing phenomenon known as polyembryony, where an embryo divides to form as many as 100 clones of itself to increase the number of offspring produced by a given colony and the chances of new colonies being established. The embryo eventually gives rise to a planktonic larva, a key stage in the life cycle of the ectoproct because this is how these otherwise sedentary animals disperse to new areas of suitable habitat.

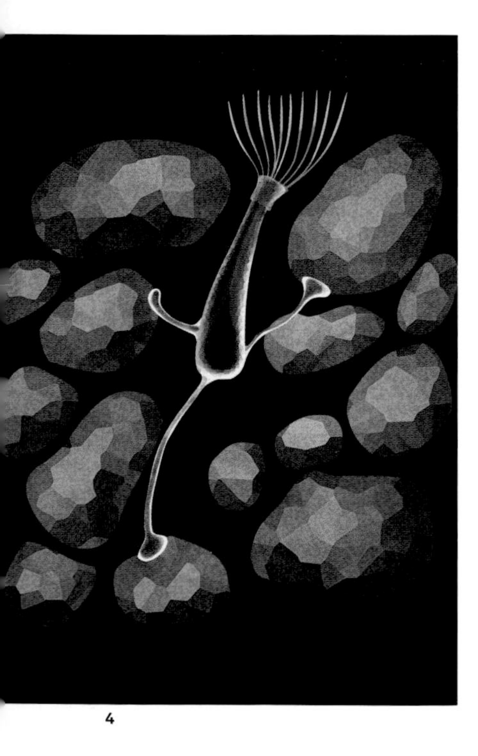

4

Some species complete all of their development in the plankton, but usually, after some time floating amongst the other minute organisms, it is time for the larva to settle and metamorphose into the founding zooid of a new colony – the ancestrula. Interestingly, inside the microscopic body of some ectoproct larva the establishment of a new colony begins early as the nascent form of the zooid takes shape and buds off copies of itself, until there is a tiny colony ready to grow when the larval housing settles.

After its metamorphosis the founding zooid is naked, but a covering of potent chemicals secreted by symbiotic bacteria affords some protection until its exoskeleton forms (some of these chemicals, notably bryostatins, may be useful in treating cancer and Alzheimer's disease in humans). Once encapsulated by an exoskeleton, the ancestrula goes about growing and making copies of itself to form the complex, multi-zooid ectoproct colony. In some of the more well-known species, the growth of the colony occurs in well-ordered rows of active units radiating away from the ancestrula with the youngest zooids toward the leading edge of the rows.

A further addition to the reproductive repertoire of the ectoprocts is the ability of the freshwater species to produce, asexually, enormous numbers of tiny, often disc-shaped and ornate resting bodies (statoblasts) to resist the ravages of cold or drought [11]. When shed, these capsules may simply adhere to the parent

5

colony, sink to the bottom or rise to the surface. Getting tangled in aquatic vegetation, snagged in the hair and feathers of animals or simply borne on currents, the resting bodies may get carried many miles from their parent colony. Eventually, the return of better conditions coaxes the little mass of dormant cells into forming a new zooid that will go on to establish a fresh colony.

Ectoprocts are diverse and abundant animals, especially in the more shallow reaches of aquatic habitats at depths of between around 20 and 80 m (65 and 260 ft). On tiny areas of suitable substrate, the shell of a long deceased mollusc, for example, it is not unusual to find anywhere between 30 and 40 different species. Regardless of their abundance and diversity, we know very little of their contribution to aquatic ecosystems, but their collective efforts in consuming planktonic organisms and edible matter must play a significant role in how nutrients and energy move through marine and freshwater habitats. The biomass they accrue in suspension feeding is greedily utilized by a huge range of other organisms that are able to breach the colony exoskeleton to get at the soft zooids within [5]. Predation of ectoproct colonies is even a speciality of many marine animals, such as some sea slugs.

ORIGINS AND AFFINITIES

The resilient exoskeleton of the ectoprocts has left its mark in the fossil record. Fossilized remains of animals in rocks around 500 million years old have been interpreted as ancient members of this lineage. However, it has been suggested the heritage of these creatures extends even further back in time, since the first ectoprocts were very likely naked, soft-bodied animals of the seabed with as much hope of resisting the rigours of fossilization as a snowflake. Only when they evolved their tough exoskeleton did they stand any chance of leaving their mark in the mineralized record of animal life on earth.

Certain aspects of the morphology and development of the ectoprocts have been a source of considerable confusion for zoologists eagerly trying to understand their relationships to the other animals. A long-held view was these animals were very closely related to the brachiopods (see pp. 266–69) and phoronids (see pp. 270–73), largely based on the fact that they all possess a similar crown of tentacles around their mouths. Today, the prevailing opinion is that the ectoprocts are more closely related to the entoprocts (see pp. 196–99) and the cycliophorans (see pp. 200–03), two equally enigmatic groups of animals.

3 A close-up of an ectoproct zooid. Note the crown of long feeding tentacles, gut (centre), mass of accumulated waste material (left), the muscles that retract the feeding tentacles, and the bud (top right) that will grow into another zooid.

4 A small number of ectoprocts are tiny, solitary animals that live between the sediment grains on the seabed. They propel themselves through the tiny spaces of this habitat using long, mobile extensions of the body equipped with sticky tips (illustration showing *Monobryozoon* sp.).

5 Ectoprocts have many predators. Here, the amphipod *Acanthonotozoma inflatum* is feeding on an ectoproct colony.

6 Zooids of the ectoproct *Flustrellidra hispida*, with their feeding tentacles extended.

7 An ectoproct colony. Many species form flat, encrusting structures that few people would recognize as an animal. Each tiny 'cell' houses a zooid (unidentified species).

6

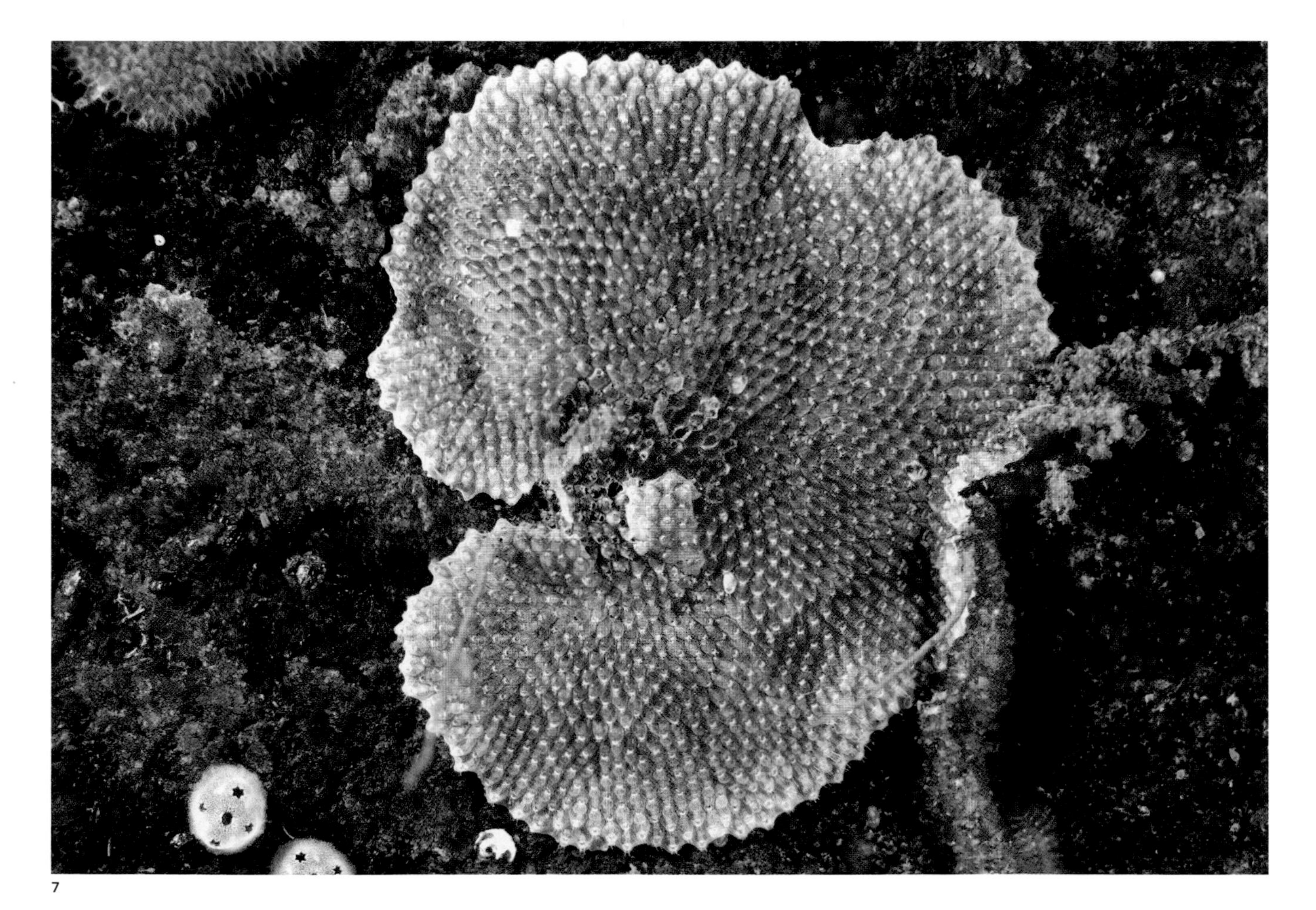

7

8

9

8 To increase the surface area for the collection of food, many species of ectoproct have evolved erect, branching forms (*Flustra foliacea*).

9 The branching forms of ectoprocts can be very elaborate (*Reteporellina denticulate*).

10 The exoskeletons of many of the zooids in this mobile colony have been modified into long bristle-like structures. These so-called vibracula sweep about, dislodging settling organisms, and can propel the colony (*Cupuladria biporosa*).

11 To survive periods of drought or freezing conditions, freshwater ectoprocts produce survival capsules known as statoblasts.

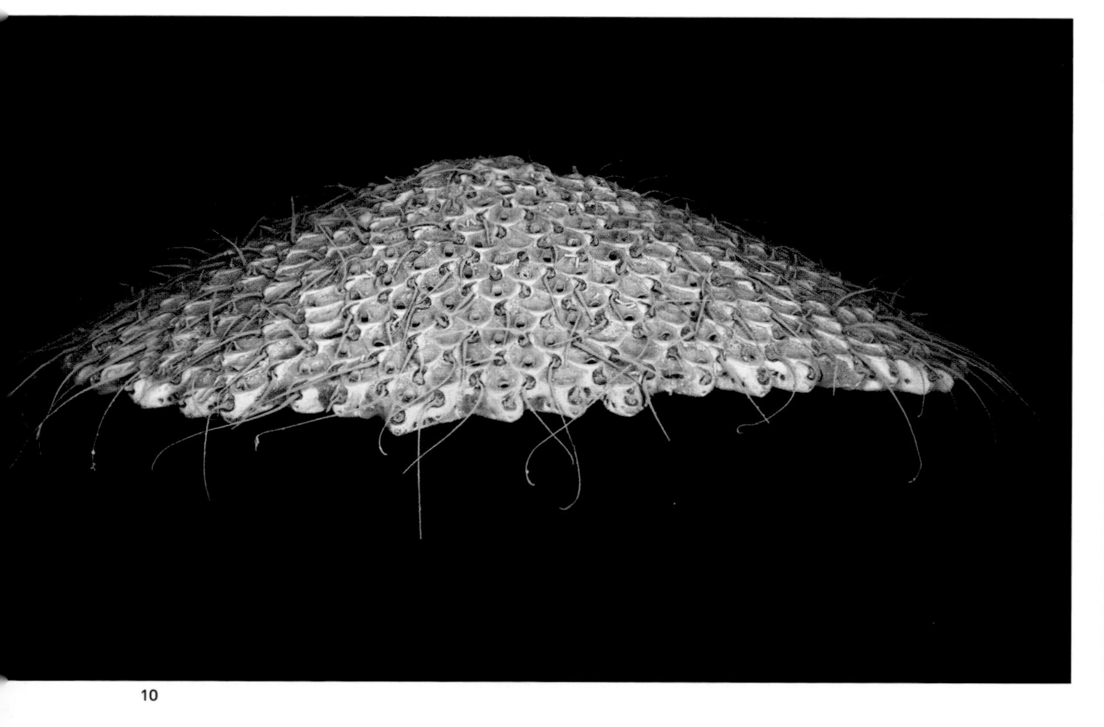

10

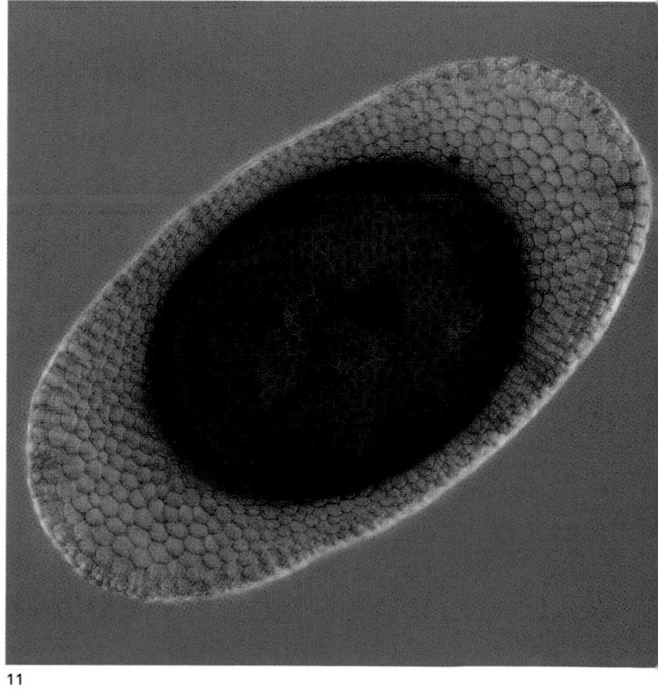

11

Entoprocta

(nodders, goblet animals)
(Greek *ento* = within, inside; *proct* = anus)

Diversity
c. 170 species

Size range
0.1 to 10 mm (zooids)
(0.004 to 0.4 in.)

1 Entoprocts can form dense colonies in suitable places on the seabed (*Loxosomella* sp.).

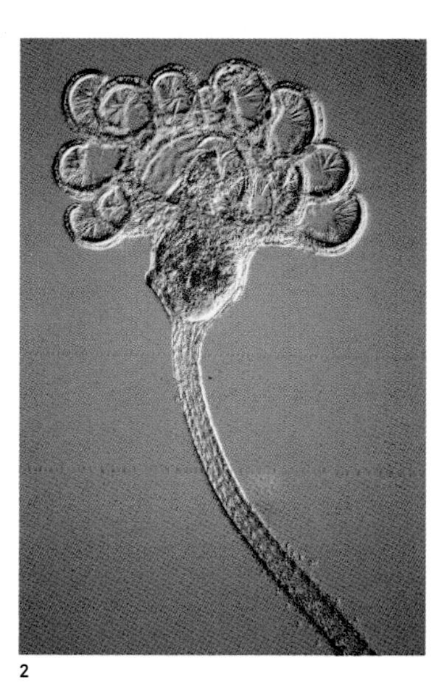

2

The entoprocts are another group of animals we know precious little about. Around 170 species have been described and all but two of these are marine. They live rather similar lives to the ectoprocts (see pp. 190–95), typically as colonies attached to all manner of underwater substrates, often the bodies of other animals.

FORM AND FUNCTION

The typical body plan looks a bit like a cup perched on a long stalk, hence the common name 'goblet animals'. The 'cup', known as the calyx, holds all the animal's organs and projecting out from it into the water is a crown of 6 to 36 solid tentacles [1, 3, 4]. They are superficially similar to the ectoprocts and even on the inside there are likenesses, such as the U-shaped digestive tract, much of which is stomach. On the other hand they also have a whole range of characteristics that differentiates them from the ectoprocts. The entoproct anus is located inside the ring of tentacles instead of outside it. There are series of spaces between the gut and body wall filled with a gelatinous matrix of cells and fluid. Also, the entoprocts have no exoskeleton and when they are disturbed they simply curl their tentacles inward rather than retracting them [2].

Most are colonial creatures, but their collectives are rather small on the whole. The exceptions to this are the colonies formed by the freshwater species, which can be substantial and conspicuous. Like the ectoprocts, the zooids in an entoproct colony are clones of the original, founding zooid, which settled on the seabed as a larva; therefore the colony *is* the animal. In a colony the zooids are all attached to a single anchor disc or attached to one another via fine, creeping outgrowths known as stolons, facilitating the sharing of nutrients [5, 7]. Their small size means there is no need for any specialized respiratory organs, but in most colonial species there is an organ akin to a heart (star-cell complex) located between the cup-like calyx and the stalk, which is thought to move nutrients between these two parts of the body. Specialized cells that have evolved into a complex network in the freshwater species eliminate waste and regulate the amount of water in the body. The brain, if you can call it that, is a small mass of nerve cells between the gut and tentacle crown. Feeding into this nerve centre are various sensory receptors all over the body. The larvae are usually equipped with simple eyes.

LIFESTYLE

Entoprocts collect all of their food by generating a current of water with the numerous cilia on their tentacles. The cilia snare microscopic organisms and organic matter and convey this edible matter to the animal's mouth in a sheet of mucus. Some species have unique glands at the base of their tentacles that discharge long, hollow, threads. Exactly what the spiral threads of these so-called lime-twig glands are for is not clear, but they are assumed to be sticky, perhaps even poisonous and therefore used to snare tiny animals.

Almost all of the entoprocts are completely sessile. It is true that they can curl their tentacles inwards, some species nod (hence the common name of 'nodders') and others twist their stalk, but that is largely it. The few solitary species are a bit more mobile, since they can somersault slowly across the substrate. In one species (*Loxosomella bifida*), the stalk has become a pair of leg-like structures on which the creature ambles along. The nodding behaviour of the colonial species may dislodge settling organisms and debris.

To make the quest for food a little easier, many species, particularly the solitary ones, live in close association with animals that generate their own feeding currents, such as burrow-dwelling worms and sponges. By positioning themselves in the flow of water they can intercept morsels of food with the minimum of effort. Some species discovered in Antarctica are only found in the colonial tubes of some ectoprocts, a microhabitat where there is not much plankton suspended in the water. It seems these entoprocts snare their food, tiny animals, with the long threads discharged from their lime-twig glands.

As is so often the way in small, easily overlooked animals, reproduction is far from simple. Asexual budding allows the colonial species to grow via the formation of nascent zooids from the calyx, stalk or stolon [8], while sexual reproduction, more common in warmer temperatures, allows sperm to be exchanged between zooids and between colonies. The solitary species may start off male but then gradually become female as they age. Zooids in the colonial forms may also change sex as they age, but others are thought to be hermaphrodite their whole life. The colonies of species that change sex may be mixed sex or single sex.

Some species discharge their fertilized eggs into the water, their development into larvae

2 An entoproct with its tentacles curled inwards; a defensive behaviour to protect these feeding structures. Note the long cilia on the tentacles (unidentified loxosomatid).

3

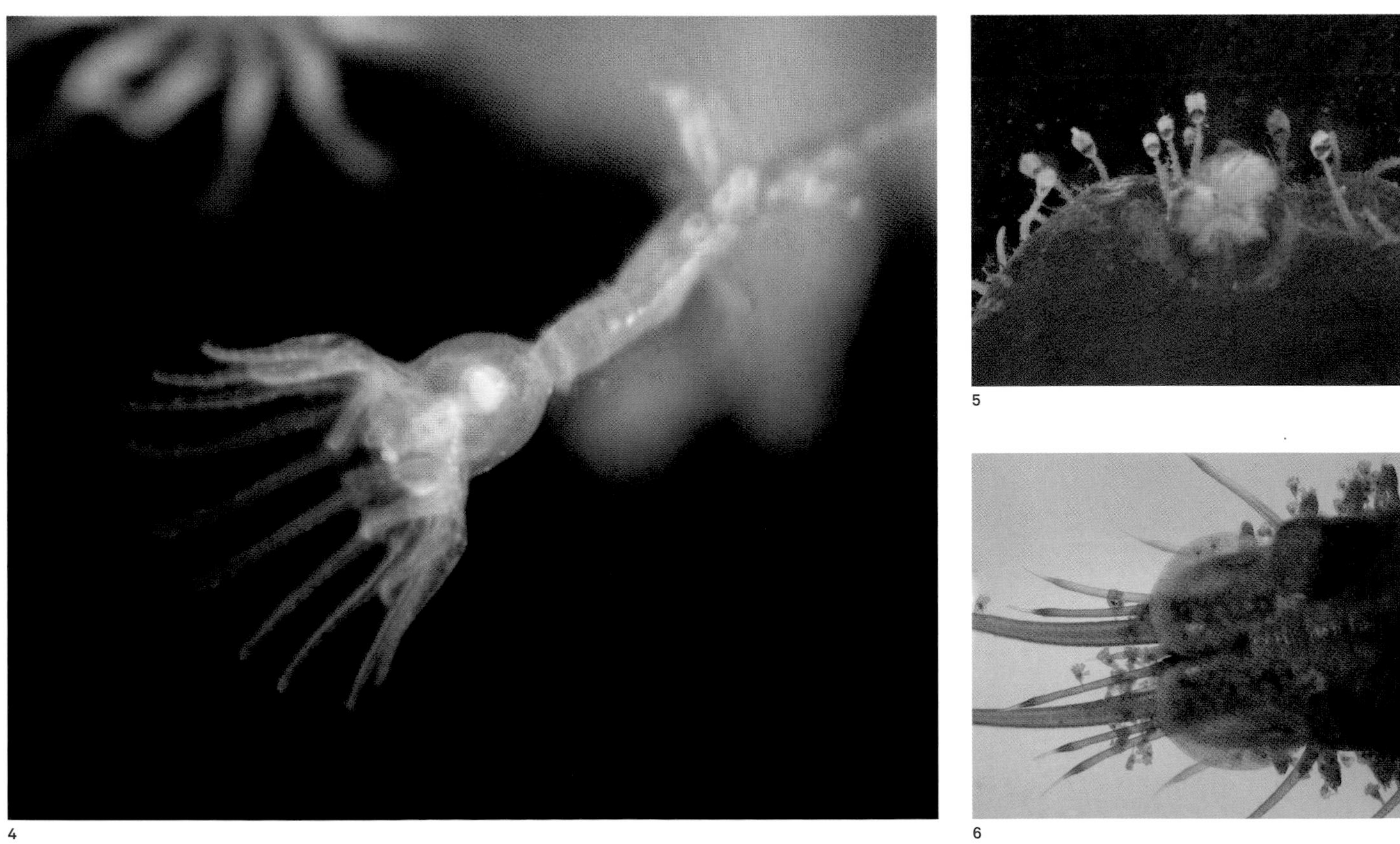

4

5

6

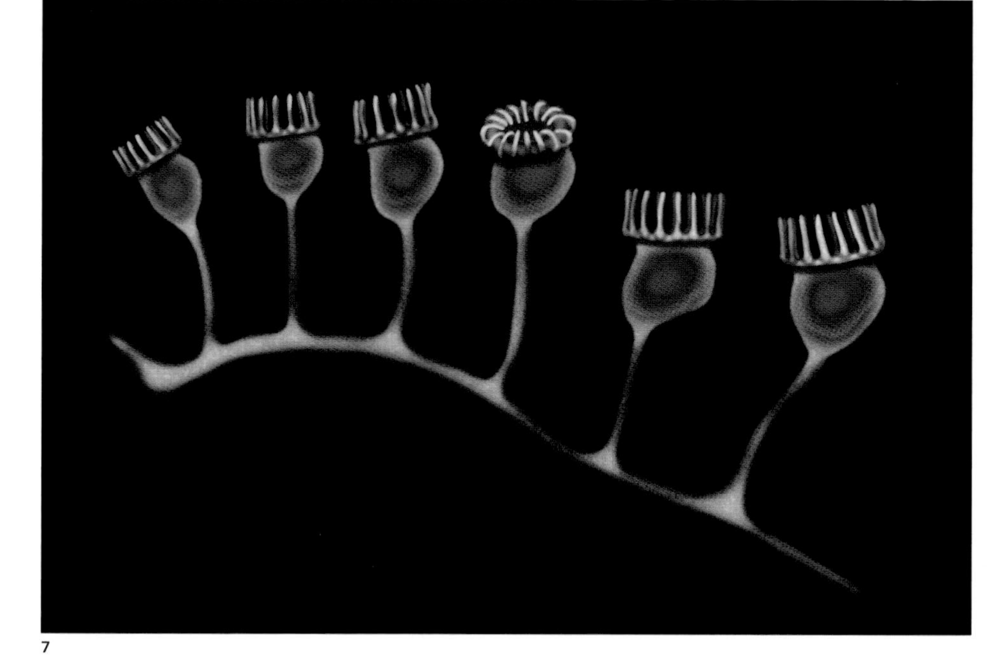

7

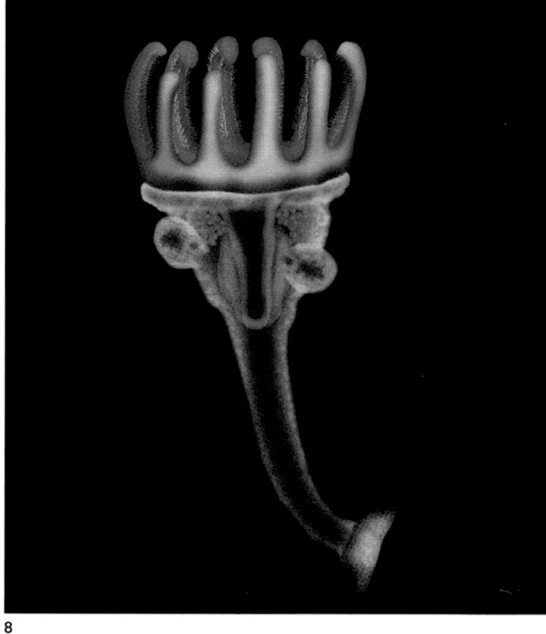

8

3 Entoprocts are superficially very similar to ectoprocts, but one of the key differences is that the entoproct anus is within the crown of tentacles (*Barentsia capitata*).

4 Being so small, entoprocts are easily overlooked animals, but in the right places they can be abundant (*Barentsia discreta*).

5 An entoproct colony on the brood sac of *Sacculina*, a parasite of crabs (see Arthropoda, p. 153). Note the almost transparent stolon connecting each of the zooids in the colony (*Pedicellina* sp.)

6 Entoproct colony on an annelid worm (*Loxosoma* sp.).

7 The zooids in an entoproct colony are connected via a network of slender, tubular outgrowths known as stolons (*Pedicellina* sp.).

8 In addition to sexual reproduction, entoprocts can also reproduce asexually. In this illustration of a solitary species two new individuals are being budded from the adult's calyx (*Loxosomella* sp.).

fuelled by a store of yolk. Most, though, brood their embryos by keeping hold of them via a short stalk and nourishing them through a placenta. The larva that develops from the embryo normally has a fleeting planktonic existence, perhaps only a few hours, before it settles on a patch of suitable substrate and begins a complex metamorphosis that will see it take on the adult form. In some species a radical transformation takes places whereby the foot of the larva, its underside, twists through 180 degrees to form the familiar calyx of the adult. In effect, the adults of these species spend the rest of their lives upside down. In another quirk of development, the larvae of some species undergo no such transformation, instead producing tiny buds that will go on to become adults.

It has been estimated that there are at least 500 species in this lineage in total, so there is still much to learn about their diversity. This, coupled with the fact that we know so little about their ecology, makes it difficult to assess the scale of the entoproct contribution to the ecosystems in which they are found. They can be common animals of the seabed, especially in shallow water, being found along the coast of every continent. Some species are even known from depths of around 500 m (1650 ft). The freshwater species are known from every continent except Antarctica and they too can often be common animals. Their relationships with other animals extend from eating tiny organisms suspended in the water to falling prey to larger, benthic animals (e.g., sea slugs and flatworms) all the way up to their complex interactions with the animals many entoproct species routinely associate with [5, 6].

ORIGINS AND AFFINITIES

Like so many of the other lesser-known animal lineages, the origins of the entoprocts are unclear and they have been moved back and forth on the animal family tree. Being soft bodied and small their fossil record is very sketchy. Fossils from Jurassic rocks, around 150 million years old, until recently represented the oldest known species, but now some Cambrian fossils around 520 million years old have been interpreted as the remains of ancient entoprocts.

When the early naturalists were getting to grips with the diversity of animal life they were fairly certain the entoprocts were simply ectoprocts. Later on, in the 20th century, the consensus was that these two lineages were not even closely related, their shared features a product of convergent evolution rather than any common, recent heritage. Now, in the 21st century, opinion has gone full circle and we are more or less back where we were in the 18th and 19th century. But the evolutionary relationships of the entoprocts are yet to be pinned down. Currently, it seems that regardless of their key differences, these two lineages are offshoots of the same evolutionary twig, along with another group of very interesting animals, the cycliophorans (see pp. 200–03), which are currently thought to be the closest relatives of the entoprocts.

Cycliophora

(cycliophorans)
(Greek *cyclo* = circle;
pherein = to bear)

Diversity
3 species

Size range
0.04 to 0.6 mm
(0.0015 to 0.02 in.)

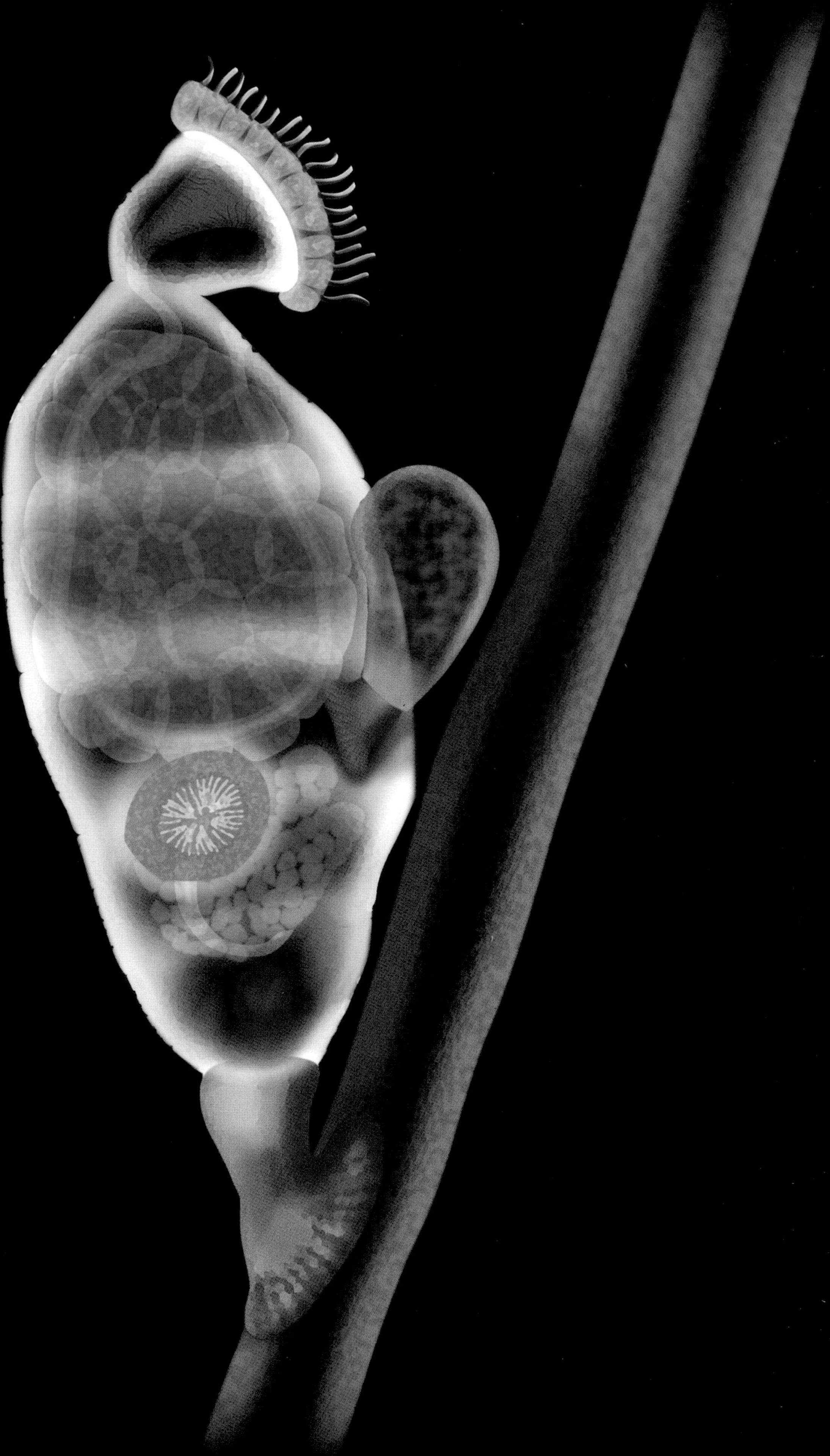

1 Illustration showing *Symbion pandora* (feeding stage). Note the feeding funnel and the U-shaped gut. A Prometheus larva containing secondary males is attached to the body of the feeding stage. In the posterior part of the animal a replacement head and digestive tract are forming.

2

2 Close-up of *Symbion pandora* feeding stages adhering to the setae of their host.

Recently discovered and living much of their lives attached to the mouthparts of clawed lobsters, the cycliophorans highlight the remarkable diversity of the more obscure animal lineages. However, what really sets them apart is the complexity of their life cycle, a bewildering cast of asexual and sexual characters in a reproductive dance set against the backdrop of their massive crustacean host. So complex is this life cycle that the zoologists who described the first species gave it the name *Symbion pandora*, because getting to grips with how it reproduces was a bit like opening Pandora's proverbial box.

FORM AND FUNCTION

The most conspicuous stage in the life cycle of the cycliophorans is the feeding stage, which spends its whole life firmly attached to the mouthparts of a lobster [1–5]. This tiny animal, around 0.3 mm (0.01 in.) long, has a feeding funnel – its mouth – lined with cilia. Containing most of the important organs is the trunk – the main part of the body – which is attached to an adhesive disc via a short stalk. In between the entrance and the exit to the U-shaped gut, with its large stomach, is the feeding stage's brain, whilst the cavity between the body wall and the gut is filled with connective tissue rather than fluid. Most curious of all is that the head (the feeding funnel) and the entire digestive tract are broken down only to grow again from an inner bud in the posterior part of the animal's trunk, a process repeated several times during the life of the feeding stage [1].

LIFESTYLE

With its cilia-lined mouth, the feeding stage collects particles of food that drift away from the lobster's mouth when it feeds. As its name suggests, it is only the feeding stage that eats anything – the other stages in the life cycle are rather short-lived and rely on energy stores in their cells.

And so to the remarkable life cycle of these animals. Alongside the tiny inner buds that replace the head and digestive system of the feeding stage are clusters of stem cells that can differentiate into three different motile life stages: the Pandora larva, the Prometheus larva and the mature female. The young Pandora larva grows until it is about one third the size of the mature feeding stage. Now, complete with its own feeding funnel and digestive system, it emerges, fixes itself to the lobster's exoskeleton and develops into a fully formed feeding stage [4]. Asexually reproducing in this way, the feeding stages can rapidly form very large populations on their hosts.

After a while building up their numbers, the feeding stages commence the sexual phase of the life cycle. In this sexual phase the stem cells in the feeding stage develop either into an immature male or a mature female. The immature male, the Prometheus larva, is a very odd little creature since he has no penis or even a gonad. He creeps from the feeding stage where he developed in search of another feeding stage on to which he clings before degenerating and spawning one to three secondary males

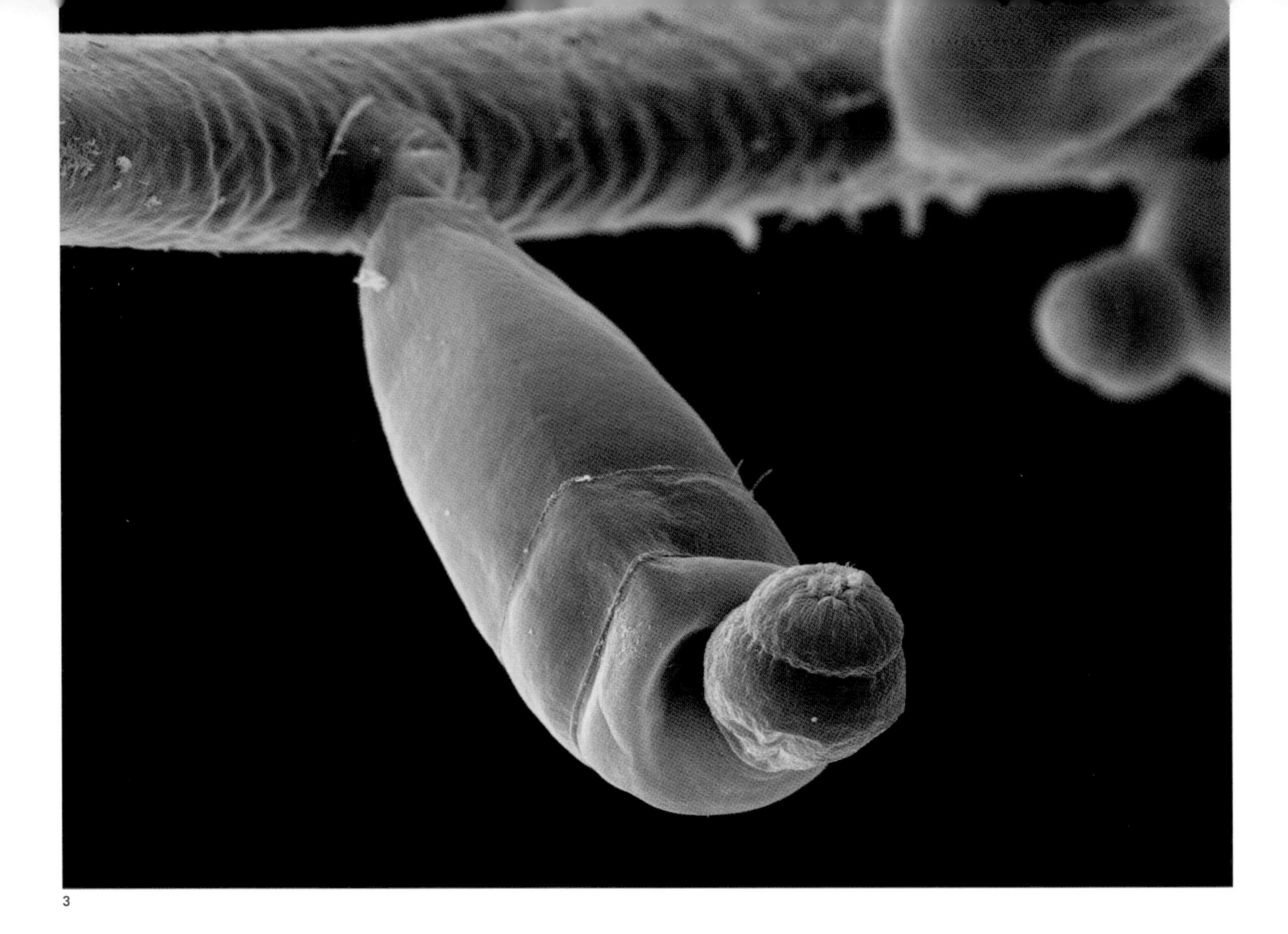

3

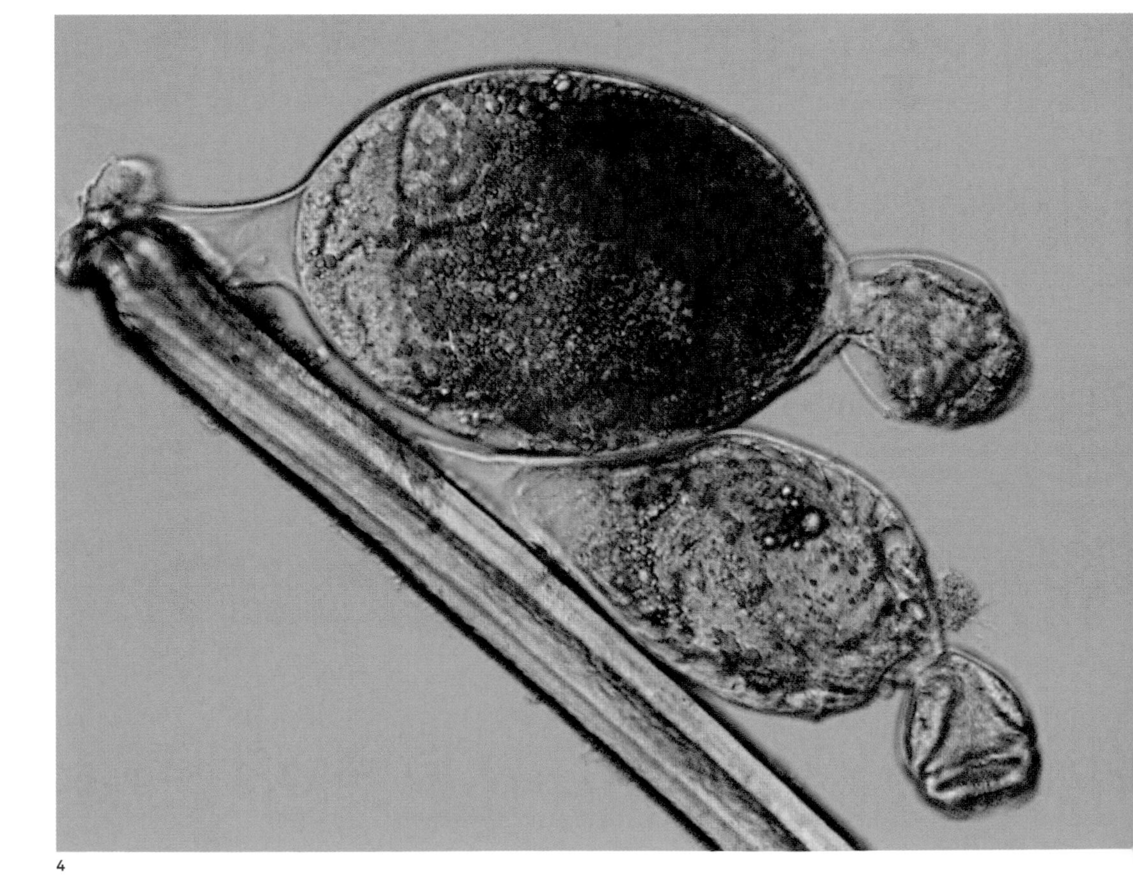

4

3 SEM of a single *Symbion pandora* feeding stage adhering to a seta of its host. Note the feeding funnel and the adhesive disc. Other individuals are visible in the background.

4 Two feeding stage individuals of *Symbion pandora* attached to a host seta. The larger individual has another stage of the life cycle in its brood chamber (most likely a Pandora larva). There is also a new feeding funnel developing at the base of the trunk in the larger individual. The smaller individual probably originated from the larger individual via a Pandora larva.

5

5 Cycliophorans (feeding stages of *Symbion pandora*) adhering to the setae on the mouthparts of their host, the Norwegian lobster *Nephrops norvegicus*.

that spring from buds within his withering body [1]. These males, equipped with a penis and gametes, are ready for action, namely fertilizing the solitary egg of the mature female before she emerges from the feeding stage. Following fertilization the mature female breaks free of the feeding stage and settles a short distance away, her tissues slowly being engulfed by a developing larva. Eventually this larva – the chordoid – with its well-developed brain and muscles hatches from the thin husk that was its mother and swims off to find a new lobster. If it is successful, it settles, a bud within it gives rise to a new feeding stage and the amazing life cycle of this animal can begin anew.

Why these animals should have such a complex life cycle is not clear. The asexual phase allows large numbers of individuals to be produced to take advantage of suitable habitat and perhaps out-compete other sessile animals that also collect edible matter suspended in the water. The sexual stages of the life cycle ultimately result in the formation of a free-swimming, motile creature that is able to leave the original host in search of new habitat. The pressures on the sessile stages in the life cycle of the cycliophorans that drive the formation of the dispersal stage are two-fold. Firstly, there is the issue of overcrowding as the feeding stages rapidly reproduce asexually to occupy the best sites on a lobster. Secondly, in order to grow, the host sheds its exoskeleton intermittently and with it any encrusting organisms, including the cycliophoran's feeding stages that are ultimately dependent on the lobster's leftovers.

It turns out the cycliophorans are rather common in the coastal waters of northwest Europe and along the eastern seaboard of North America. At least three species are now known from three species of lobster and it is highly likely there are many more out there.

It is not uncommon to find thousands of cycliophorans clinging to the mouthparts of their crustacean host, and in some areas these tiny passengers are found on as many as three quarters of the lobsters. Interestingly, the relatively conspicuous feeding stages were known as far back as the 1960s, but they were not described. Working out the life cycle of the cycliophorans with its cast of tiny, microscopic characters was a labour of love, but even now, several years after the discovery of these animals first made the headlines, many of the finer details of their remarkable lives are still to be resolved.

ORIGINS AND AFFINITIES

The zoologists who described *Symbion pandora* suggested it had certain affinities with the entoprocts, but a few years after this the rotifers were touted as the closest relatives of the cycliophorans. Now, in the light of other discoveries, it does indeed seem that the entoprocts are the closest relatives of these very strange, lobster-lip-dwelling animals.

Dicyemida

(dicyemids)
(Greek *di* = two; *kyēma* = embryo)

Diversity
c. 125 species

Size range
0.1 to 5 mm
(0.004 to 0.2 in.)

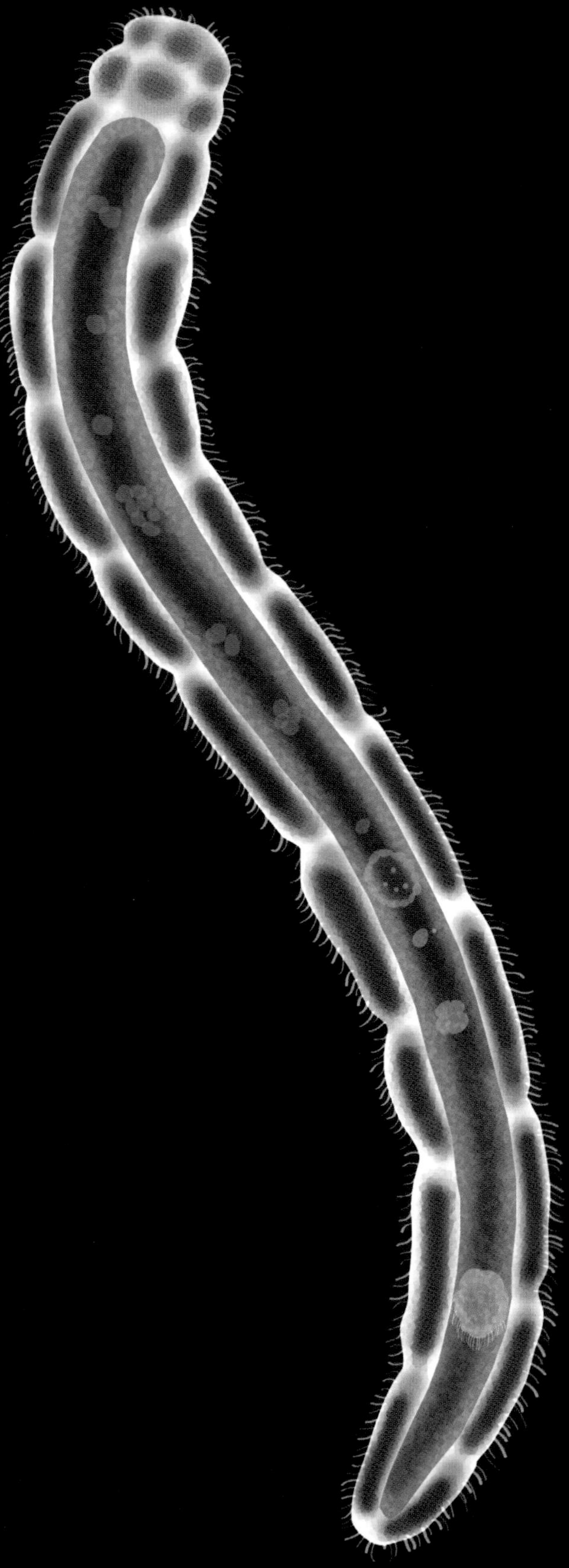

1 Illustration showing an adult dicyemid. Within the long, central cell new individuals are forming. At the anterior end (top) is the calotte.

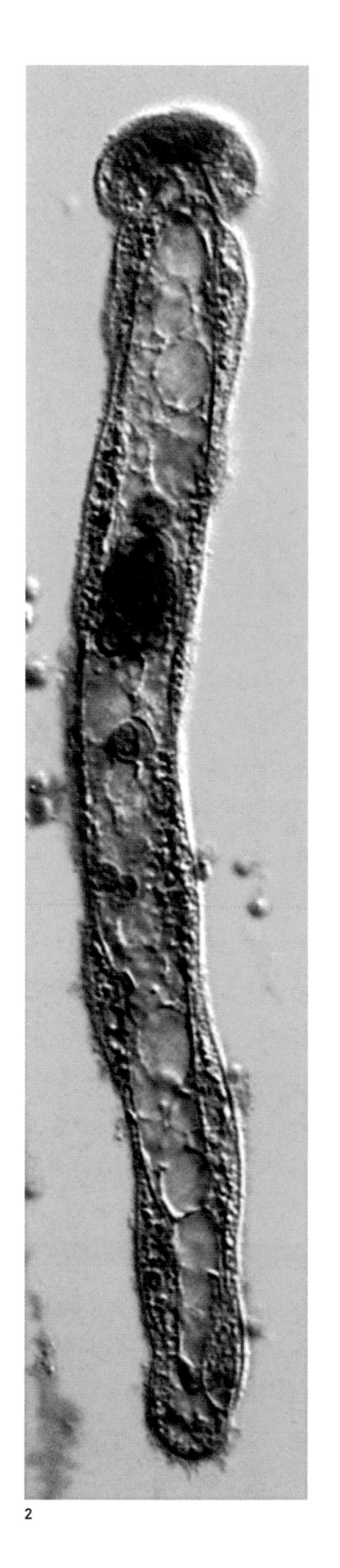

2

2 An adult dicyemid. These extremely simple animals live inside the kidneys of cephalopods, typically octopuses and cuttlefishes. It seems as though their relationship with the host may have extended beyond parasitism to become a symbiosis.

Larger than the orthonectids and slightly better understood, the marine dicyemids also live in the bodies of other sea animals, but the nature of their relationship with their hosts is unique: the most conspicuous stage in the life cycle spends its time attached to the inside of a cephalopod's kidney. Here, submerged in urine, these tiny, worm-like creatures actually benefit the host and are therefore symbiotic rather than parasitic.

FORM AND FUNCTION

Like the orthonectids (see pp. 206–07), the dicyemids have a very simple body plan. There are no organs, no body cavities and no gut. The thin body of one of these animals is instead composed of a long, cylindrical cell surrounded by between 8 and 30 ciliated peripheral cells. The calotte, a bulb of cells at one end of the animal, sprouts cilia, affording a loose grip on the wall of the cephalopod kidney, the animal's home.

LIFESTYLE

These worm-like forms can be either asexual or sexual. Cells contained within the long, central cell divide, simultaneously giving rise to several asexual individuals that squeeze out from between the peripheral cells of their mother and latch on to the wall of the kidney. These mature and go on to produce yet more clones, allowing large populations to build up quickly in the host. But when things start to get a little too crowded the animal somehow switches to sexual reproduction. Small clusters of cells in the long central cell give rise to tiny hermaphroditic bodies with an outer layer of eggs surrounding a cluster of tail-less sperm. The sperm fertilize the surrounding eggs, which then go on to develop into tiny larvae, the release of which is achieved by the rupturing of the parent. Squirted out in a jet of urine, the larvae quickly sink to the seabed, weighed down by a pair of dense cells. How these larvae get inside another cephalopod is not clear, but it is possible they infect the host's embryos on their way out as the stream of urine mixes with the water the female octopus or cuttlefish uses to ventilate her developing young. There is also the possibility that once on the seabed the larvae somehow find their way into an animal the cephalopod predates. Once inside the host the larvae develop into the worm-like adult form to start the cycle all over again.

The dicyemids have no sign whatsoever of anything resembling a digestive tract, so it is assumed that they absorb all the nutrients they need from the urine they are bathed in. Their ruffled epidermis greatly increases the surface area through which molecules can be absorbed. Being submerged in food, even if it is another animal's urine, seems like quite an easy way to live, but even here, in the confines of a cephalopod's kidney, there is the potential for competition. Cohabiting dicyemids minimize this by each developing a distinctly different anchoring calotte, allowing them to cling on to specific parts of the host's kidney.

Octopuses and cuttlefish are often found with huge numbers of these animals in their kidneys, but even when they are present in very high population densities they do not cause any damage. On the contrary, they seem to benefit the host by helping it get rid of harmful metabolic waste products, as well as improving the flow of urine through the kidney. Perhaps a very long time ago the ancestors of the dicyemids were true parasites, ravaging the internal organs of their molluscan hosts and causing disease. However, as parasite and host became ever more attuned to one another the negative effects slowly disappeared and a symbiosis gradually emerged. With even more time, far into the future, maybe the distant descendants of the dicyemids will be to their host what the mitochondria are to a eukaryotic cell: an integral part of its inner workings.

ORIGINS AND AFFINITIES

For a long time the dicyemids were grouped together with the orthonectids as the Mesozoa, which is roughly translated as 'middle animals', alluding to the belief that they were a sort of halfway house between the single-celled Protozoa and the more complex animals. This has since been proved to be wholly inaccurate and is another example of how morphology alone can confuse our attempts at deciphering the evolutionary relationships of the animals.

Now, with new technologies at our disposal, it is known that the dicyemids are actually close relatives of annelid worms and molluscs. The various complex features that typify these animals were, it seems, an extravagance for a parasitic and eventually symbiotic existence, and they were gradually lost in the dicyemids. This pared down body is more than enough to obtain nutrients and ensure the continuation of the lineage.

Orthonectida

(orthonectids)
(Greek *orthos* = straight;
nektos = *swimming*)

Diversity
c. 45 species

Size range
0.05 to 0.8 mm
(0.002 to 0.03 in.)

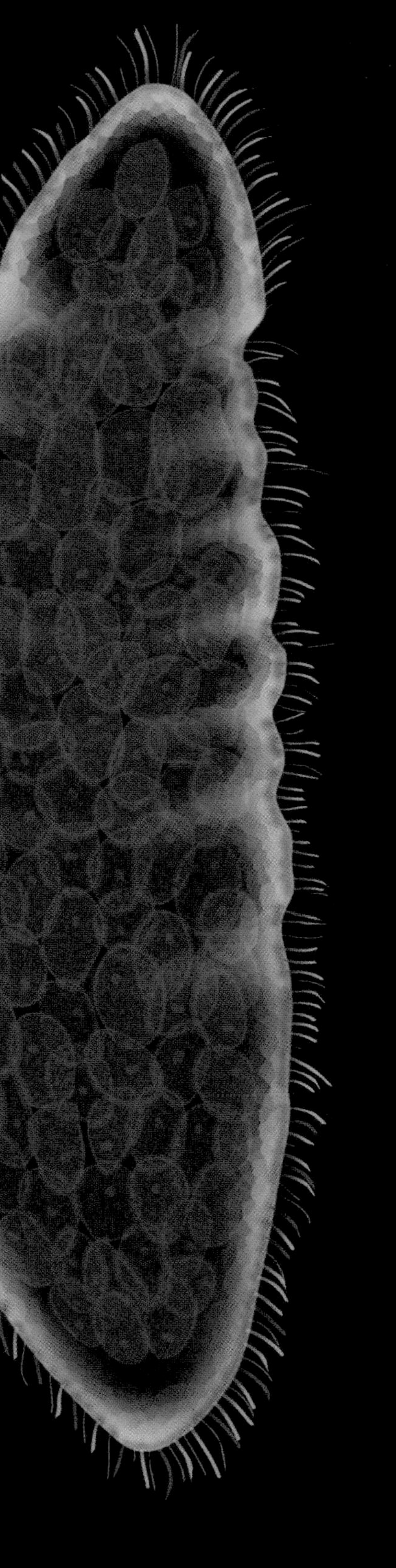

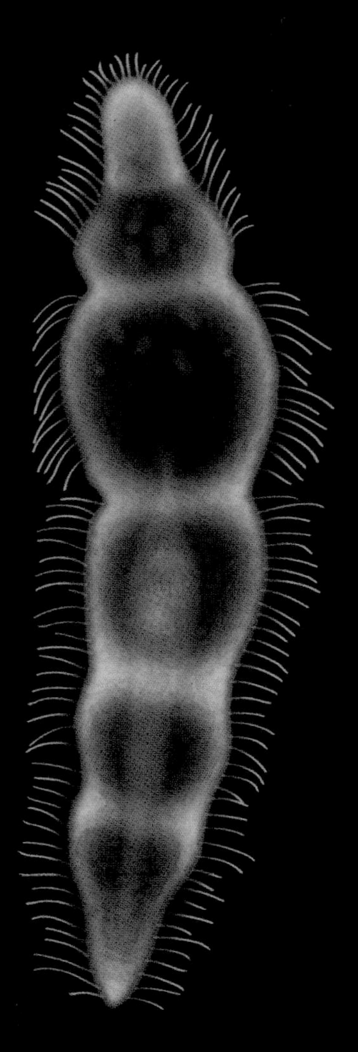

1 An illustration of an adult female (left) and male (right) orthonectid. The body of an adult orthonectid is little more than a mass of reproductive tissue encircled by ciliated and nonciliated jacket cells.

Very little is known about this lineage of rarely seen marine animals whose scientific name – Orthonectida – alludes to the propensity of the adults to swim in straight lines. They are typified by their microscopic dimensions, very simple body plan and parasitic lifestyle. All of them live out a portion of their lives inside the bodies of various other sea creatures, including brittle stars, free-living flatworms, annelid worms, ribbon worms, sea squirts, ectoprocts and some molluscs.

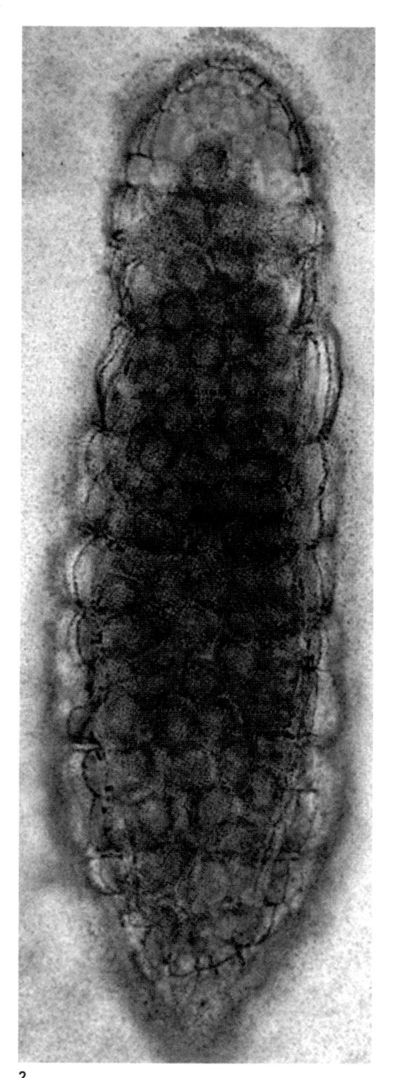

2

FORM AND FUNCTION

The largest and most conspicuous stage in the life cycle is the adult, but even these are little more than swimming gonads. The body of the adult male and female is bounded by many rings of ciliated and non-ciliated jacket cells and a muscle sheath, beneath which there is a burgeoning mass of sperm or egg cells. This is all there is to these creatures, although it has recently been suggested that they may have a rudimentary nervous system and a sense organ, the function of which is unknown.

Adult females are much larger than males and oddly they come in two distinct forms: a larger, elongated female and a smaller, ovoid female, both similarly packed with egg cells. These different forms are never found in the same host. Why there should be two distinct female forms and why they do not occur together is unknown.

LIFESTYLE

In some ways the complex life cycle of the orthonectids is reminiscent of various single-celled organisms. Adult males and females are ephemeral, free-living creatures. After completing their development they exit the host, often via its genital pore, and swim off propelled by the cilia adorning their outer surface. The male locates a female and pulls alongside her so that his genital opening is lined up with that of his relatively massive mate, and the sperm make a rush for the female to fertilize the eggs she is packed with. Some species, however, are hermaphrodite and appear to fertilize themselves. The fertilized eggs develop into tiny, ciliated larvae, the release of which is the end of the road for the female since her body must rupture to set them free.

The larvae seek out and gain entry into a new host to begin the parasitic phase of the life cycle, but exactly how they do this is not fully understood. A germinal cell from a larva may enter the unlucky host and take up residence inside one of its gonad cells. Within the hijacked cell, the parasite divides to form a diffuse structure known as a plasmodium that draws nutrients from the host to fuel the growth of berry-like clusters of cells, which will eventually become the short-lived adults that leave the host. In doing this some species completely obliterate their host's reproductive tissues and stimulate changes in other parts of its body. In some hermaphrodite hosts this parasite-induced damage and eventual castration is restricted to the female gonads, while the male gonads are left alone.

The orthonectids appear to be rather rare parasites of various marine animals, but how many scientists are actively looking for them? The answer is hardly any. These tiny creatures have been found in many locations around the world, but typically in the coastal reaches of cold and temperate waters. And we simply do not know how they affect the populations of their hosts by influencing reproductive success. Also, as with many other internal parasites, it is not inconceivable that they tweak the inner workings and behaviour of their hosts to suit their own ends.

ORIGINS AND AFFINITIES

The affinities of the orthonectids are a real puzzle. On their first discovery they were considered to be at or very near the base of the animal family tree, perhaps a transitional stage between single-celled protozoa and outwardly 'simple' animals, such as sponges. This idea persisted for a long time, mainly because the orthonectid body plan is so simple. Today, this theory has fallen out of favour and it now seems the orthonectids are closely related to more complex animals, their simplicity a result of their parasitic lifestyle. This certainly would not be the first time in the history of the animals that evolution has pared down a body plan to the bare basics as a consequence of living inside another animal (a perfect example of this evolutionary degeneracy is seen in the mucus animals (see p. 50), which are extremely simple, yet unequivocally Cnidaria).

Unfortunately, we do not yet know what branch of the animal family tree the orthonectids belong to. A possible affinity with the flatworms has been touted, but this has largely been discredited, so until the DNA of the orthonectids has been more fully compared with that of other animals, their affinities will remain a tantalizing mystery.

2 An adult female orthonectid. Adult males and females exit their host for a brief, free-living existence. In adapting to a parasitic way of life many of the the more complex characteristics of their ancestors have been lost.

Annelida

(bristle worms, beard worms,
spoon worms, peanut worms,
earthworms, leeches)
(Latin *annellus* = little ring)

Diversity
c. 18,950 species

Size range
~0.1 mm to 3 m
(~0.004 in. to 10 ft)

1 Segmentation, a distinguishing feature of the annelids is clearly visible here.

We are all familiar with the earthworms that abound in soils around the globe, as wells as the infamous leeches and the common polychaetes (bristle worms) of the seashore, such as ragworms and lugworms. However, the species we commonly come into contact with are but a fraction of annelid diversity. Animals in this lineage are found in a huge range of habitats, from sediments in the deepest reaches of the oceans to the moist soils of alpine valleys and the aquatic microcosms amongst the leaves of epiphytic plants high in the rainforest canopy.

FORM AND FUNCTION

Segmentation (see Introduction, p. 16) is central to the success and diversity of the annelids. Between the head and tail of an annelid there are multiple, repeating units (or segments), each with its own body cavity, gonads, appendages and excretory organs [1, 21, 22]. Nerve fibres, blood vessels and muscles run the length of the body passing through thin membranes partitioning each segment. In some of the annelids the outward signs of division into segments have been secondarily lost, but their organs still retain a degree of segmentation [BOXES 1 AND 2].

The segmented annelid body is typically a long thin cylinder enveloped in a cuticle that includes collagen. In many tube-dwelling species this cuticle has become extremely thin and delicate, its protective role assumed by the tube [23, 24]. The annelid body often sports an array of appendages, including long, sensory structures and fleshy outgrowths [25–28]. These outgrowths, known as parapodia, are extremely important in locomotion; the way they are moved is reminiscent of how a millipede uses its legs to walk. Parapodia are typically short and

BOX 1 ***Spoon worms (echiurans)***

These oddities of the seabed, numbering around 240 known species, have gone on something of an evolutionary tangent and so bear little resemblance to the 'typical' polychaete annelids from which they derive.

Segmentation is limited to the internal organs and the body cavities. The body consists of a squat trunk and a non-retractable proboscis, and the largest species can be 1.5 m (5 ft) long. Fond of burrows and crevices, they normally feed by using their proboscis to collect organic matter from the surface of the sediment, although some species, notably the fat innkeepers (*Urechis* spp.), secrete a mucus net to filter feed.

Many spoon worms exhibit an incredible degree of sexual dimorphism with numerous dwarf males, a little over 1 mm (0.04 in.) long, taking up residence inside the body of a female whose total length may approach 1 m (3 ft). In some species gas exchange occurs across the lining of the hind-gut and/or the cloaca when water is drawn in and out of these spaces via the anus.

2 3

4

2 A spoon worm (unidentified thalassematid). Note the curved chaetae, which are important in burrowing.

3 *Urechis* spp. are the largest spoon worms. They are sometimes called fat innkeepers because their U-shaped burrows are home to at least three specialist commensals (a crab, a scaleworm and a clam) (*Urechis caupo*).

4 A spoon worm (*Bonellia* sp.). The body is normally tucked away in a crevice and the proboscis extends out on to the seabed for the collection of edible particles. The green colour is due to a toxic pigment, bonellin, which may deter predators or kill settling organisms.

BOX 2 ***Peanut worms (sipunculans)***

Like the spoon worms, the inner workings of the peanut worm single them out as annelids, even though superficial similarities are few and far between. Around 1500 species are known, and they range in length from 2 mm to 70 cm (0.08 to 27 in.).

The mobile, anterior end of these animals (the introvert), can be retracted into the squat body. They are typically sessile, inhabiting burrows or crevices, although some species are capable of boring into rock. Peanut worms can be filter or deposit feeders, and to prevent their digestive enzymes being diluted by ingested water their gut is equipped with a bypass that sends water straight to the end of digestive tract.

Unique to the peanut worms are waste-processing microorgans known as ciliated urns. Some of these float free in the large body cavity, entangling foreign bodies and moribund cells in trailing strings of mucus that they secrete.

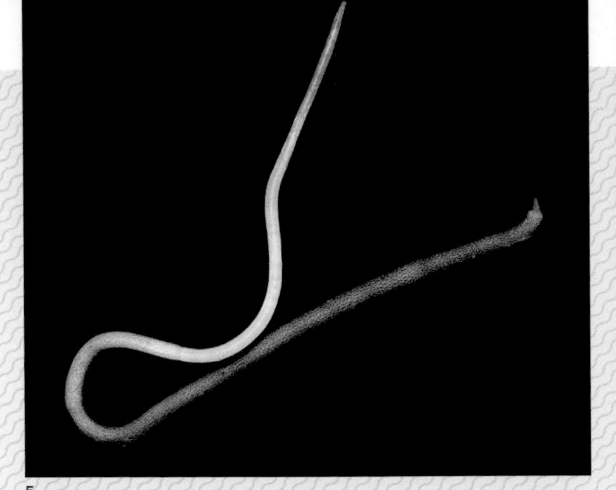

5

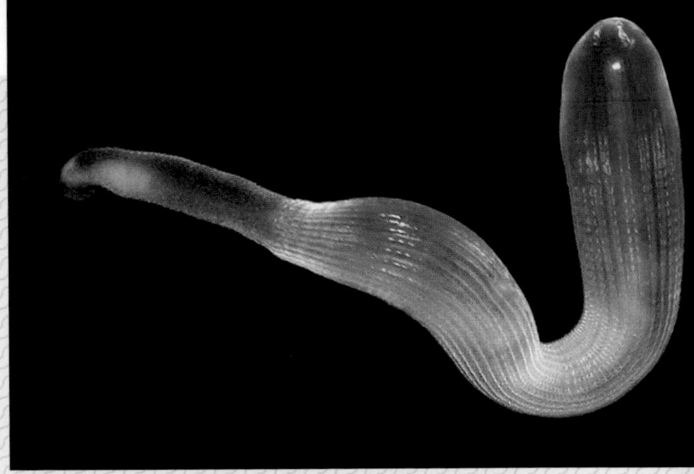

6

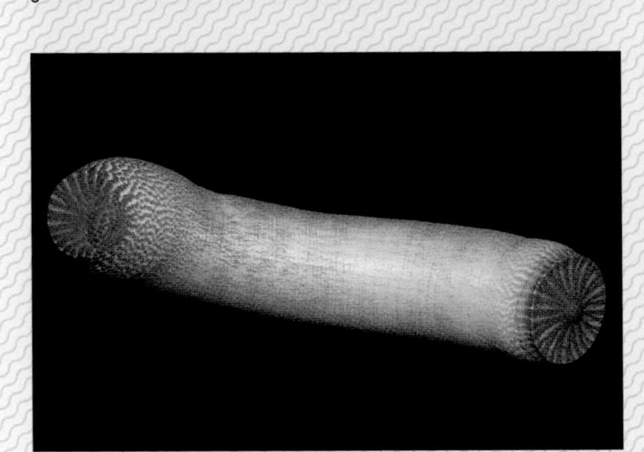

7

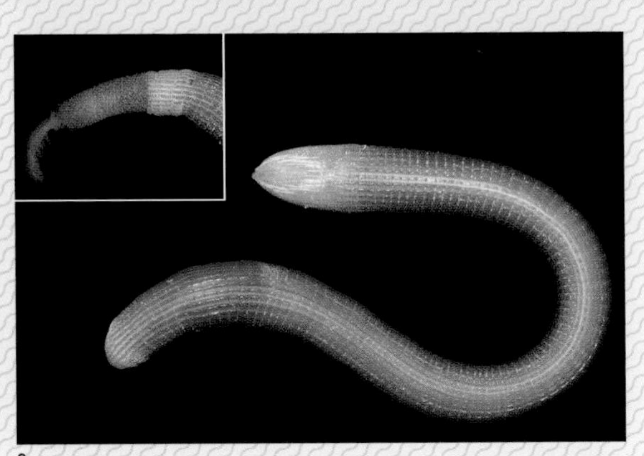

8

5 A peanut worm (*Aspidosiphon gracilis*).

6 Unidentified deep-sea peanut worm.

7 Some peanut worms, such as this *Aspidosiphon* sp., bore into rock and have calcareous shields on the front (left) and rear of their body. The former caps the entrance to their retreat, while the latter bores and anchors them in place.

8 Other peanut worms, such as this *Sipunculus* sp., burrow in sand. The inset shows the extended introvert.

stubby, but in some swimming species they are long and slender [45]. Annelids also have chitinous bristles known as chaetae, which are important in locomotion, traction and anchorage. They come in a huge variety of shapes and sizes, from the paddle-like structures of some swimming species to the inconspicuous hooks of tube-dwellers [2, 34, 35, 36].

Controlling these appendages and processing information from the environment is a complex nervous system consisting of a well-developed brain, a ladder-like arrangement of nerve fibres and ganglia – in effect, mini-brains serving each segment of the body. In some species, the nerve fibres running the length of the body are the thickest of any animal (around 1.7 mm (0.07 in.) in diameter). Being so broad means there is less resistance, and electrical signals can travel from the brain to the muscles very rapidly, as much as 40 times faster than if the nerve fibres were of average thickness. This translates into a lightning-quick escape response – evident when a startled earthworm retreats down its burrow. Annelids, especially the polychaetes, also exhibit an array of complex sensory structures, including eyes, statocysts (for balance and orientation) and nuchal organs (for chemoreception).

Gases are able to move in and out of tissue across the thin, moist annelid body wall, although some forms, especially tube-dwellers, have elaborate gills to increase the surface area for gas exchange. The process of gas exchange is further enhanced by a discrete circulatory system consisting of blood vessels and even numerous hearts to pump the blood around the body. Within this blood gases are bound to a respiratory pigment, often haemoglobin, although other pigments are common, lending the blood green, pink or violet hues. Respiration in the earthworms is thought to be made more complicated by the high concentration of carbon dioxide in the soil (emitted by subterranean bacteria), which essentially prevents other molecules of the same gas from leaving the body of the worm. It is thought these soil-dwelling annelids overcome this problem using unique organs – the so-called calciferous glands – that combine carbon dioxide gas with calcium to form calcium carbonate crystals, which are voided from the body with the faeces.

Around parts of the circulatory system as well as the gut there is a specialized tissue that fulfils a comparable function to the liver of other animals, namely the production of energy storage molecules (glycogen, lipids) and the detoxification of harmful substances. The gut is usually a simple straight tube: at the head of this digestive tract there is often a muscular pharynx, and some species even have a series of gizzards for grinding their food. The pharynx, sometimes equipped with zinc- or copper-reinforced jaws or venom-dispensing fangs, can be squeezed out of the mouth to engulf food [29, 31–33]. Defecation is not an issue for free-living species, but tube- and burrow-dwellers have evolved a range of adaptations to prevent their homes being sullied. The posterior half of the U-shaped mason worms is reduced to little more than a tube for conveying faeces into the sea, while feather duster worms [36, 37] have a long, cilia-lined groove to transport faecal pellets to the tube entrance.

The majority of annelids have formidable powers of regeneration. The burrow- and tube-dwellers often have their tentacles and heads nibbled off by predatory animals, but the lost parts simply grow back. Parchment worms and others are even able to regenerate their whole body from a single segment. This ability to regenerate also comes into its own in asexual (clonal) reproduction, as an adult worm can bud off other individuals (see below).

LIFESTYLE

Annelids have evolved to fill a huge range of niches. Filter feeding is common, as is the seemingly unsavoury practice of eating sediment and soil to extract edible morsels. Some species are active predators with a range of adaptations for catching and subduing prey. However, extreme environments are their forte. Notable examples include the beard worms that thrive around hydrothermal vents [BOX 4], *Osedax* worms digesting the marrow in whale skeletons on the sea floor and glacier-dwelling ice worms (*Mesenchytraeus* spp.), which simply 'melt' if they are exposed to temperatures as high as 5°C (41°F). Neither have annelids overlooked the opportunities offered by other animals: leeches [BOX 3], however unpleasant to us, are superbly adapted to a parasitic way of life and there are numerous other annelids for whom the exterior or interior of other animals has become home [38].

It is in the oceans where the lifestyles of the annelids are most diverse. Many of the benthic species sport a huge variety of structures for the filtering of seawater or the collection of organic matter from seabed sediment [36, 37, 39–41]. Christmas tree worms, for example, use their spiral of multicoloured tentacles to filter edible particles from the water [41], while the tangle of 1-m- (3-ft-) long tentacles emanating from the head of a spaghetti worm collect edible morsels from the sediment surrounding its tube [30]. The lugworms are more inconspicuous, spending their adult lives in the horizontal part of an L-shaped tube, eating sand and digesting edible morsels. So as not to foul its home the lugworm shuffles back to the entrance of its tube and leaves a mound of defecated sand on the surface – a familiar seashore worm-cast. *Armandia* and related species swim like fish through loose sand and water (see illustration on p. 21), while other annelids have completely adapted to a pelagic existence, spending their entire adult lives swimming gracefully through the water column [45].

Parchment worms [42] live in tubes around the same size and shape as a banana, though the tube is buried in the sand with only the two ends protruding. The tube they make is an extension of their very fragile body and serves to protect and feed them. By rhythmically beating some of their appendages they drive water through the tube and trap suspended organic matter in a mucus mesh bag secreted from a hoop formed from other, hugely modified segmental appendages. A small ciliated cup grips the bag and feeding continues until it is laden with food particles, at which point the cup rolls the bag and its contents into a ball, which is projected forward and deposited in a ciliated groove along which it trundles to the mouth.

The earthworms and their aquatic relatives are typically scavengers of plant and animal matter, much of which consists of very resilient substances that are not digested easily [18]. To help them release the energy contained in this food, they have formed an alliance with

BOX 3 ***Leeches***

Predominantly freshwater animals, the 680 or so species of leech are parasites or active predators. They have evolved a suite of unique adaptations for their way of life, while many of the typical annelid characteristics have been secondarily lost.

To grip on to their hosts, most leeches have well-developed suckers on their front and back ends, while all but one species have lost all traces of chaetae. The typical, segmented annelid body cavity is replaced by a single open space, much of which is filled with tissue for storing nutrients. The loss of these partitioned compartments has allowed the evolution of different forms of movement, from sinusoidal swimming to inch-worm-type locomotion.

Although they do not have large, conspicuous sense organs, leeches are well equipped to respond to any stimuli that may betray the presence of prey or hosts. Terrestrial leeches, the bane of Southeast Asian rainforests, wave their front ends frantically in the air when they sense an approaching host (perhaps detecting vibration, heat or carbon dioxide). When they locate a host their blade-like jaws or digestive secretions are quick to breach the skin. To avoid alerting the host they secrete a potent anaesthetic as well as a cocktail of complex chemicals, including blood vessel dilators, anticoagulants and antibiotics.

Feeding opportunities are rare to say the least, so leeches are able to imbibe ten times their own weight in food in one sitting. Unusually, they produce very little in the way of digestive enzymes, instead primarily relying on an army of symbiotic bacteria to convert the meal into usable nutrients. It is these symbiotic bacteria that produce the antibiotics to inhibit the growth of opportunistic bacteria that may otherwise make short work of the food themselves. This may be one reason why the digestion process can take as long as 200 days, meaning that there is no need to feed more than twice a year.

A group of 150 annelid species, closely related to the true leeches, scour the exoskeleton of their crustacean hosts for microorganisms and organic matter. Rather than parasitic, these annelids can actually be beneficial to their hosts in some situations.

9

10

11

12

9 The medicinal leech, *Hirudo medicinalis*. Leeches are usually tapered toward their anterior end and large posterior sucker.

10 A tiger leech, a terrestrial species from the rainforests of Southeast Asia (*Haemadipsa picta*).

11 A marine leech using its posterior sucker to grip onto a fish (unidentified species).

12 Most species of leech have three blade-like jaws in their mouth (note the serrations on the jaws). It is these that slice through the skin of the host, leaving a Y-shaped wound (*Hirudo medicinalis*).

BOX 4 ***Beard worms (siboglinids)***
Beard worms surely rank as some of the most unusual of all animals. Many are large (up to 1.5 m or 5 ft long) and they live in very deep water, often in close proximity to the superheated water being discharged from hydrothermal vents. Adult beard worms have no mouth, anus or even a normal gut; instead, where you would normally find a digestive tract, there is a solid mass of symbiotic bacteria. The worm supplies these symbionts with copious quantities of oxygen and sulphur containing compounds so that they can synthesize more complex molecules, some of which the worm utilizes as food (alongside some of the bacteria themselves).

Osedax is a genus of small beard worms, commonly known as zombie worms. They too depend on symbiotic bacteria, but instead of indirectly using the chemical energy from hydrothermal vents, they bore into whale bones on the sea floor. Their symbiotic bacteria digest the fats and oils in the bone marrow, releasing nutrients that the worm can use. If this is not unusual enough, they also display one of the most extreme examples of sexual dimorphism: the males are microscopic and a 'harem' of between 50 and 100 of them live inside the female's transparent mucus tube.

13

14

15

16

13 Beard worms, such as these *Lamellibrachia* sp., build chitinous tubes in dense aggregations on continental slopes and near hydrothermal vents or cold seeps of hydrocarbons.
14 *Lamellibrachia* sp. emerging from its tube. The tentacles are bright red because of the haemoglobin-loaded blood running through them.
15 Beard worms are a dominant component of a complex community of animals that ultimately depend on chemical energy rather than energy from the sun. In amongst the beard worms (*Lamellibrachia* sp.) here are decapod crustaceans and bivalve molluscs.
16 An adult female zombie worm (*Osedax* sp.).

bacteria. These symbionts dwell in the wall of their intestine, where they secrete enzymes to break down cellulose and chitin. Earthworms are found throughout the world, but they are not fully terrestrial because the soil in which they live needs to be moist. During periods of drought they burrow down, perhaps 3 m (10 ft) or more, and enter a state of suspended animation until conditions improve. Other species confronted by dry conditions also enter a period of inactivity, but they envelope themselves in a layer of mucus that hardens into a tough cocoon.

Annelid reproduction is also wonderfully varied. In the marine forms there are typically separate sexes and the gametes from the diffuse gonad tissues are either shed through the same pores that excrete waste or, more brutally, by rupture of the body. A reproductive phenomenon

BOX 5 ***Epitoky***

Many polychaetes display an interesting reproductive approach known as epitoky. For much of the year the worm exists as an atoke, a benthic asexual creature that cannot breed. However, as the breeding season approaches, the animal undergoes a radical transformation. Depending on the species, complete new sexual individuals or segments bud from the rear of the atoke forming a long chain. These reproductive individuals or chains of gamete-laden segments are known as epitokes.

The palolo worm (*Eunice viridis*) is one of the more well-known epitokous polychaetes. Huge numbers of these worms, with their epitokes fully formed, wait in their refuges for a specific cue to spawn – the moon. In October or November, at the beginning of the last lunar quarter, all the worms release their epitokes at the same time. Free of the atoke, they swim to the surface where the light of the rising sun, detected by the eye spot, triggers their synchronous bursting to release eggs and sperm, turning the sea, close to shore, into gamete soup. The eggs are fertilized rapidly and by the next day tiny larvae have formed. After drifting for two or three days these offspring begin to settle and find rocky hideaways of their own, where they will develop into adult worms in preparation for the next year's mass spawning. The atoke, still in its burrow, will regenerate a new epitoke for the following year's breeding season.

Epitoky and swarming is quite common among the polychaetes. This phenomenon makes it possible for a dispersed population of worms to come together briefly, thus increasing the chances of successful fertilization.

17

17 *Myrianida pachycera*. This polychaete, like the palolo worm, reproduces by epitoky, but the atoke (the largest individual in the image) buds off a chain of smaller epitokes containing either eggs or sperm, depending on the sex of the atoke. These epitokes meet in the water to reproduce.

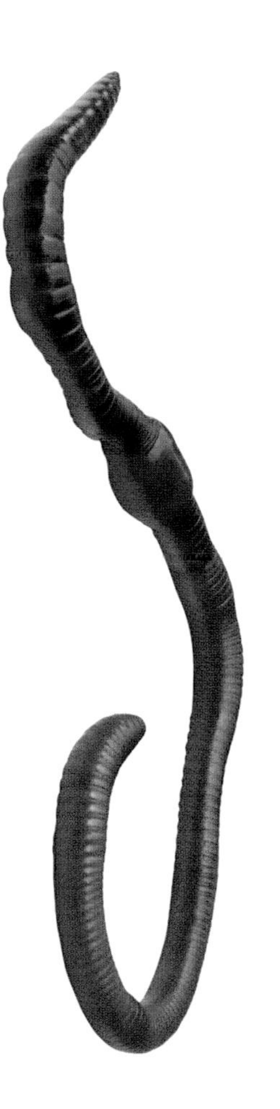

18

unique to many types of polychaete is known as epitoky [BOX 5]. Swarming is also a key element of reproduction in other marine polychaetes. Males and females of the Bermuda glow worm, *Odontosyllis enopla*, gather in frantic, reproductive swarms, a behaviour triggered by light intensity and the cycles of the moon. In the summer, 50 to 60 minutes after sunset and for up to 12 days following a full moon, the female worms glide to the surface and swim in circles while emitting a steady green glow (bioluminescence). This soon attracts the males whose green glow blinks off and on as they swim towards the females. When males and females meet there is a flash of green light and the gametes are released.

The earthworms and their relatives are hermaphrodites. During a reproductive embrace that can last several hours the individuals exchange sperm and store it for later use. Unique to these animals is a girdle toward the head end of the body, conspicuous in many common species [18]. This structure, the clitellum, is responsible for forming a cocoon that will protect and nourish anywhere between 1 and 20 eggs. The cocoon begins life as a thick layer of mucus secreted by the girdle that slides forward over the pores where the sperm and eggs are stored. The gametes are released into the mucus sheath, fertilization takes place and the sheath eventually slips off the worm's head. The ends of the sheath constrict to form a protective cocoon loaded with nutritious albumen.

The leeches are also hermaphrodites, but they start off male and then swap gender as their ovaries mature. Sperm are normally introduced into the female with the help of a penis; but in those species lacking a penis copulation is a bit barbaric as a little packet of sperm (spermatophore) is simply forced through the body-wall of the partner.

Following fertilization, the annelid norm – but secondarily lost in the earthworms and leeches – is for the egg to develop into a characteristic planktonic larva – the trochophore [44, 47]. This feeds using girdles of cilia around its middle, eventually metamorphosing into a juvenile worm with a small number of segments [30, 43]. New segments are added from a band of cells towards the read end of the young animal.

From an ecological point of view the annelids are one of the most important animal lineages. In many marine habitats it is not unusual for annelids to reach staggering densities of more than 1.5 million individuals per sq. m (140,000 per sq. ft) of seabed, accounting for anywhere between 40 and 80 per cent of the sediment fauna. Collectively these marine burrowers are capable of churning vast quantities of sediment – as much as 4500 tonnes per ha (11,000 tonnes per acre) of sea floor each year. On land there can be as many as 700 earthworms per sq. m (65 per sq. ft) of soil. In scavenging organic debris from the surface, earthworms are the main drivers of nutrient recycling, a by-product of which is the formation of rich, deep, well-drained and aerated soils. Drawing on Darwin's earthworm treatise from 1881, we can estimate that these relentless burrowers bring as much as 40 tonnes of soil to the surface per ha (or 100 tonnes per acre) each year. Even the leeches can reach huge population densities, with as many as 10,000 individuals per sq. m (950 per sq. ft).

Being relatively large, soft-bodied and hugely abundant, it is not surprising that larger animals relish annelids; in the ocean and on land annelids are a major component of the diet of arthropods, fish, mammals and birds. This has exerted a selection pressure for the evolution of an array of defensive adaptations [BOX 6].

ORIGINS AND AFFINITIES

The annelids have a rich fossil record. The oldest unequivocal annelids date to the middle Cambrian (around 520 million years ago), and the lineage appears to be well established in the seas of the late Cambrian. Based on more recent fossils and what we know about how the annelids have diversified, we can speculate that the ancestor of the lineage was a marine animal with a segmented body and long chaetae – a body plan that many living species retain and which is suggestive of a surface-dwelling form. Some of the ancestors may have gradually transitioned to life in freshwater sediments and then progressively drier substrates, i.e. soil.

A long-held view was that annelids were the progenitors of those other conspicuously segmented animals, the arthropods. Zoologists from the 18th to the 20th centuries were seduced by this idea, such are the superficial similarities between these two major lineages. To confound things even further the velvet worms (see Onychophora, pp. 144–47) were hailed as the 'missing-link' between the annelids and the arthropods. Today, these theories have been debunked and it is accepted that the annelids and arthropods belong on separate branches of the animal family tree. The closest relatives to annelids are thought to be molluscs.

18 Earthworms are crucially important in soil formation. The swollen girdle is the clitellum, which secretes the egg cocoon (*Lumbricus* sp.).

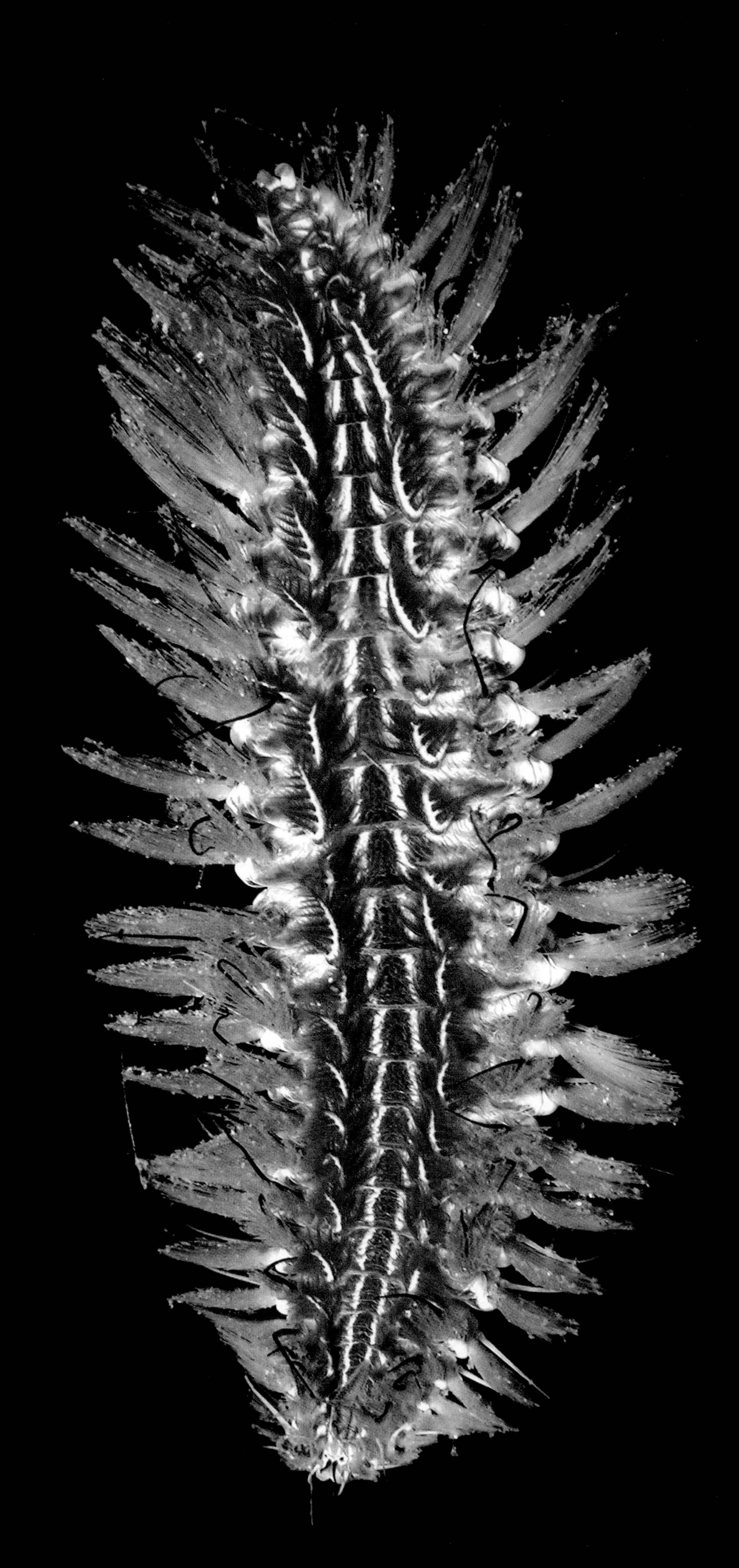

BOX 6 ***Annelid defences***

Annelids can protect themselves in a splendid variety of ways. The green bomber worm (*Swima bombiviridis*) and its relatives are able to cast off bulbous, modified gills that give off a green bioluminescence, which is thought to distract or confuse predators. The sharp, brittle chaetae of other polychaetes, commonly known as fireworms, are loaded with toxins. These break off in the skin of an attacker and cause a very painful burning sensation.

Not to be outdone, some earthworms are capable of squirting fluid from their body cavity as much as 30 cm (1 ft) to repel an attacker. There are even some leeches that can change colour dramatically, but it is not yet known if this is important defensively.

19 *Chloeia* sp., one of the infamous fireworms.

20 *Hermodice cf. carunculata*, a fireworm.

21

22

21 *Glycera capitata* and its relatives have a large, eversible pharynx equipped with four fang-like jaws through which venom is discharged into the prey. These polychaetes live in complex burrow systems in marine sediments.

22 An *Autolytus* sp. polychaete.

23 Many annelids build beautiful, very elaborate tubes, which are an extension of their body. The sand grains forming the outer surface of this tube are embedded in layers of organic material secreted by the worm, the structure and composition of which are very similar to the animal's cuticle (*Pectinaria* sp.).

24 *Pectinaria* sp. emerging from its elaborate, sand-grain tube. The animal builds this structure by using its mouthparts too add grains of the appropriate size and shape to a secreted tube.

23

24

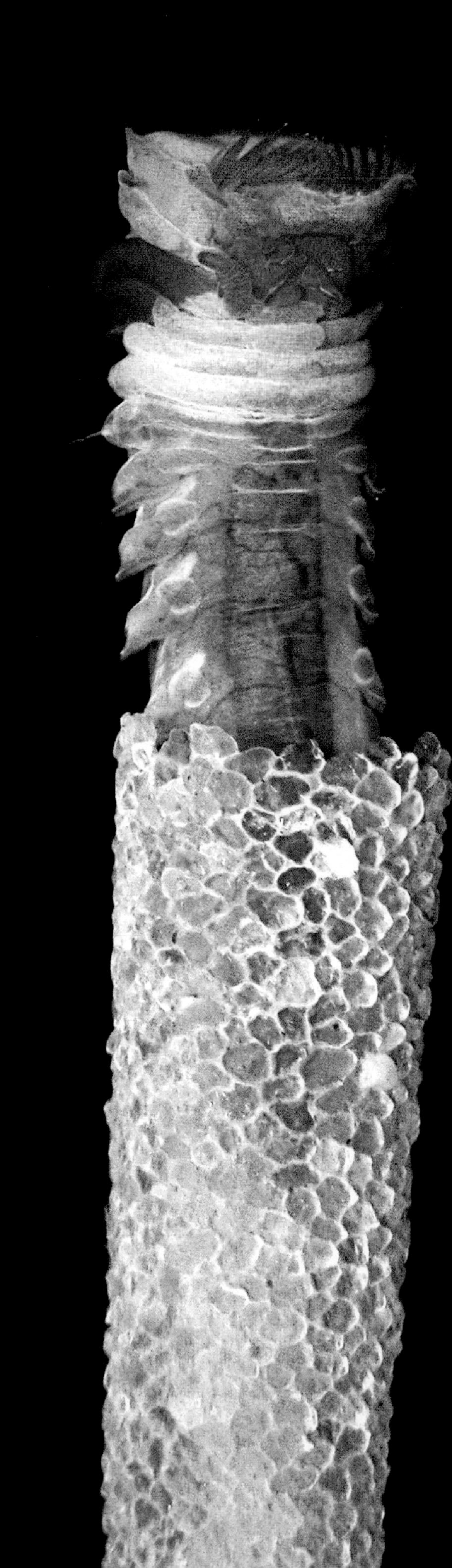

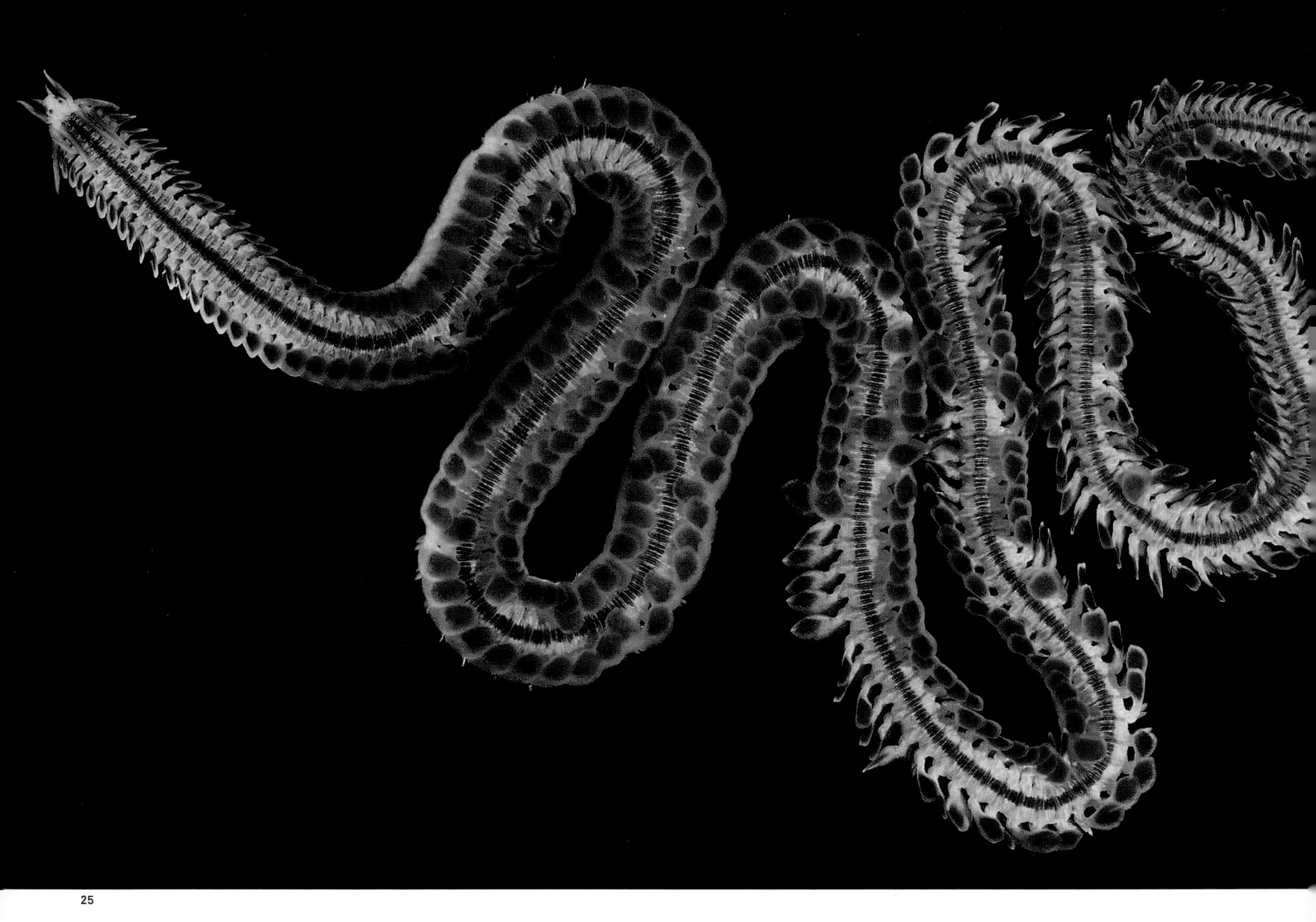

25

26

25 The fleshy outgrowths of the body segments are clearly visible in this colourful annelid, *Phyllodoce citrina*.

26 Sensory appendages on the head of the polychaete annelid, *Alitta virens*.

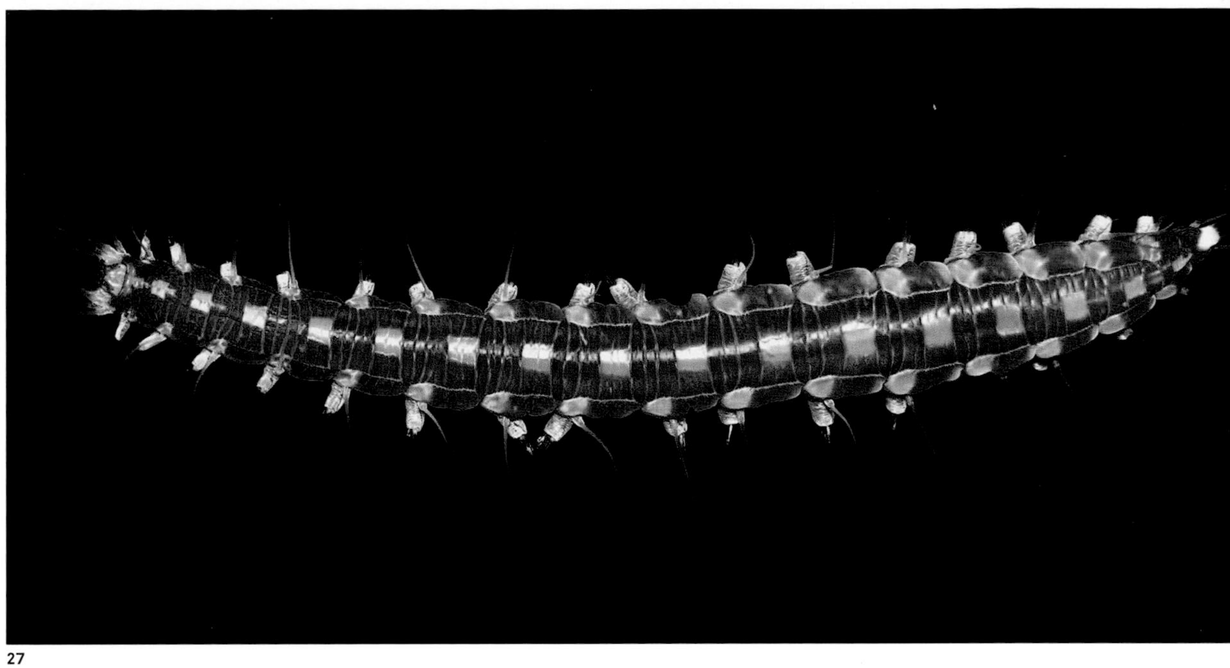

27

27 Chaetae-bearing parapodia are clearly visible in this polychaete (unidentified hesionid).

28 *Pterosyllis finmarchica* has long tentacles sprouting from its segments. These annelids use a pharyngeal tooth to pierce their prey.

28

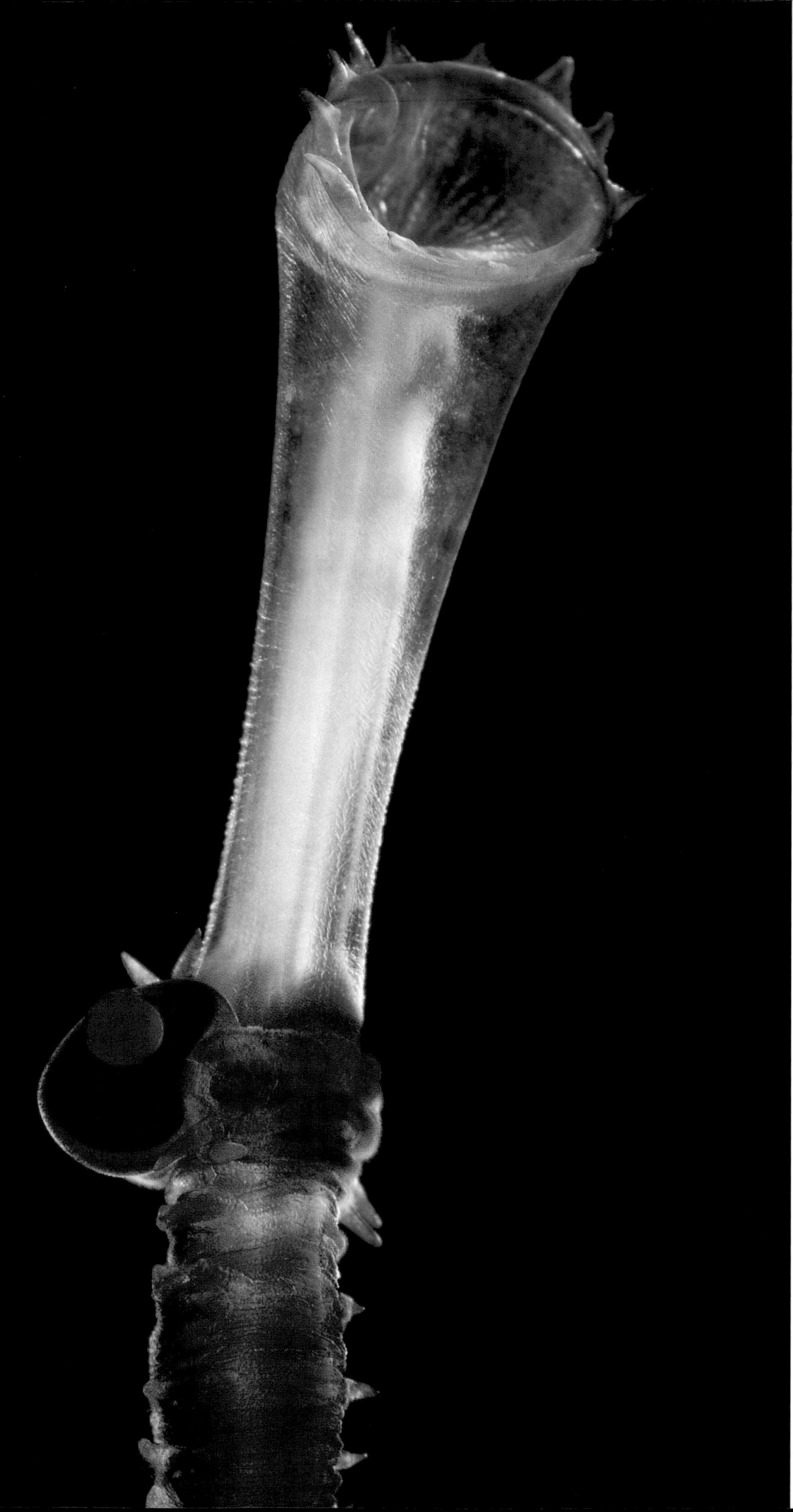

30

29

29 This *Vanadis* sp. polychaete uses huge, bulbous eyes and a capacious, eversible pharynx to locate and engulf prey.

30 A spaghetti worm (*Terebellides stroemi*). These annelids extend their mass of mobile tentacles from the mouth of their tubes onto the surrounding sediment. Edible particles stick to the mucus on the tentacles and are conveyed to the mouth by cilia.

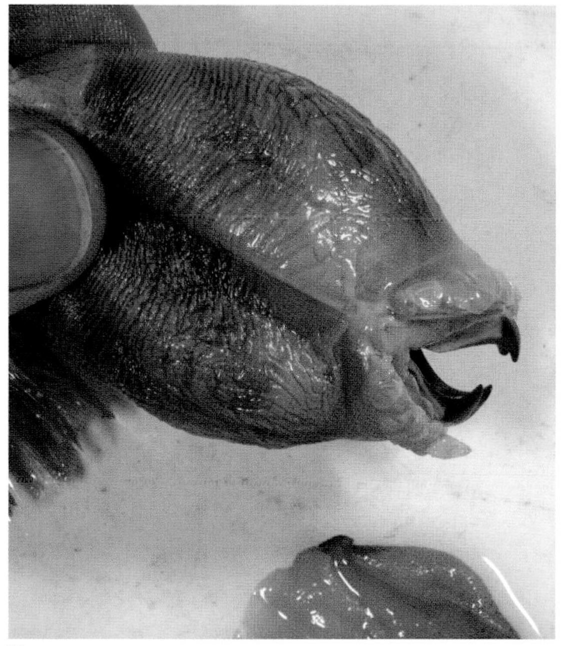

31

32

31 The fearsome jaws of a giant Antarctic scale worm, *Eulagisca gigantea*.
32 *Eunice* spp. are among the largest annelids. They use trap-like jaws to snare prey.
33 A polychaete, *Nereis virens*, everting its pharynx and revealing its impressive jaws.
34 This freshwater annelid (*Vejdovskyella comata*) uses long, paddle-like chaetae to swim through the water.
35 The long, thin bamboo worms (maldanids) live upside down in their tubes, where they ingest the sediment to form cavities. These protective retreats are occupied by various commensal species, including bivalve molluscs and amphipod crustaceans.

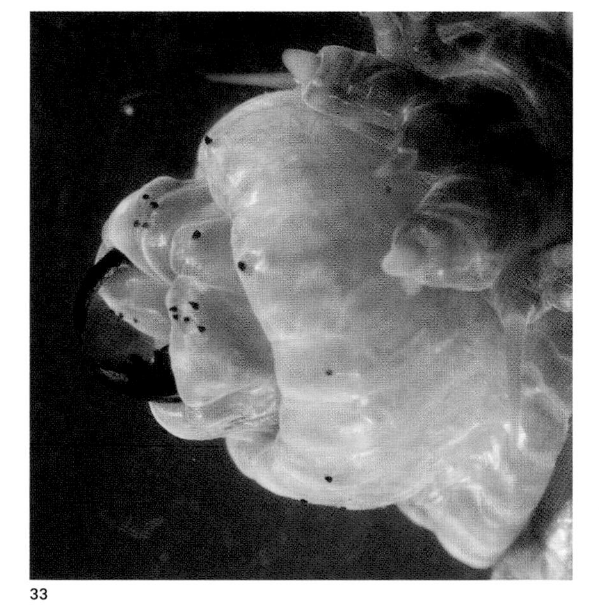

33

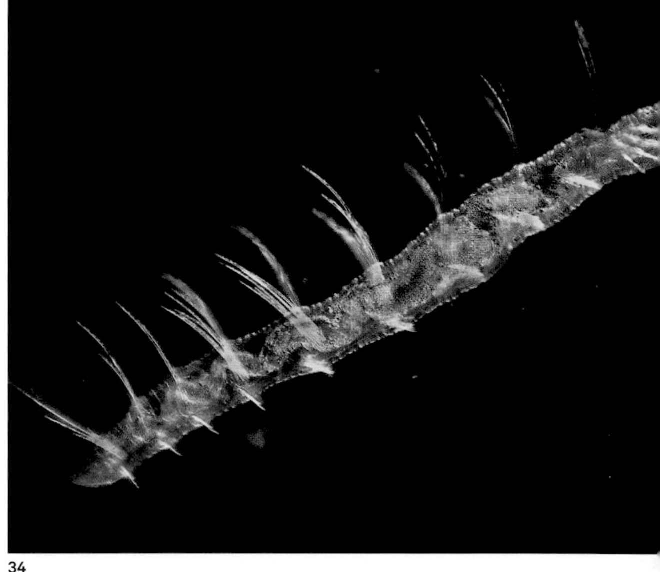

34

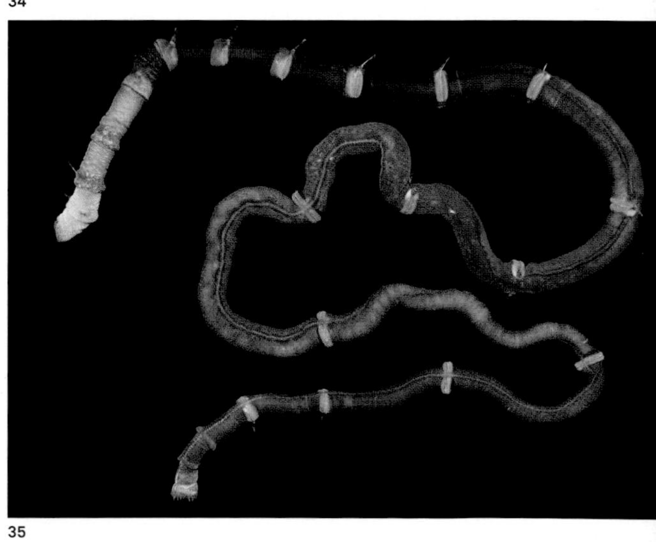

35

Overleaf
36 Feather duster worms, like this *Branchiomma arctica*, are among the most beautiful annelids. The feathery structures on the anterior end are specialized tentacles for feeding and gas exchange.
37 Feather duster worm (*Branchiomma* sp.).

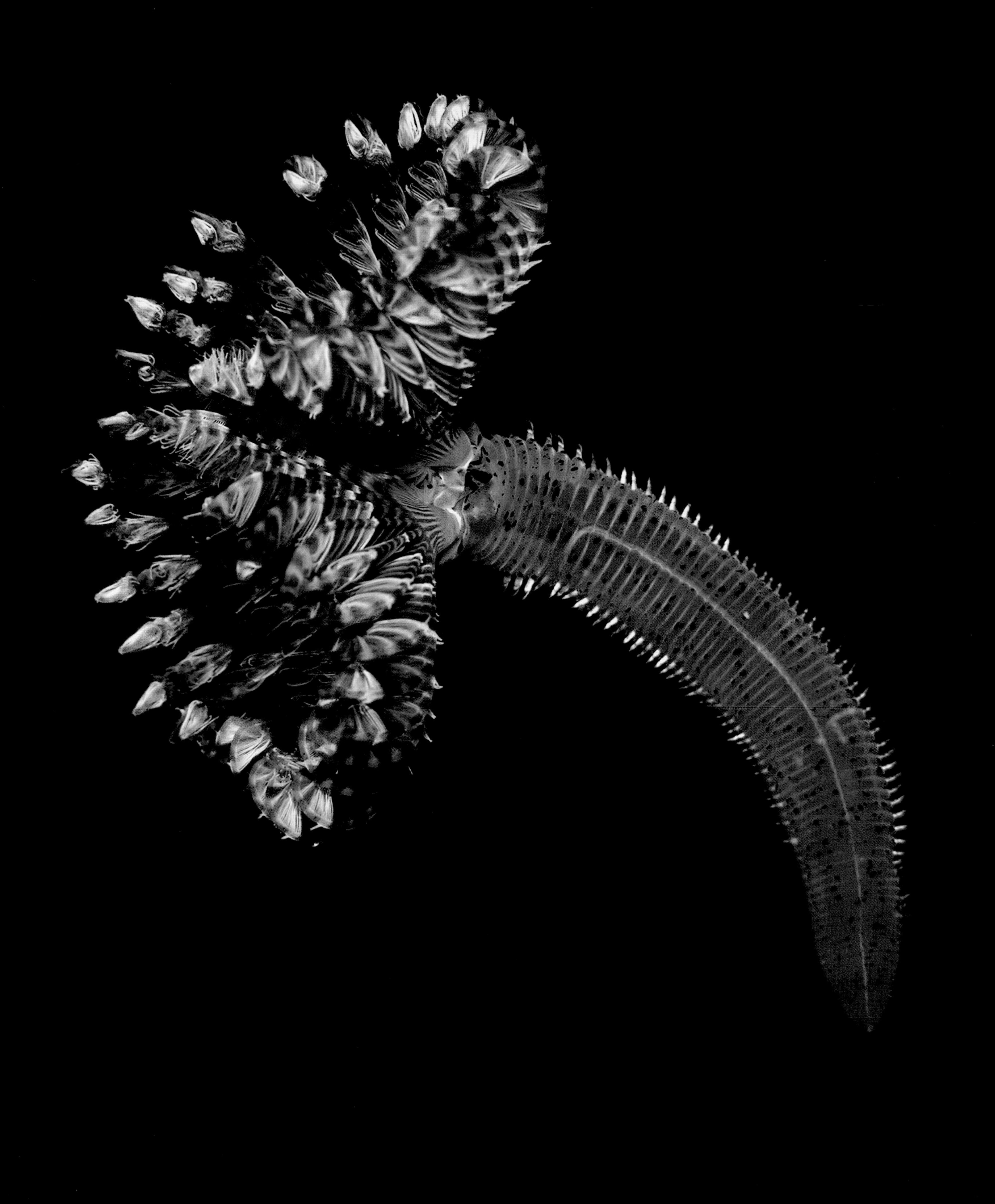

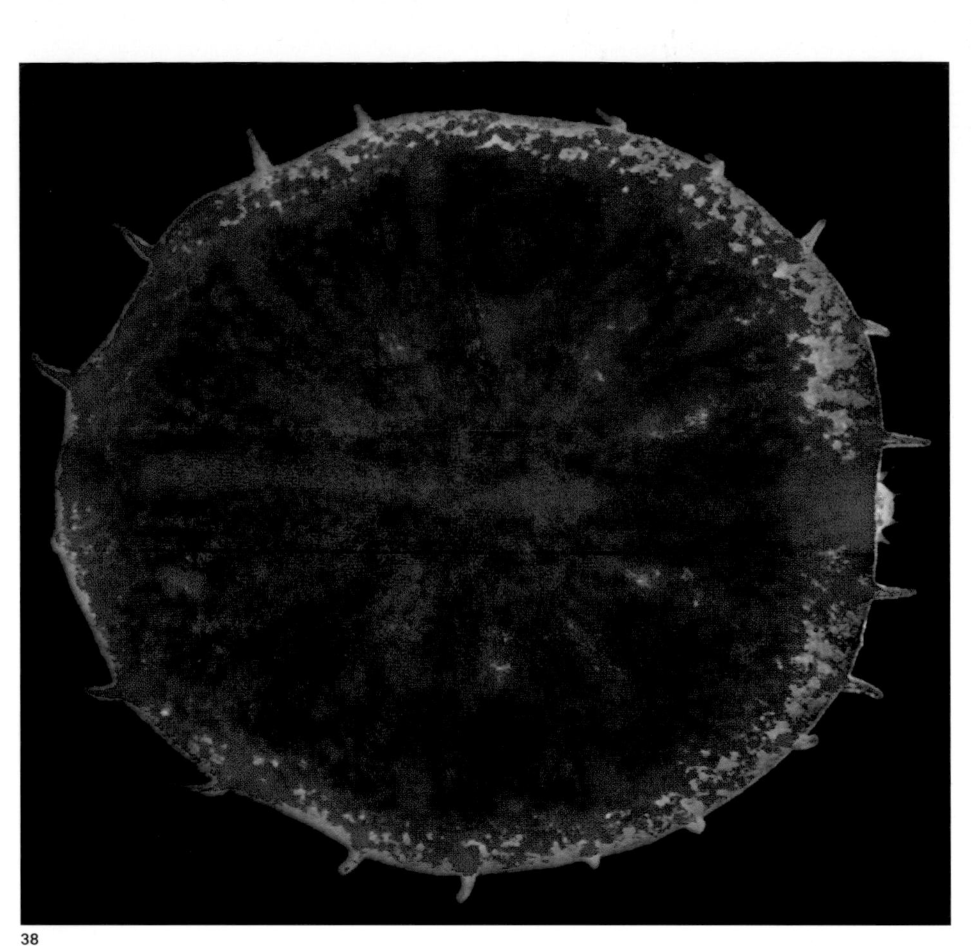

38

39

40

41

38 Myzostomes are very peculiar annelids that live as commensals or parasites on various echinoderms (*Myzostoma seymourcollegiorum*).

39 Annelids have evolved some incredible ways of harvesting edible particles from the water. Windmill worms add radiating 'spokes' to their tubes as a framework for a mucus web to trap edible particles in the water, much like a spider uses a silken web. Once the mucus web is loaded with food the worm eats it and secretes another (*Praxillura maculata*).

40 Many tube-dwelling polychaetes have elaborate, colourful tentacles for filter feeding and gas exchange. The funnel-shaped structure (operculum) seals the tube when the animal retreats inside (unidentified serpulid).

41 The tentacles of a Christmas tree worm (*Spirobranchus spinosus*). The rest of the body is out of sight in a tube.

42

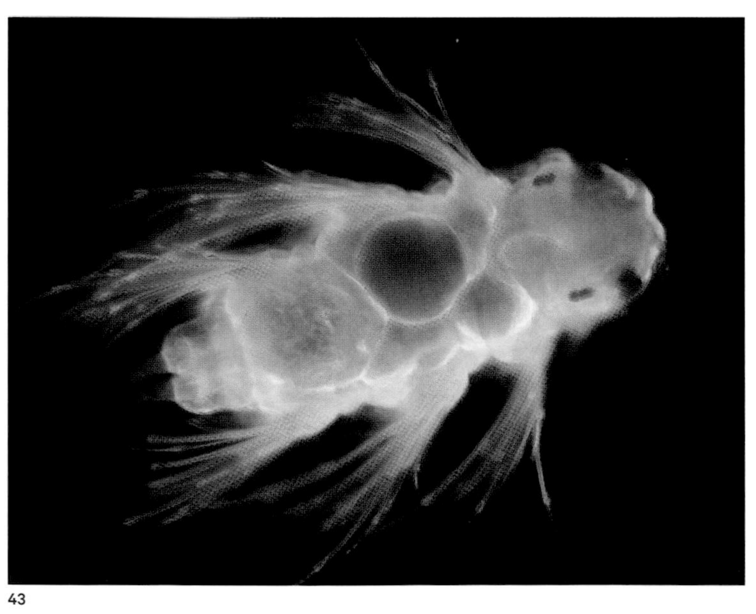

43

42 The extremely fragile parchment worm lives in a banana-shaped burrow in marine sediment.

43 A juvenile nereid polychaete.

›44 A trochophore larva of the shingle tube worm, *Owenia fusiformis.*

‹45 In adapting to a pelagic existence, some annelids have evolved long parapodia for swimming (*Tomopteris* sp.).

46 Annelids increase in length via the growth of new segments (unidentified polynoid polychaete).

47 Illustration showing a generalized trochophore larva. This planktonic stage in the life of many annelids swims using its ciliated girdles. The mouth (left), stomach (large central cavity) and anus are visible.

47

46

Mollusca

(snails and slugs, octopuses and squids, chitons, bivalves, scaphopods, solenogasters, caudofoveates)
(Latin *molluscus* = soft)

Diversity
c. 117,350 species

Size range
<1 mm to 20 m
(<0.04 in. to 65 ft)

1 The majority of the molluscs are aquatic, and this lineage comprises almost a quarter of all described marine organisms. The variety of body forms is extremely diverse. This sea angel, *Clione limacina*, about 5 cm (2 in.) long, feeds on other pelagic molluscs by using eversible tentacles, chitinous hooks and brute force to haul unfortunate victims from their shells. The prey is then swallowed whole.

From microscopic, interstitial sea slugs to enormous squid and a dizzying array of forms in between, the molluscs are second only to the arthropods in terms of diversity. Many species in this lineage are familiar to us from everyday life: most of us would be able to name at least a few, whether the slugs berated by gardeners, the oysters coveted by foodies or any number of others that come into our lives on a regular basis, normally – unfortunately for the molluscs – as food.

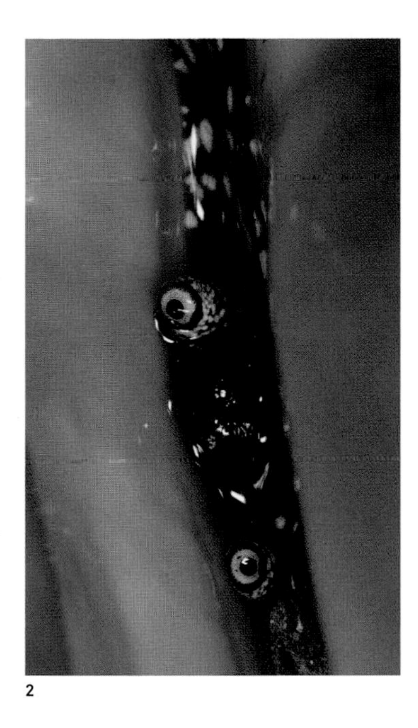

2

2 The eyes of molluscs are incredibly diverse. From the simple, pigment cups of limpets to the very complex lens eyes of cephalopods and an array of variations in between, it seems these sensory organs have evolved independently on a number of occasions in this group of animals. The eyes of this spider conch are equipped with lenses and can probably form detailed images (*Lambis* sp.).

FORM AND FUNCTION

What makes a mollusc a mollusc? This is quite a tricky question because they are such diverse animals [1, 14–18, 29]. With that said, there are a number of generalizations we can make about their body plans, unique characteristics uniting these creatures under the molluscan banner. Depending on the species in question some of these traits are more evident than others, and in some aberrant groups they may have been reduced to mere vestiges.

The general mollusc is an unsegmented, soft-bodied animal with a muscular foot on its underside [20]. Most of the remaining body is covered by skin forming a protective coat, the so-called mantle, which in many molluscs produces calcium carbonate arranged into small scales and spines or – more often – a shell. This was a real innovation when it evolved many hundreds of millions ago as a solution to the diversification of predators in the oceans, especially those with jaws that could make short work of soft-bodied animals. Over time, the simple scales and often multiple shell-plates of the ancestral molluscs evolved into single shells. The mollusc shell is a strong composite of protein and calcium carbonate layers that, across this group, has evolved into some weird and wonderful shapes, from the flattened internal shell of squid (the familiar cuttlebones given to birds) to the incredibly elaborate and seriously strong retreats of snails [19, 21–23].

The mantle is typically folded to create a substantial and unique cavity, often connected to the outside world via a pore (pneumostome) or a tubular extension of its surface (siphon) [22, 28]. This mantle cavity houses the gills/lung, a very sensitive smell/taste organ, and the outlets of the gonads, digestive tract and excretory organs. In aquatic species the water in this cavity is circulated by the beating of cilia.

Another characteristic feature of molluscs is the radula; a flexible ribbon of chitinous teeth, sometimes hardened with iron or silica [25, 27]. This elaborate structure is poked out of the mouth and used like a rasp to graze or rend whatever food the mollusc is interested in. In the filter-feeding species, namely bivalves (think mussels and oysters), the radula is surplus to requirements, so it has been lost. Edible matter and any inedible particles swallowed by the mollusc are conveyed to the stomach in a mucus string. To separate the food from the inedible bits and pieces, the molluscs have evolved an elaborate sorting mechanism in the stomach consisting of ciliated grooves and ridges. In bivalves, some of the inedible particles are consolidated into a stiff, rod-shaped mass, which is rotated by the cilia to reel the mucus string with its embedded food into the stomach.

To keep the tissues well oxygenated, a muscular heart pumps blood loaded with respiratory pigments through short stretches of vessels and a series of blood-filled cavities. The respiratory pigment (haemocyanin) is based around copper, lending oxygenated molluscan blood a blue tint. However, the blood of some molluscs – for example the mud snails often found in garden ponds – also contains haemoglobins, red respiratory pigments similar to ours. In the cephalopods (squid, octopuses, nautiluses) the blood system is even more sophisticated, being a completely closed series of arteries, capillaries and veins and smaller, accessory hearts to pump blood through the gills. Well-developed kidneys remove waste from the blood and regulate the amount of water in the body.

Typically, molluscs have a nervous system consisting of a simple brain (cerebral ganglion) and two to four pairs of connected nerve cell clusters (ganglia), which receive input from a range of sensory organs, including eyes [2], statocysts (for orientation) and osphradia (akin to noses). Notable exceptions to this general set up are some snails and the cephalopods, which have an extremely well-developed central nervous system hooked up to a range of very finely tuned senses [BOX 1].

LIFESTYLE

The diversity of the molluscs is at its most staggering in the oceans. A considerable number of snail species are to be found on land [24, 28], but they occupy a rather narrow ecological niche and have not adapted to a terrestrial existence in the same way as some of the arthropods and craniates. Molluscs have diversified into an

BOX 1 ***Cephalopod convergence***

The evolution of the cephalopods is convergent with that of the craniates, especially the fishes. Both groups of animals are highly specialized aquatic predators with complex brains, sensory organs and hydrodynamic forms. The architecture and performance of the average cephalopod brain compares well with what we see amongst the craniates, even though the evolutionary paths of these lineages diverged hundreds of millions of years ago.

The cephalopods, once thought of as rather lowly, are now known to be very intelligent beings. They can form memories and are even capable of complex feats of learning, made all the more impressive by the fact that they are relatively short-lived animals. Experiments and observations have shown that they can solve problems, use tools and cooperate with one another via a complex range of subtle signals. Their abilities are at least on a par with those of some craniates.

Their eyes are also among the most sensitive that have evolved in any animal, with an elaborate structure that is convergent with, but structurally superior to the eyes you are reading this book with.

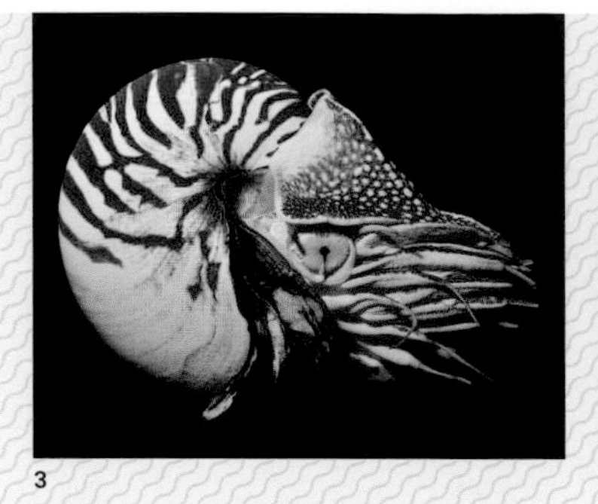

3

4

3 *Nautilus* spp. are shelled cephalopods. They have been around, superficially unchanged, for at least 500 million years (*Nautilus belauensis*).

4 Octopuses are typically benthic cephalopods. They are known to be highly intelligent animals (unidentified species).

5 Squid and octopuses have very well-developed eyes that are structurally superior to those of the craniates (*Ilex* sp. squid).

5

enormous range of niches, from slow, marine grazers rasping algae from stones, through tiny, worm-like predators and deposit feeders, to sessile filter-feeders and large, agile, open-water predators. Take the three groups thought to represent the most ancestral body plan among the living molluscs: the chitons and the worm-shaped solenogasters and caudofoveates. The chitons (polyplacophorans) are flattened marine animals with eight shell plates surrounded by a so-called girdle bearing small scales and spines [30, 31]. They can be found from the shoreline downwards and spend their days rasping away on microscopically small algae growing on rocks and stones. The solenogasters and the caudofoveates (aplacophorans) are typically small animals covered in a cuticle that is pierced by tiny scales or spines [17]. The colourful solenogasters are carnivorous, engulfing mostly cnidarians and polychaetes [32, 34], while the generally well-camouflaged caudofoveates burrow in soft sediment and take their fill of nutrient-rich mud containing small algae and other microorganisms [33].

When we think of molluscs the word 'slow' springs to mind. It is true that most of them glide around on a mucus-lubricated foot and that these are not exactly swift or agile, but there are lots of species that have evolved other, more rapid forms of locomotion. The muscular foot of the sea angels and sea butterflies, diaphanous pelagic animals, has evolved into wing-like fins [1, 35–37], while the violet snail clings to a raft of bubbles produced by its foot. The cephalopods, particularly squid, have become masters of the open water, using jets of water from their siphon and large fins to propel them at speed through the ocean. Even some of the bivalves (such as scallops), normally sedentary animals at best, can use jet propulsion to evade predators and disperse.

Many of the herbivorous molluscs live a simple and measured life. Consider the snails that dwell on rocky coastal shores: these hole up in their shells during low tide and emerge to graze on algae and other encrusting organisms when the tide comes in. There are lots of variations on the theme of herbivory though. The marsh winkle, for example, has taken to farming, long thought to be the preserve of humans and insects. This little mollusc scrapes saltmarsh cordgrass with its radula and smears the wounds with its own faeces to encourage the growth of a fungus on which it feeds.

Instead of simply grazing, some of the sea slugs (a term that is commonly used to describe a huge variety of marine molluscs, including nudibranchs, sacoglossans and sea hares) [11, 13, 16, 41–47] have evolved to drain algal cells of their nutritious contents using needle-like radula teeth. Some of these algal-feeding sea slugs are remarkable in how they make use of the algal cell's contents once they have ingested them. As a baby the emerald green sea slug (*Elysia chlorotica*) feeds on algal cells, digests their contents, but somehow spares the chloroplasts, the organelles where photosynthesis takes place. The chloroplasts are taken up by cells lining the digestive tract and here they go on photosynthesizing, supplying the mollusc with nutrients so that it can survive for long periods before feeding again.

Some snails have also taken to a predatory way of life with impressive flair. Many of these predatory snails have an extendable, tubular proboscis, which is used to great effect to make short work of their prey. Others, commonly known as drills, use their radula and acidic secretions to bore through the shells of other marine molluscs. Once the prey's shell has been breached, the drill slips its proboscis inside and sets about rending the flesh of the hapless victim. Some simply use their enlarged foot to smother their prey or burrow through the sand in search of small animals they subdue with acids or toxins in their saliva. There are even swimming snails that patrol the open water in pursuit of other small, swimming creatures. Perhaps the most remarkable predatory snails are the cone snails and their relatives: the radula teeth of this species have evolved into long, barbed harpoons laden with some of the most potent neurotoxins known. A cone snail lies in wait in the sand, a harpoon at the tip of its very mobile proboscis ready to be thrust into a polychaete, another mollusc or even a fish. Once stabbed the prey is very rapidly immobilized by the venom before being engulfed by the cone snail's capacious proboscis.

Many of the sea slugs have also become adept predators. In a similar way to how *Elysia chlorotica* exploits algal cells, some other sea slug species sequester components of their prey for their own use [38–40]. The prey in question are various cnidarians and the components are the cnidocysts (see Cnidaria, pp. 48–67). The sea slug consumes the prey's tissue, but preserves the cnidocysts and prevents them from discharging. The appropriated cnidocysts

6

are stored in special sacs in the sea slug's skin where they are primed to repel its own enemies, but exactly how it controls these weapons is unknown. Some open-water sea slugs [51] that move around on the underside of the water surface predate and use the cnidocysts of colonial cnidarians, such as the Portuguese man-of-war. Many sea slugs advertise their formidable acquired defences to potential predators using bright, bold, and often psychedelic colours and patterns (known as aposematism) [11, 13, 16, 41–47].

Many snails have evolved to scour the seabed for edible matter, including tiny organic particles amongst the sediment and the remains of other animals. There are also snails that filter feed either using modified gills or secreted mucus nets. Amongst the most remarkable of these are some of the open-water sea butterflies [36, 37, 48], which secrete a mucus net that can be as much as 2 m (6 ft) across. They dangle from the bottom of this filter by their proboscis, the cilia in which create a current to draw food particles into the mouth.

By far the most specialized filter-feeding molluscs are the bivalves [6, 9, 49, 50, 52], well known animals because we eat so many of them. The bivalves come in a huge range of sizes, from tiny clams, barely 2 mm (0.08 in.) across and inhabiting the burrows of worms, to centenarian giant clams more than 1 m (3 ft) wide and 300 kg (650 lbs) in weight. The whole mollusc body plan has been reconfigured for this sessile, filter-feeding lifestyle. The soft body, with only the vestige of a head and no radula, is protected by a unique, hinged two-part shell, which is kept closed for extended periods by strong muscles (think of scallops and their disk-shaped shell muscles). The gills of bivalves are hugely modified in order to filter food (normally phytoplankton) from the water. Cilia on the surface of these elaborate structures generate a current that ventilates the gills, delivers suspended material and separates it into edible and inedible particles. Typically, bivalves also have a long siphon that pokes out into the water from the sediment [8].

These adaptations allowed the ancestral bivalves to filter water from the safety of the sediment, and because they went so far down the road of specialization this is still the lifestyle of most living bivalves. Some, however, have used these adaptations to live a life on the surface of the seabed, often attached tenaciously to rocks with extremely strong protein threads (byssus) or organic cement that can resist the full force of storms. The scallops also dwell on the surface, but they have evolved the ability to swim using jet propulsion by forcibly clapping their shells together to force water out of their mantle cavity. This is by no means an energy efficient way of getting around so it is reserved for evading predators.

Some of the bivalves have even taken to boring into various substrates, such as coral or wood. They start the tunnelling process when they settle as larvae and the burrow they excavate will be the only home they'll ever know;

6 *Argopecten irradians*, a bivalve mollusc. Note the numerous, well-developed eyes (blue) along the margin of the mantle.

8 In some burrowing bivalves the body is no longer contained within the shell valves. Siphons contained within an outgrowth of their body wall allow these animals to draw in water to filter out edible particles while most of the body is safely buried in the sediment. This example, *Panopea abbreviata*, can reach a considerable size and live for more than 150 years.

9 Some bivalves have adapted to living on the bodies of other animals. Here, *Pseudopythina rugifera is* attached to the abdomen of a Pacific mudshrimp, *Upogebia pugettensis*.

BOX 2 ***Colour changing***

Cephalopods make the colour changing abilities of chameleons look very ordinary. This is made possible by chromatophores, pigment skin cells that can be stretched and squeezed by muscles around their perimeter. When the muscles are relaxed the cell is contracted and the pigment spot is inconspicuous; however, when the muscles contract, the cell is drawn out and the area of pigment expands. The pigment can be yellow, orange, red, blue or black, all of which may occur in groups or layers. Underlying reflective cells known as iridocytes enhance the visual effect by lending an iridescent or metallic edge to the overall colour. With this elegant system these animals can communicate with a riotous display of shimmering colours or else blend into the background, either to hide from predators or remain undetected by prey.

7 The spots visible on the body of this squid are its chromatophores (*Sepioteuthis australis*).

8

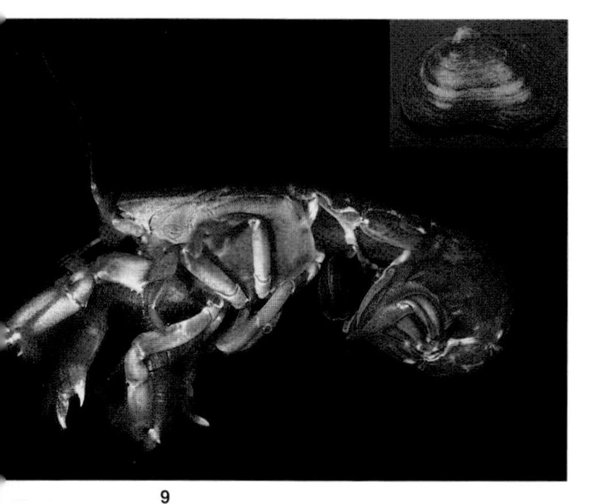

9

their only communication with the outside world a siphon they can extend to the tunnel entrance. The best known borers are the shipworms, worm-like bivalves capable of excavating tunnels more than 2 m (6 ft) long in submerged wood with shells that have evolved into small rasps. Shipworms are alone amongst the bivalve borers because they consume the substrate in which they tunnel. They enlist the help of symbiotic bacteria to digest and enrich this tough and nutritionally impoverished food.

The most accomplished predators amongst the molluscs are the cephalopods. Their agility, acute vision, big brains and very mobile tentacles are used to locate and catch a huge range of marine animals. Suckers, often toothed, stud the tentacles and are important in keeping a firm grip on struggling prey. At the centre of the tentacles is a pair of hefty jaws reminiscent of a parrot's beak, and it is these that tear the prey asunder [26]; the radula behind drags the resultant morsels into the mouth. Some of the cephalopods, such as the blue-ringed octopus, have a venomous bite, the potency of which is due to tetrodotoxin (see Chaetognatha, pp. 122–25). Squid and nautiluses are typically predators of open water, but the octopuses have specialized as benthic hunters, using their brains and brawn to winkle out prey from nooks and crannies in the seabed.

Commensalism is relatively common in the molluscs, especially the bivalves, many species of which have adapted to living in the burrows of other animals, where they presumably feed on leftovers – although the exact nature of these relationships is poorly understood. Parasitic molluscs are not that common, although they have evolved some interesting takes on this way of life. Apart from a bivalve that lives in the gut of sea cucumbers, all the parasitic molluscs are snails. Some of these use an elongated proboscis to suck blood and other fluids from the bodies of other molluscs. When young *Stilifer* snails burrow into the body wall of starfish and become embedded in a thickening of the host's tissues, which is stimulated by their presence. From this gall they extend their proboscis into the tormented echinoderm's body cavity to feed on its fluids and tissues. There are also snails that have so taken to a parasitic way of life that you would be hard pressed to identify them as a mollusc at all. These live in the body cavities of sea cucumbers, and can reach lengths of 1.3 m (4 ft). Absorbing all the nutrients they need from their host, these snails have become degenerates: often their body is nothing more than a hollow sac filled with reproductive organs, identifiable as a mollusc only because a pregnant individual gives birth to hundreds of tiny snails (see Dicyemida and Orthonectida, pp. 204–07, for other examples of how parasitism can lead to the loss of morphological complexity).

As you have probably already guessed, reproduction in the molluscs is also very diverse. There is a general theme – separate sexes, the shedding of gametes via the excretory system, external fertilization, an egg mass deposited by the female and a planktonic, trochophore

BOX 3 *Freshwater pearly mussels*

The freshwater pearly mussels have evolved some astonishing ways of using fish to nurture and disperse their offspring. Their eggs hatch into tiny, parasitic larvae called glochidia, which remain inside a pouch derived from their mother's gill until a host can be attracted. In some species the edge of the female's mantle develops into a lure – an astounding mimic of a small fish.

A real fish, mistaking the lure for a tasty snack, lunges to attack, and in doing so ruptures the gill pouch and gets doused in the tiny larvae, which clamp on to the fish's own gills. A successfully attached glochidium stimulates the tissue of the fish to produce a small cyst in which it will be protected and nourished. After 10–30 days it will break out, sink to the bottom and metamorphose into a juvenile mussel.

Other species release their glochidia in tethered gelatinous masses that look like tasty worms or insect larvae, or clamp onto the head of a bold, hungry fish and pump glochidia into its mouth.

10 The edge of a freshwater pearly mussel's mantle mimics a small fish (*Lampsilis altilis*).

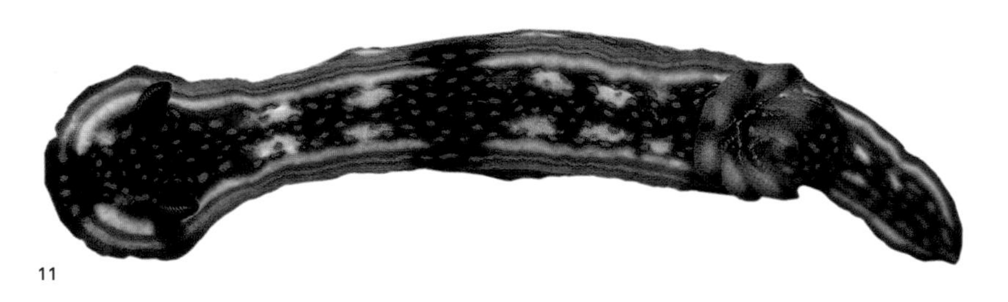

11

larva (see Annelida, pp. 208–31) – but this has been tweaked to produce myriad strategies. For example, hermaphrodites, internal fertilization and a combination of both are well known in these animals. Reproduction in the humble, land-dwelling slugs and snails can be extremely complicated. The familiar members of this group are hermaphrodites and often a long, tender courtship terminates in the exchange of sperm. In some slugs, such as the leopard slug (*Limax maximus*), the act of copulation is extravagant to say the least. The slug couple take to the branch of a tree and relinquish their grip, but instead of hitting the ground a mucus thread fixed to the branch arrests their fall and there they dangle, wrapped in an embrace, ready to share sperm. Steadily a penis extends from each slug – sometimes reaching a length of 80 cm (30 in.) – until their tips meet, intertwine, flare and exchange packets of sperm that will be carried back to the slugs on the retracting penes.

Breeding is also very elaborate in the cephalopods, which are typically dioecious and display internal fertilization. Often there is a lengthy courtship display where the males and females communicate using their unparalleled colour-changing abilities [BOX 2]. Following the pleasantries of courtship the male must get one of his packets of sperm into the body of the female, and he does this with a modified tentacle that functions a bit like a very mobile penis. In some cases this tentacle detaches from the male during courtship allowing him to mate only once in his life. Baby cephalopods hatch from the egg as a minute version of their parents.

In contrast, most bivalves, as well as chitons, scaphopods, caudofoveates and some marine snails, are free-spawning – in other words the eggs and sperm are released directly into the open water. The fertilized eggs develop into larvae that propel themselves through the water with cilia.

In all molluscs where the eggs are fertilized internally, the female deposits an egg mass [53], which comes in a variety of forms, including grape-like bunches, calcareous cases or long strings. Many species of snail retain their eggs in a brood sac and eventually give birth to lots of baby snails. A planktonic larval stage, or sometimes two, is common, especially in the marine species, but these immature stages are often completed in the egg before hatching as juveniles. Larvae are essential in the dispersal of marine species that are more-or-less sessile as adults, but in river bivalves a larva could easily get carried into an unsuitable habitat; therefore alternative, shorter-range forms of dispersal have evolved, some of which are truly brilliant [BOX 3]. The larval mollusc may go through a radical metamorphosis to take on the adult form, and in the snails this can involve an extreme twisting of the body through 180 degrees.

11 The use of colour and striking patterns to warn prey of a secondary defence system (aposematism) is at its most extreme in the nudibranchs (*Hypselodoris agassizii*).

13 The tentacle-like structures at the anterior end of this nudibranch are rhinopores, which are used to detect (smell/taste) chemicals in the water. The feathery structures on the dorsal surface are gills (a branchial plume) encircling the animal's anus (*Nembrotha kubaryana*).

12 One of the very few photos of a living monoplacophoran (*Laevipilina* sp.).

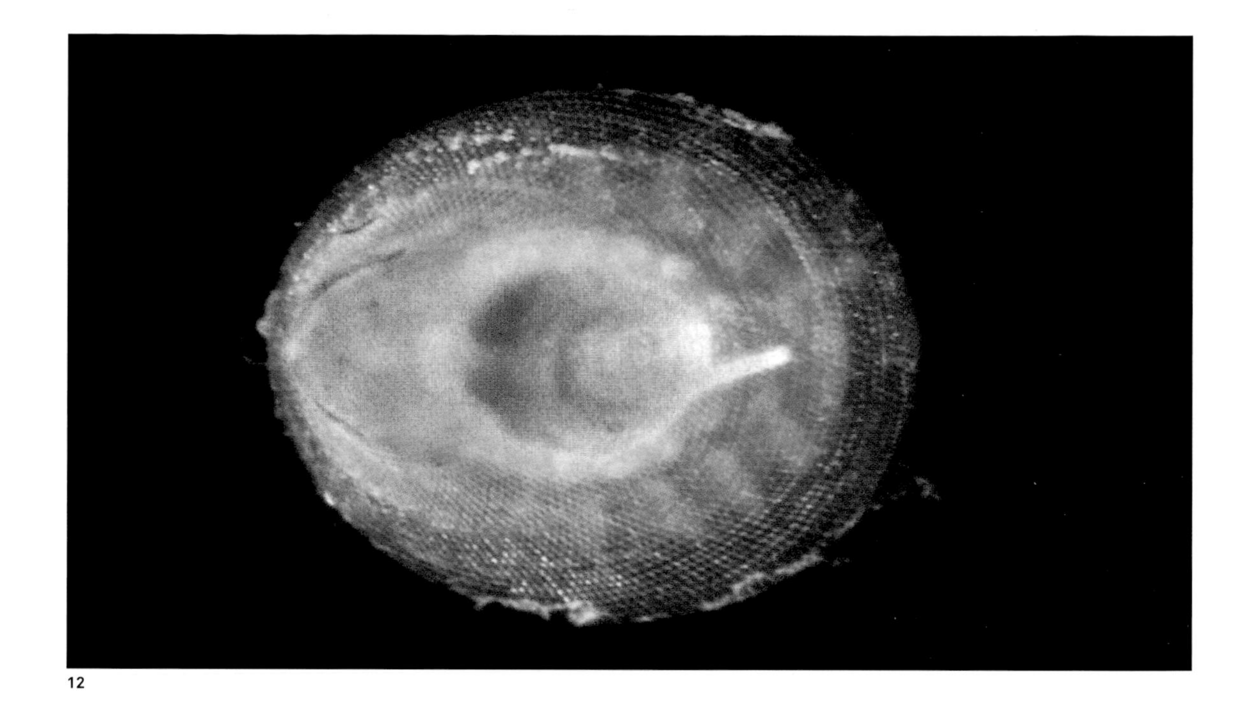

12

13

Ecologically, the molluscs are enormously important. In the oceans they occupy a huge range of niches and are commonly to be found in enormous numbers, from hordes of burrowing bivalves in mud-flats to prodigious shoals of squid in open water. In grazing, filtering, predating and getting eaten they are integral in marine food webs. Many larger animals rely on molluscs for food and their burrowing contributes enormously to the mixing of marine sediments and the consumption of organic matter that would otherwise litter the seabed. Aside from their ecological importance, human activities have conspired to turn some molluscs into pests. The planktonic larvae of many molluscs make it all too easy for them to be accidentally introduced into ecosystems where they do not belong, often in the ballast tanks of ships.

ORIGINS AND AFFINITIES

As so many of the molluscs have a hard shell of some kind, their fossil record is fantastic, one of the finest there is. Lots of the forms we are familiar with today were also flourishing some 500 million years ago in the Cambrian seas – and even the very specialized cephalopods with their complex anatomy were well established in these ancient times. Many, such as the well-known ammonites and belemnites, have become extinct, but some of these ancient animals have clung on – more or less unchanged – into the modern era, their ancestors somehow surviving the cataclysms that have laid low countless other species. Perhaps the most intriguing of these are the monoplacophorans, small, limpet-like creatures that were dredged from a depth of 3590 m (11,750 ft) off the coast of Costa Rica in 1952 [12]. Previously, similar animals were only known from fossils 375 million to 530 million years old: the remarkable discovery of living individuals is testament to the biological treasures no doubt still to be discovered in the oceans of the world.

For a long time the diversity and specialization of the molluscs made it very difficult to trace the threads of relatedness. The monoplacophorans, which display conspicuous repetition of their gills and some muscles indicative of segmentation, were touted by some as evidence of an evolutionary link between the molluscs and the annelids (as the latter have very obvious segmentation). Although this particular theory is now largely discredited, other lines of evidence, such as developmental and molecular biology, do indeed suggest that the molluscs and annelids are closely related, even though the superficial similarities between adult animals from these lineages are few and far between.

14

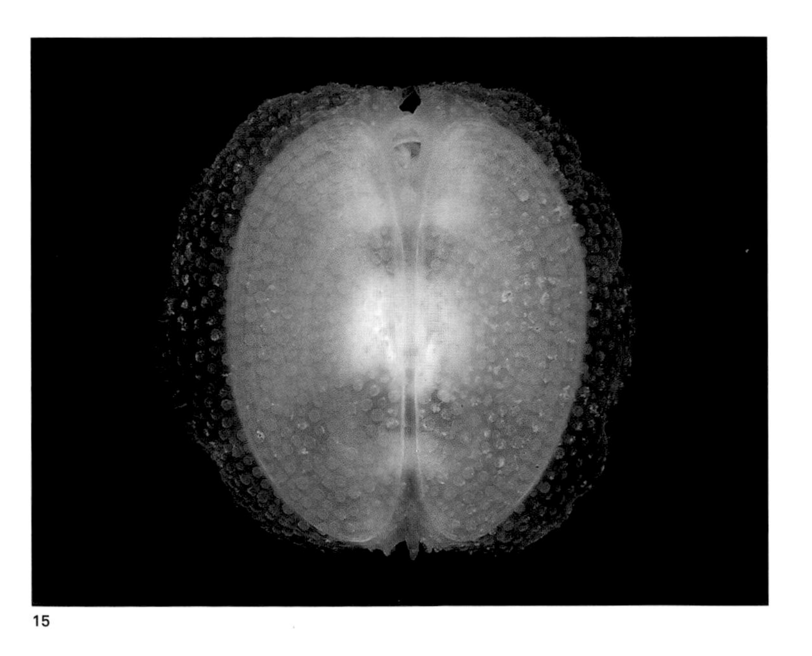

15

14 Squid have adapted to a similar ecological niche as pelagic fish and have therefore convergently evolved a very hydrodynamic shape, superb swimming abilities and acute senses (*Sepia* sp.).

15 The body of a bivalve mollusc is enclosed in a pair of calcareous shell valves. This particular bivalve, *Ephippodonta cf. gigas*, is unusual in that the mantle covers the valves.

› 16 Nudibranchs, together with a huge variety of other marine molluscs, are commonly known as sea slugs (*Coryphella polaris*).

17 Aplacophorans are peculiar, often worm-like molluscs. The mantle covers the entire body and instead of a shell they have calcareous spicules, which are microscopic in this species (unidentified species from Grand Cayman).

18 Scaphopods are a distinct type of burrowing mollusc. The shell they secrete looks a bit like a tooth or tusk, hence their common name, tusk shells (unidentified species).

19 Mollusc shells are elaborate, secreted structures that are one key to the group's relative success, as exemplified here by a marine snail (*Calliostoma annulatum*).

17

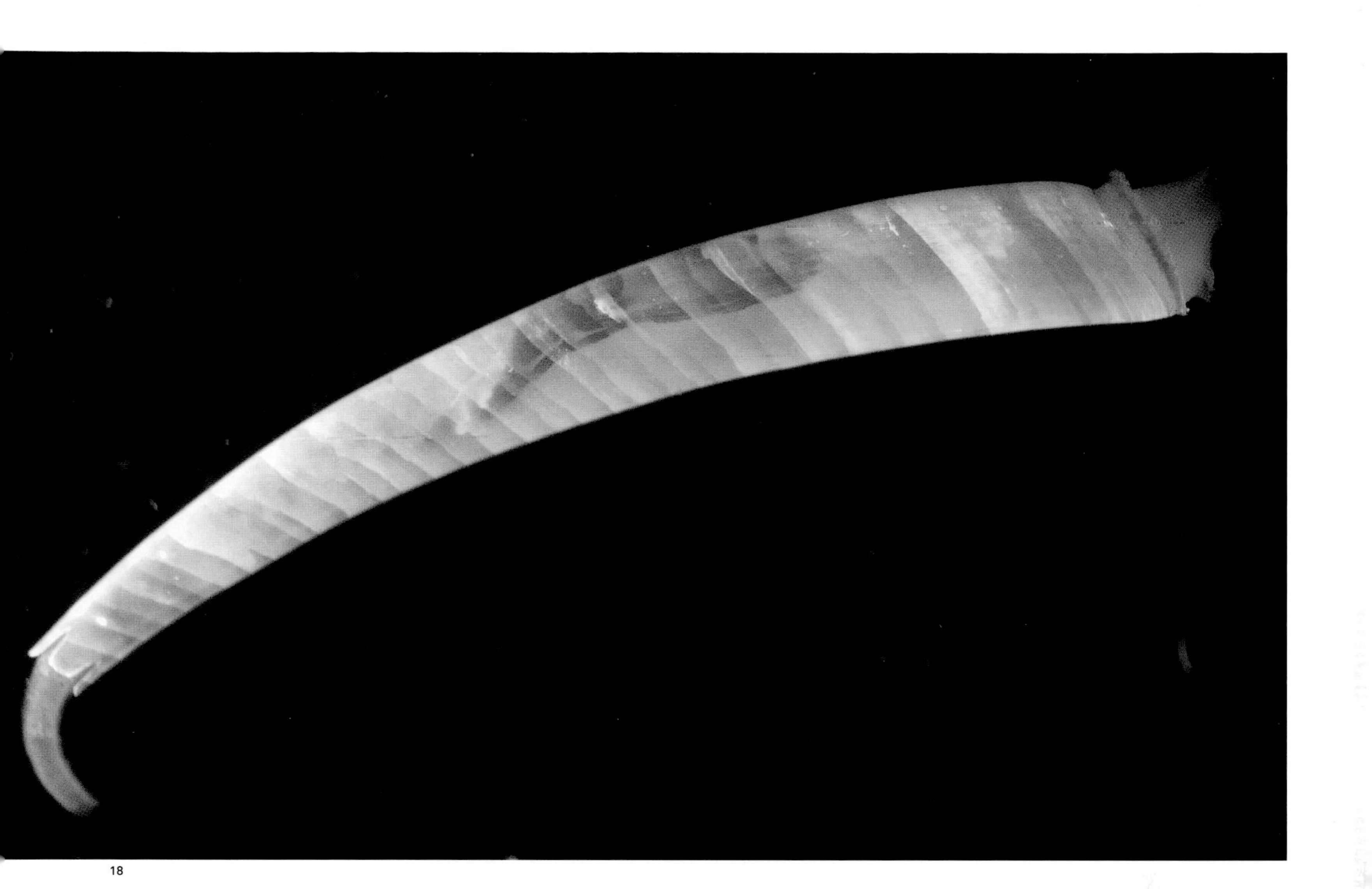

18

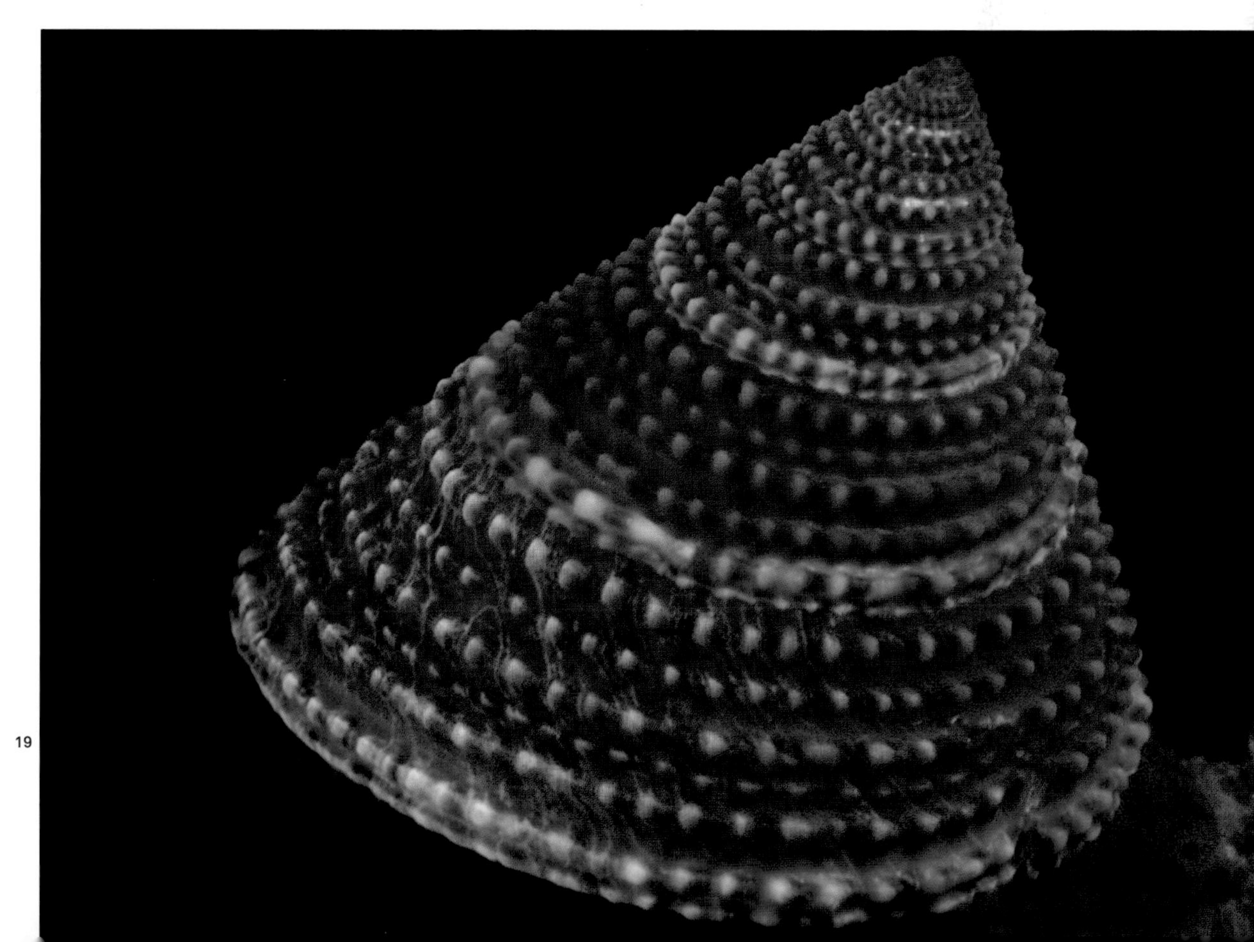

19

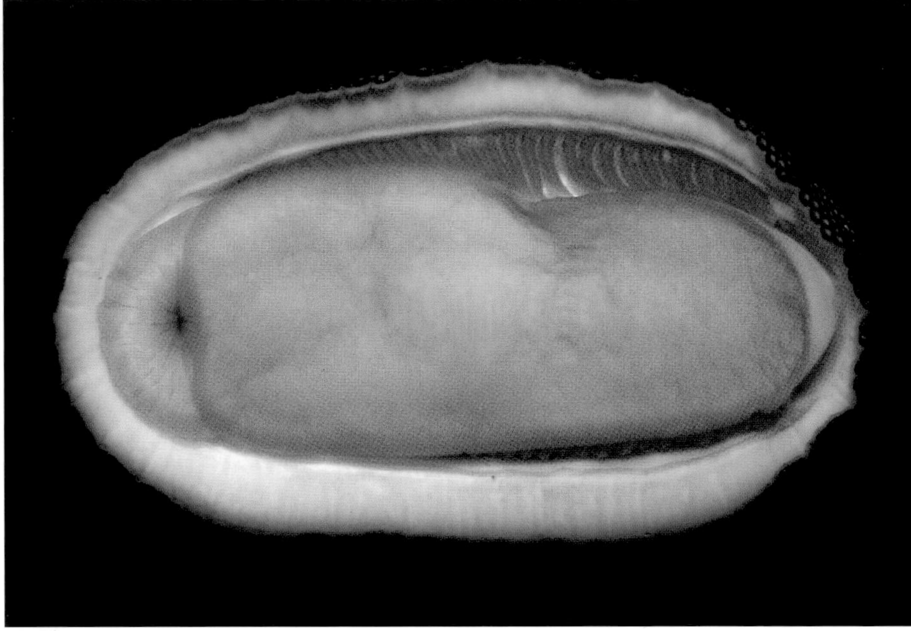

20

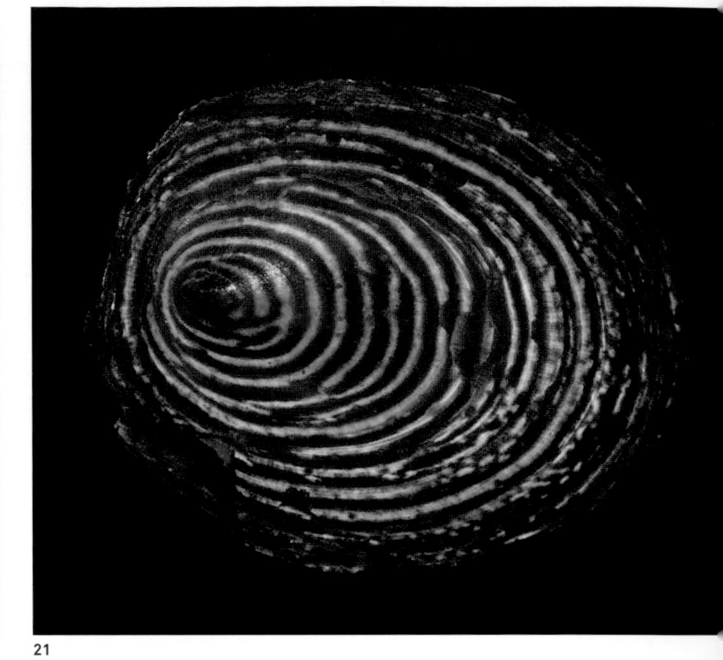

21

22

20 The muscular foot is a characteristic feature of the molluscs. In this image you can also see the gills (brownish structures at the edge of the body) and the mouth (pinkish aperture at left) (unidentified chiton from Chile).

21 Limpets are highly successful benthic animals with a very well-developed foot and shell that are able to withstand being battered by waves on the rocky shore (*Atalacmaea fragilis*).

22 The short, curving tube near the front of this marine snail's shell is the siphon, an elongated outgrowth of the mantle cavity (*Terebra felina*).

23 Many land-dwelling molluscs, including the archetypal snails, have a calcareous shell, which prevents desiccation and affords a degree of protection from predators (*Oxychilus* sp.).

24 The small shell of this terrestrial snail is almost completely covered by the mantle (unidentified helicarionid).

23

24

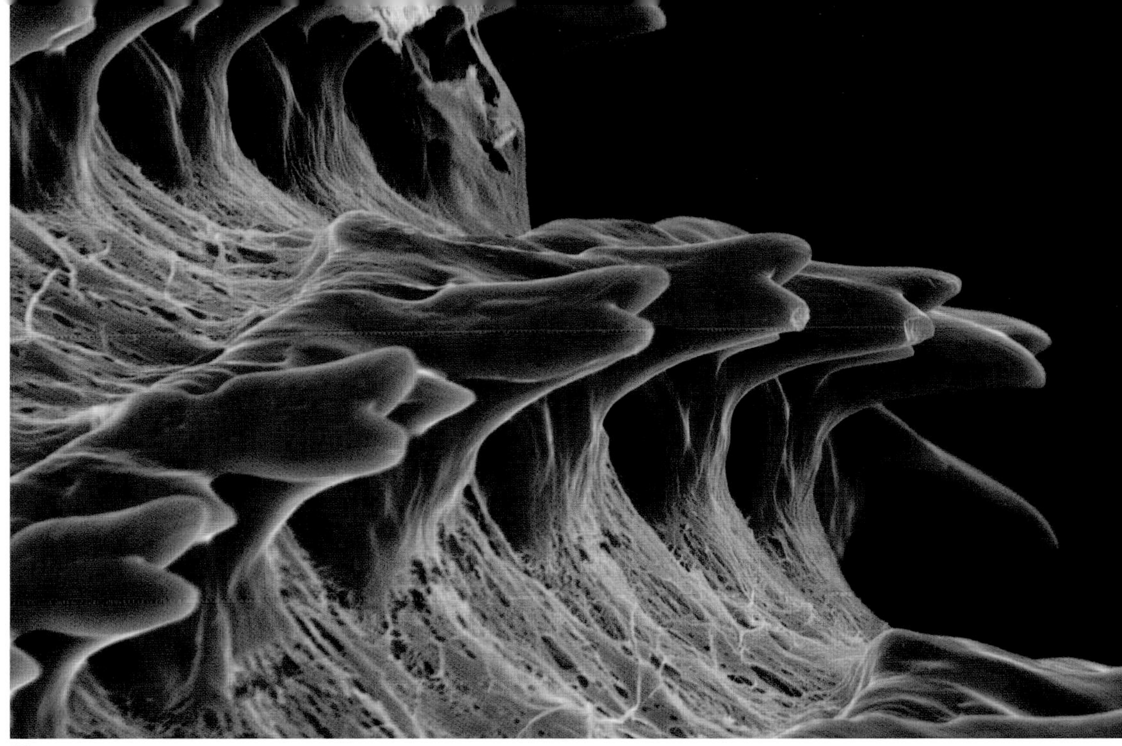

25

26

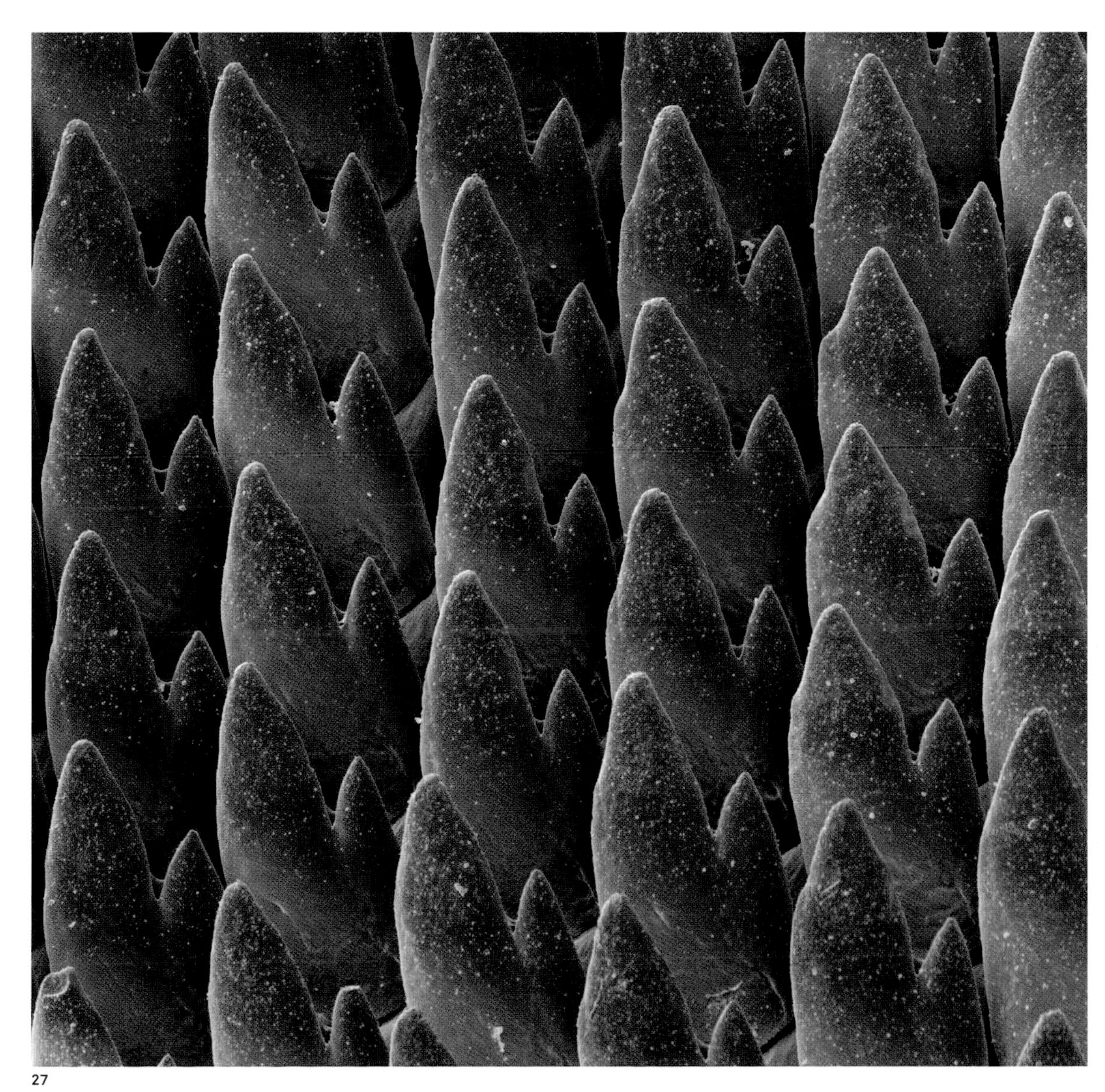

27

25 The radula of *Helix aspersa* showing the rasping teeth.

26 Cephalopods, such as this Humboldt squid (*Dosidicus gigas*) have very robust jaws reminiscent of a parrot's beak. Note the rasping radula between the jaws.

27 The radula is unique to the molluscs. Typically, it is a flexible ribbon of chitinous teeth. This is the radula of a common garden snail (*Helix aspersa*).

28

28 Slugs are very well-known molluscs that live on land, but they have been unable to fully exploit the opportunities offered by the terrestrial environment. Clearly visible on this European black slug (*Arion ater*) are the long, optical tentacles, the short, smell/taste tentacles and the pore (pneumostome) that opens into the mantle cavity, which houses the animal's lung.

29 Chitons (polyplacophorans) are superbly adapted to life on the rocky shore, where they cling tenaciously to rocks and shells. Their shell consists of eight overlapping plates.

30 In this chiton the eight separate plates of the shell are much reduced (*Cryptoplax* sp.).

31 A chiton (*Acanthochitona crinita*). Note the shell plates and spine-bearing girdle.

29

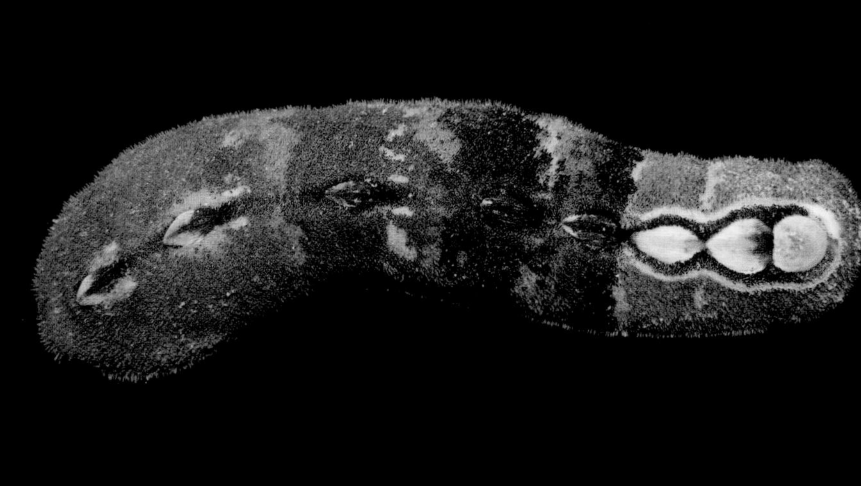

30

32 Most solenogasters are worm-like animals, but this species (*Plawenia* sp.) is rather egg-shaped. It feeds on polychaete worms that are seized and engulfed with the help of a large radula.

34 A more typical solenogaster. Note the pedal groove running the length of the ventral surface. This is thought to be the vestige of a muscular foot. At the far anterior end (right) is the mouth, below which is a sensory pit. The body is covered in very short, calcareous spicules (unidentified species).

31

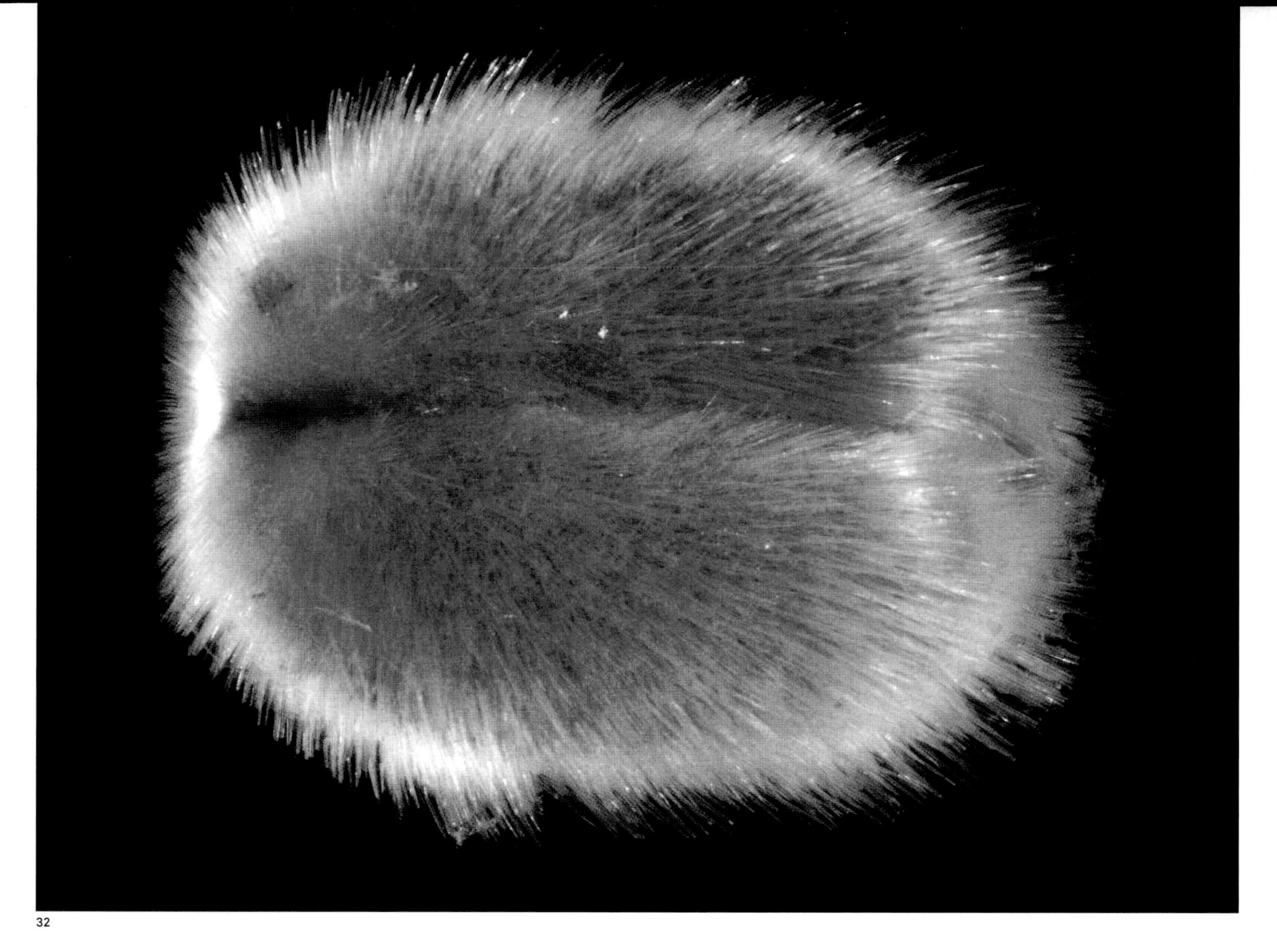

32

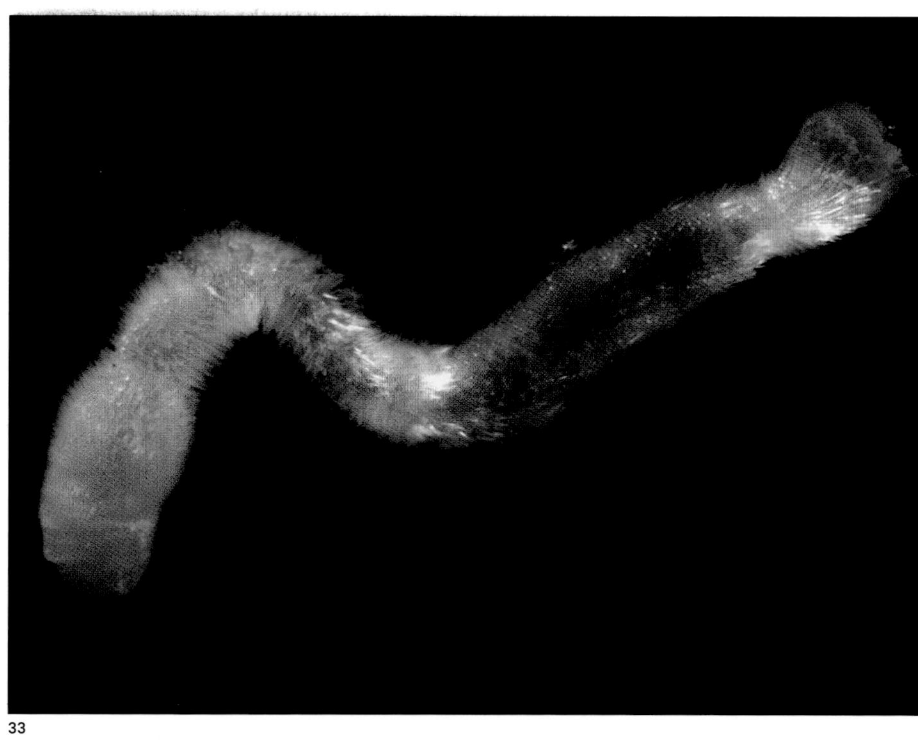

33

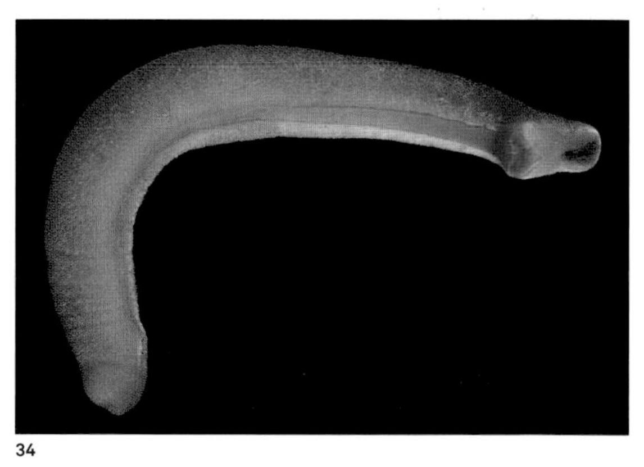

34

33 A caudofoveate. Note the long calcareous spicules covering the body of this species (*Chaetoderma* sp.).

35

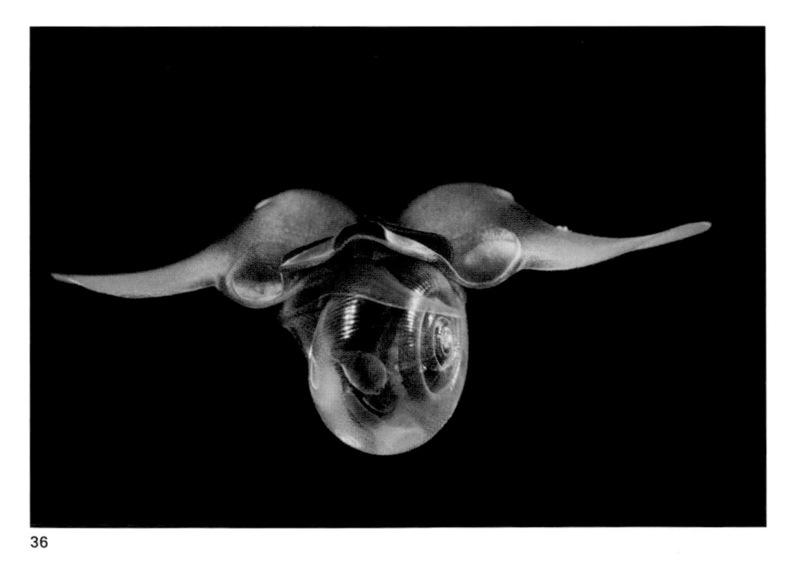

36

35 A sea angel, *Clione limacina*. In this image the grasping tentacles and chitinous hooks are retracted (compare to [1]).

36 This pelagic sea butterfly, *Limacina helicina*, collects suspended edible matter with a mucus net. It is preyed upon by *Clione limacina*.

›37 *Limacina helicina*.

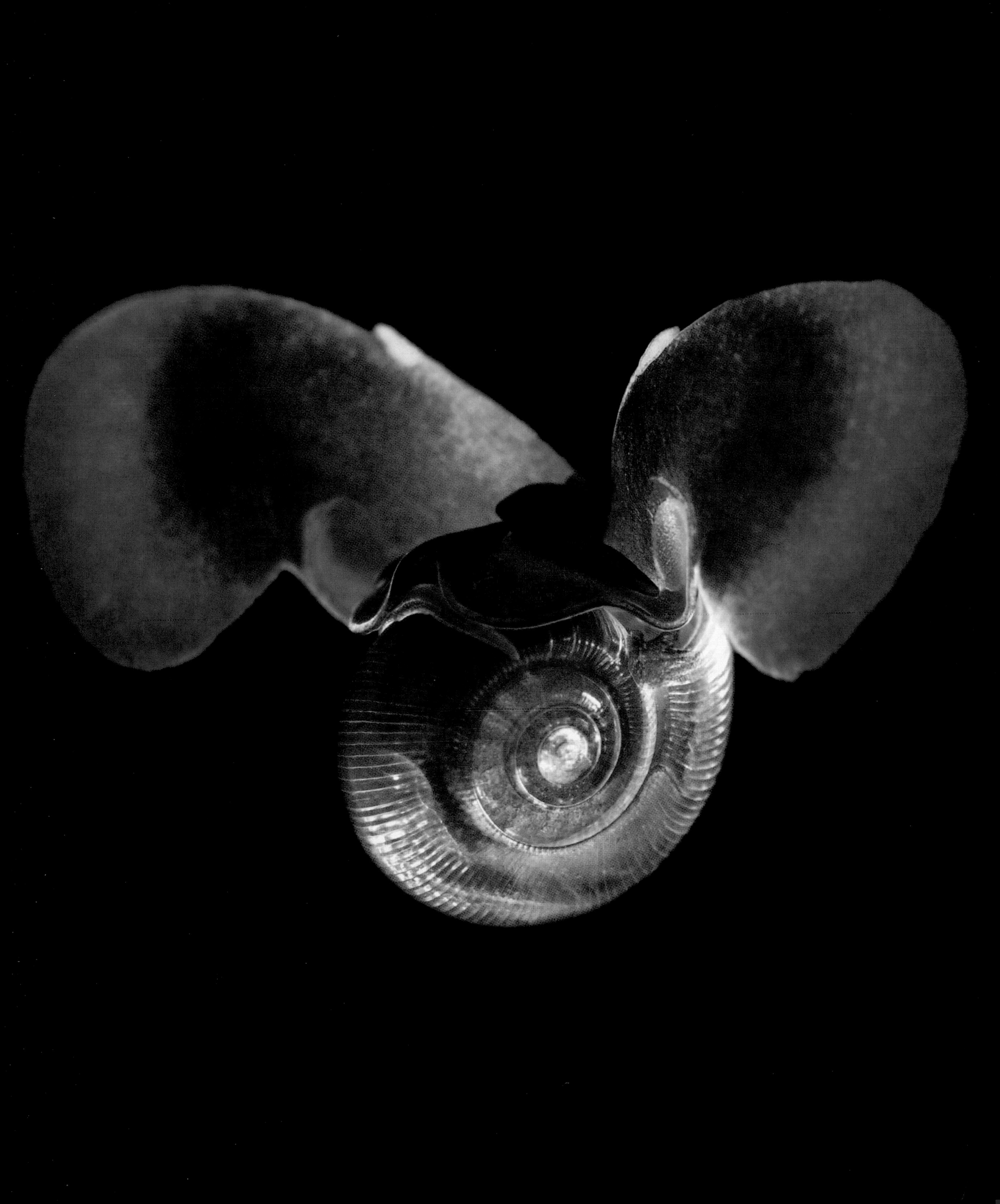

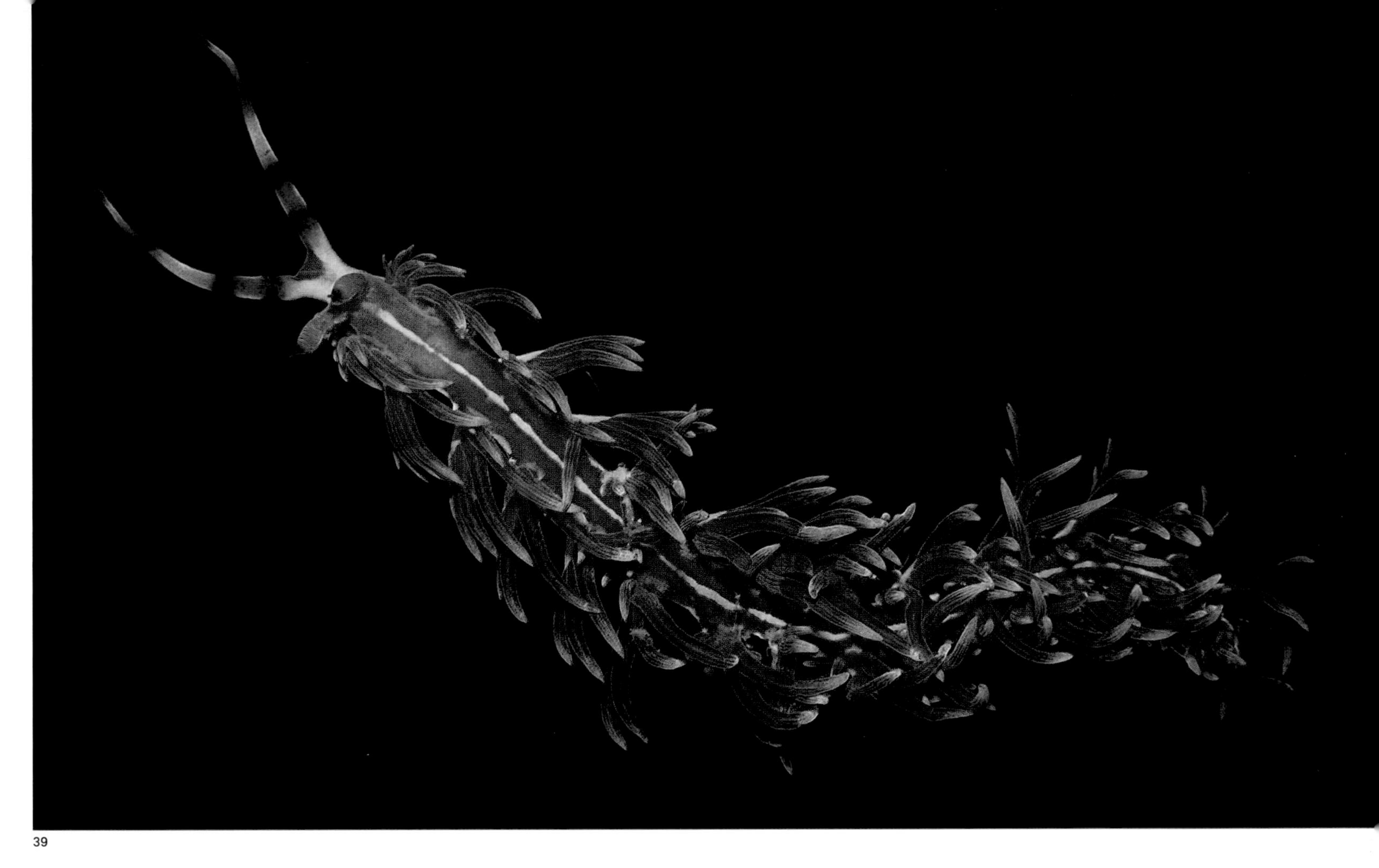

39

40

‹ 38 Many nudibranchs prey on cnidarians. Here, *Coryphella polaris* consumes a hydroid, *Tubularia indivisa*. The mollusc swallows the prey's stinging components (cnidocysts), but somehow saves them from digestion and uses them for its own protection.

39 In some nudibranchs the mantle is extended into finger-like projections (cerata). These increase the surface area of the body surface for gas exchange and in some species house extensions of the digestive system. It is in the cerata where the commandeered cnidocysts of the mollusc's cnidarian prey are stored (*Pteraeolidia* sp.).

40 A nudibranch with long cerata (*Spurilla cf. neapolitana*).

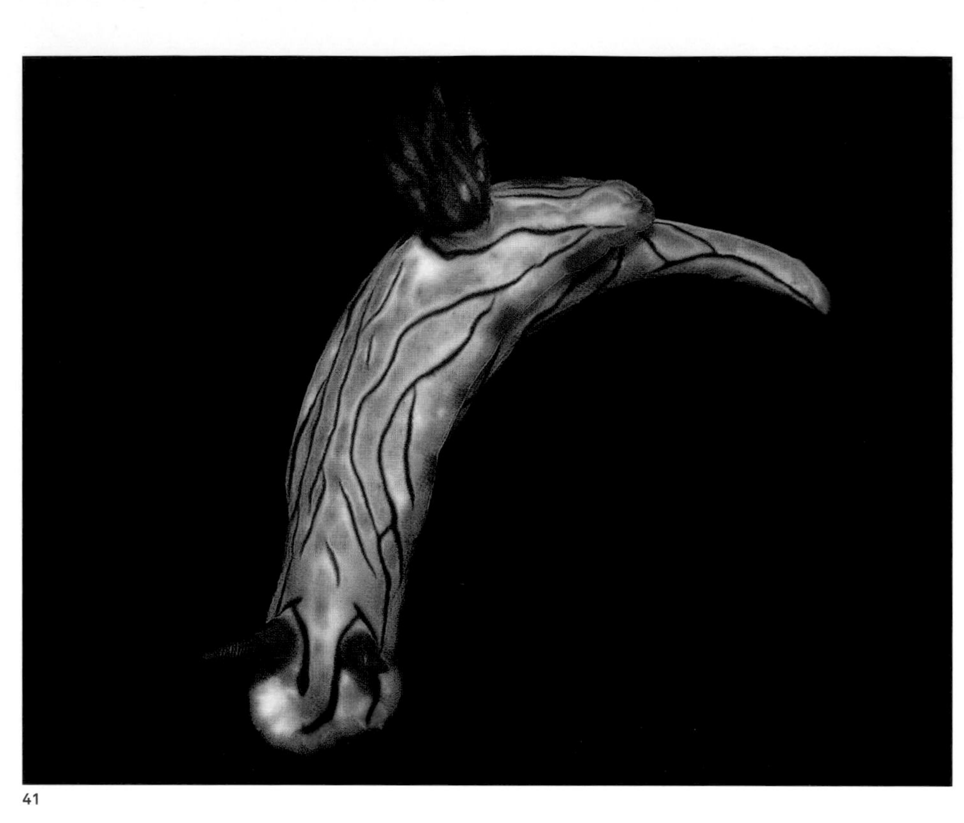

41

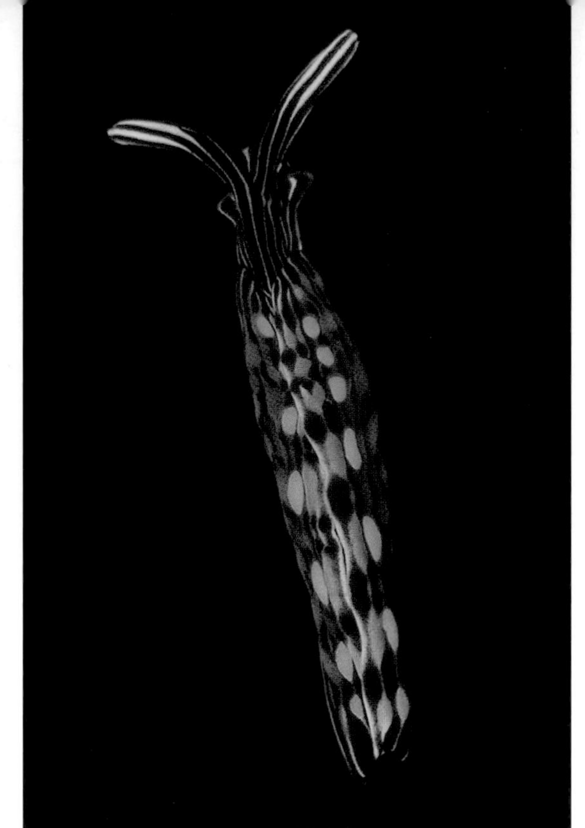

42

41 The term sea slug covers a number of superficially similar groups of animals, including nudibranchs and sacoglossans. This strikingly patterned example is a nudibranch (*Chromodoris* sp.).

42 A sacoglossan (*Elysia* sp.).

43

43 The colours and patterns of the sea slugs warn predators of their toxicity. This nudibranch is *Chromodoris annulata*.

›44 A nudibranch (*Glossodoris edmunsi*).

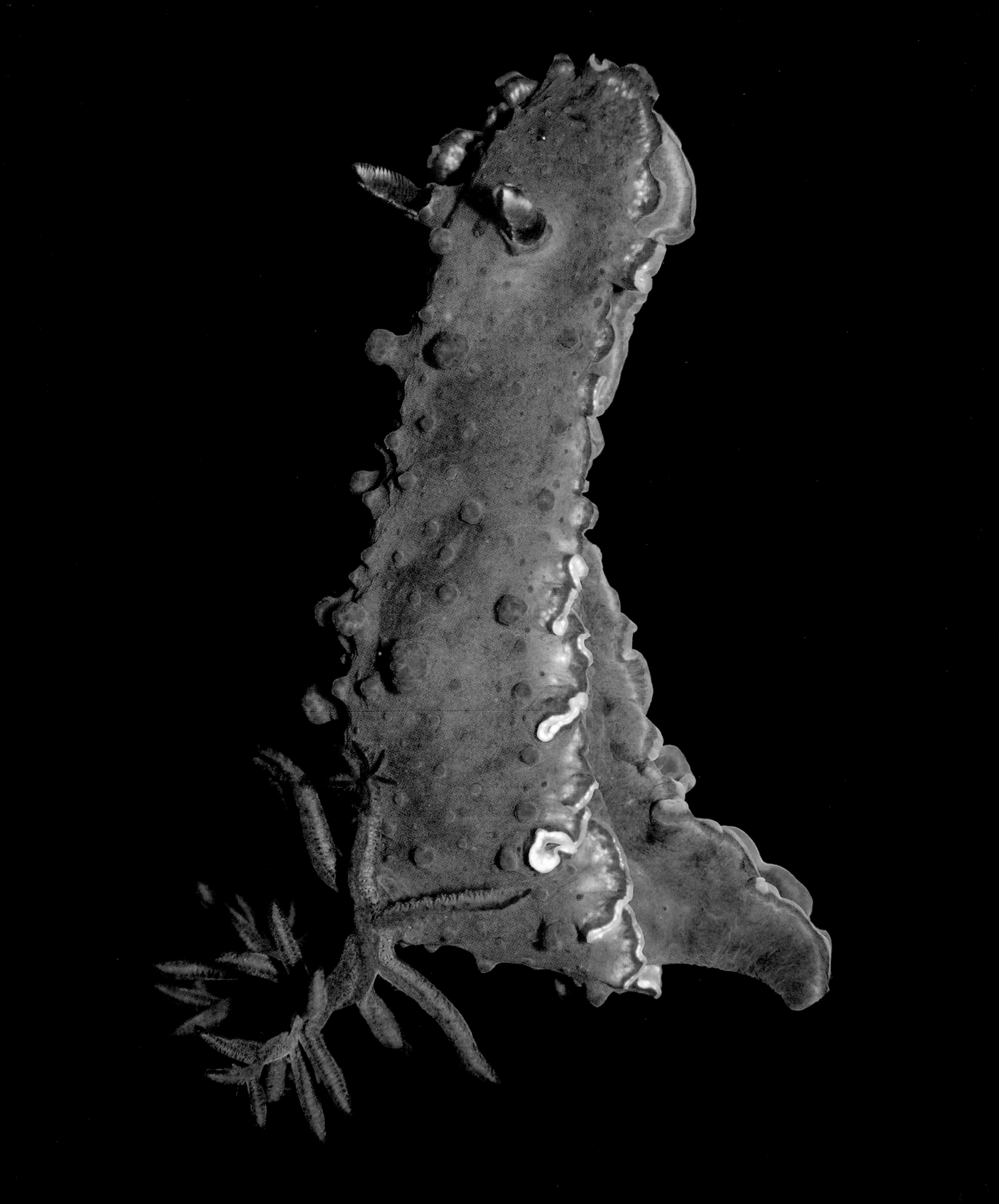

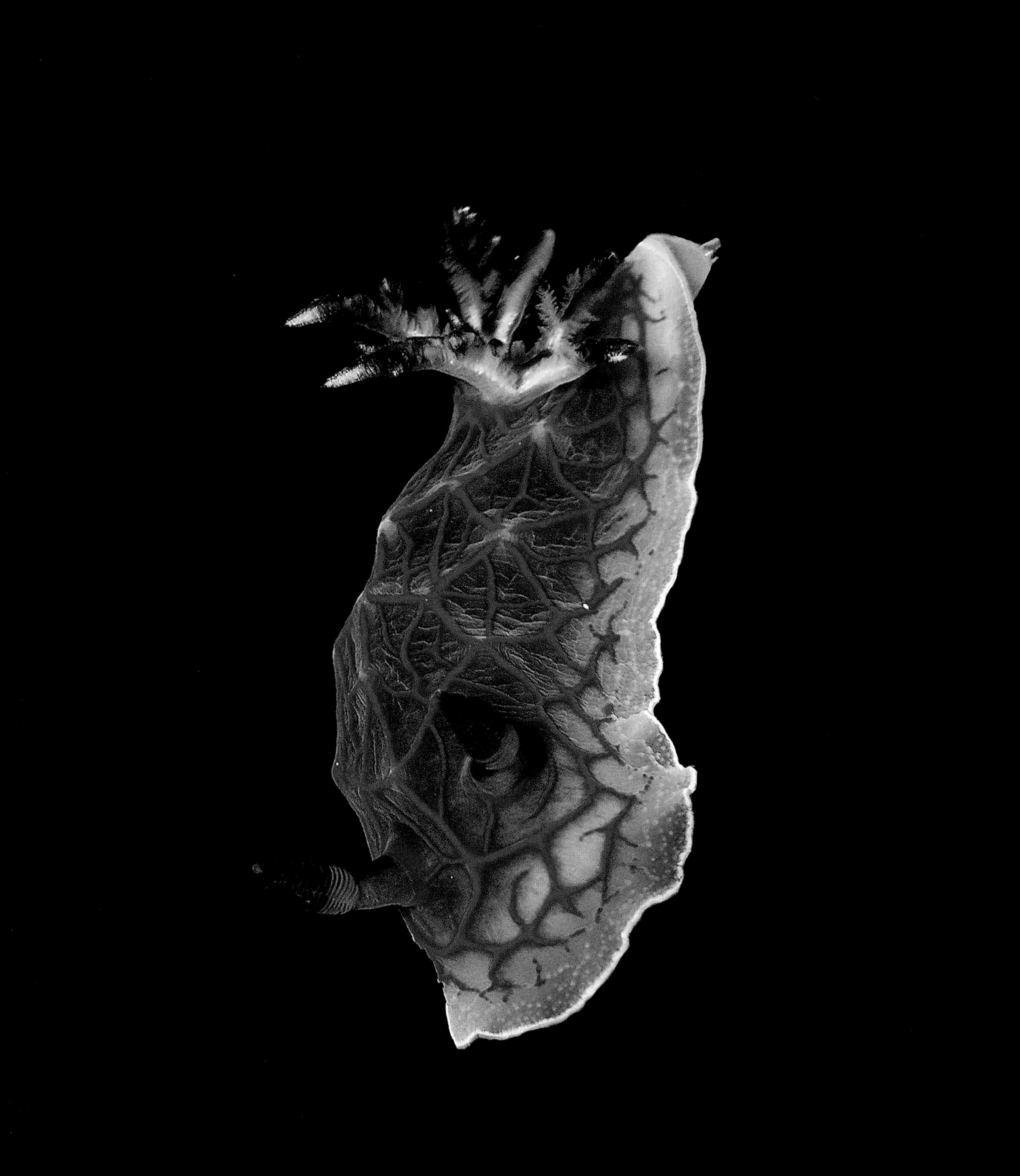

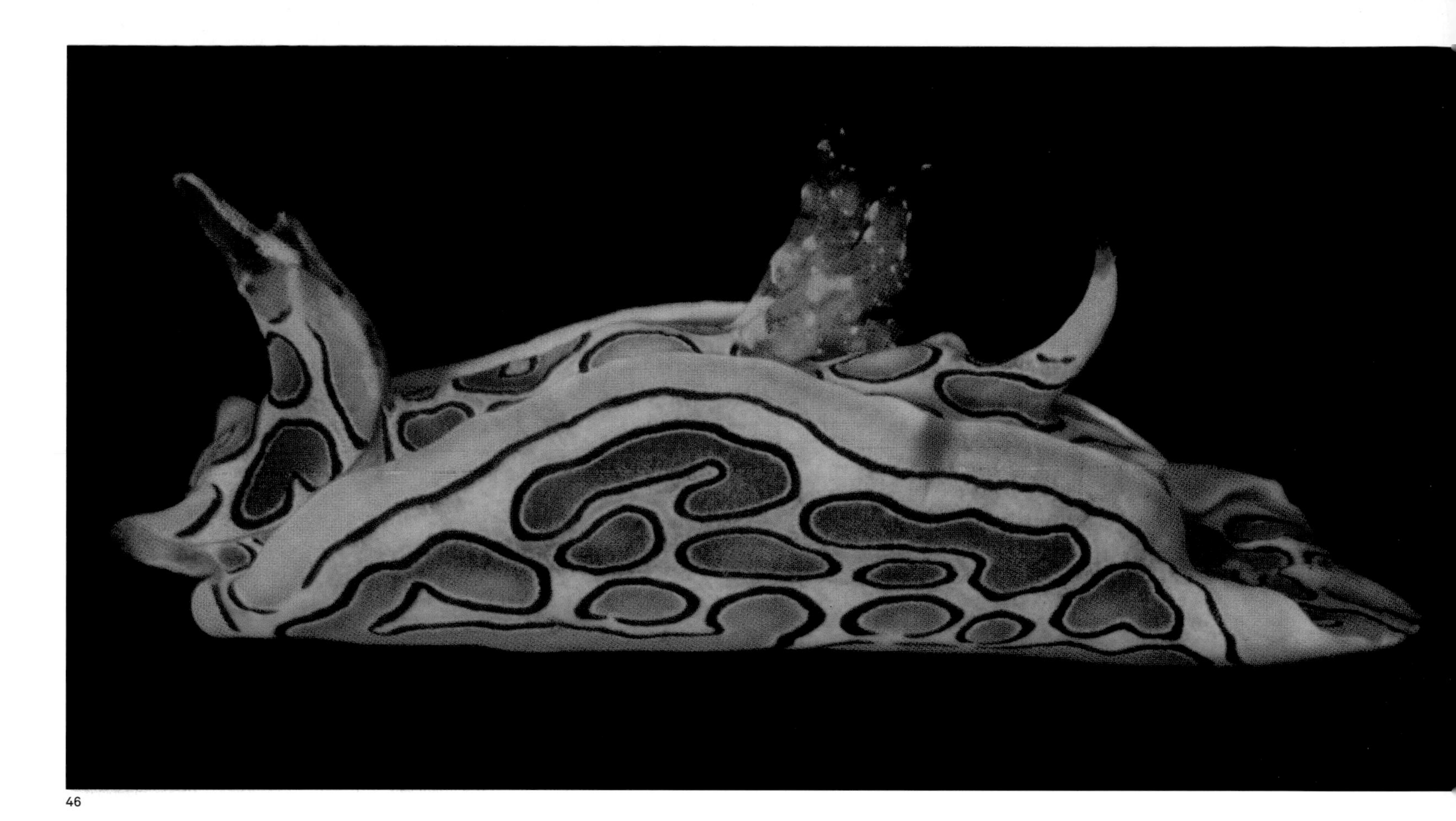

46

Boldly coloured and patterned nudibranchs:

‹45 *Halgerda* sp.
46 *Sagaminopteron psychedelicum.*
47 *Chromodoris coi.*

47

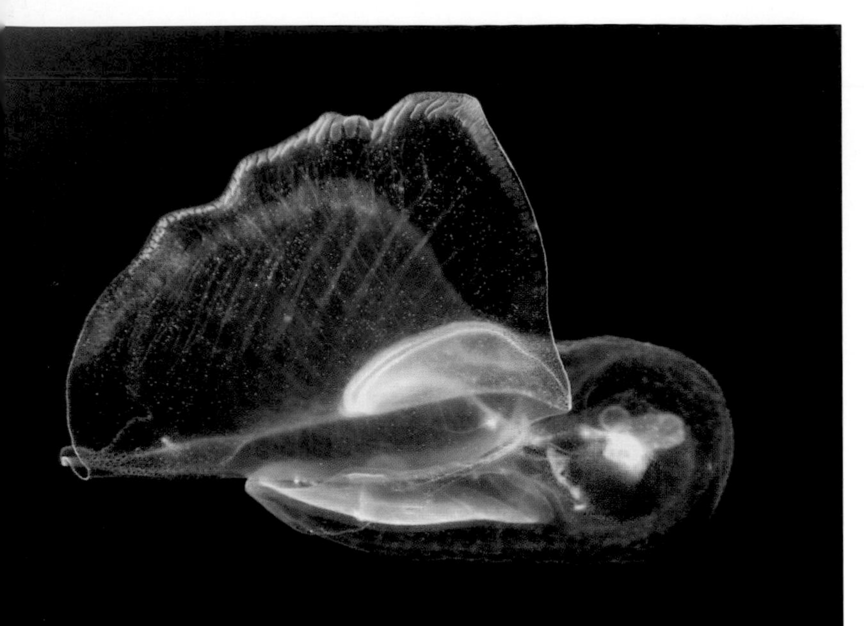

48

49

48 *Corolla spectabilis*, another type of sea butterfly. The fins are outgrowths of the muscular foot.
49 In this bivalve the mantle overgrows and largely obscures the shell valves (unidentified galeommatid).
50 Bivalve molluscs (*Scintilla* sp.) with a thin layer of mantle tissue covering the shell valves.
›51 *Glaucus atlanticus* moves around upside down, clinging to the underside of the water surface. It preys on pelagic cnidarians, such as *Physalia physalis* (Portuguese man-of-war).

50

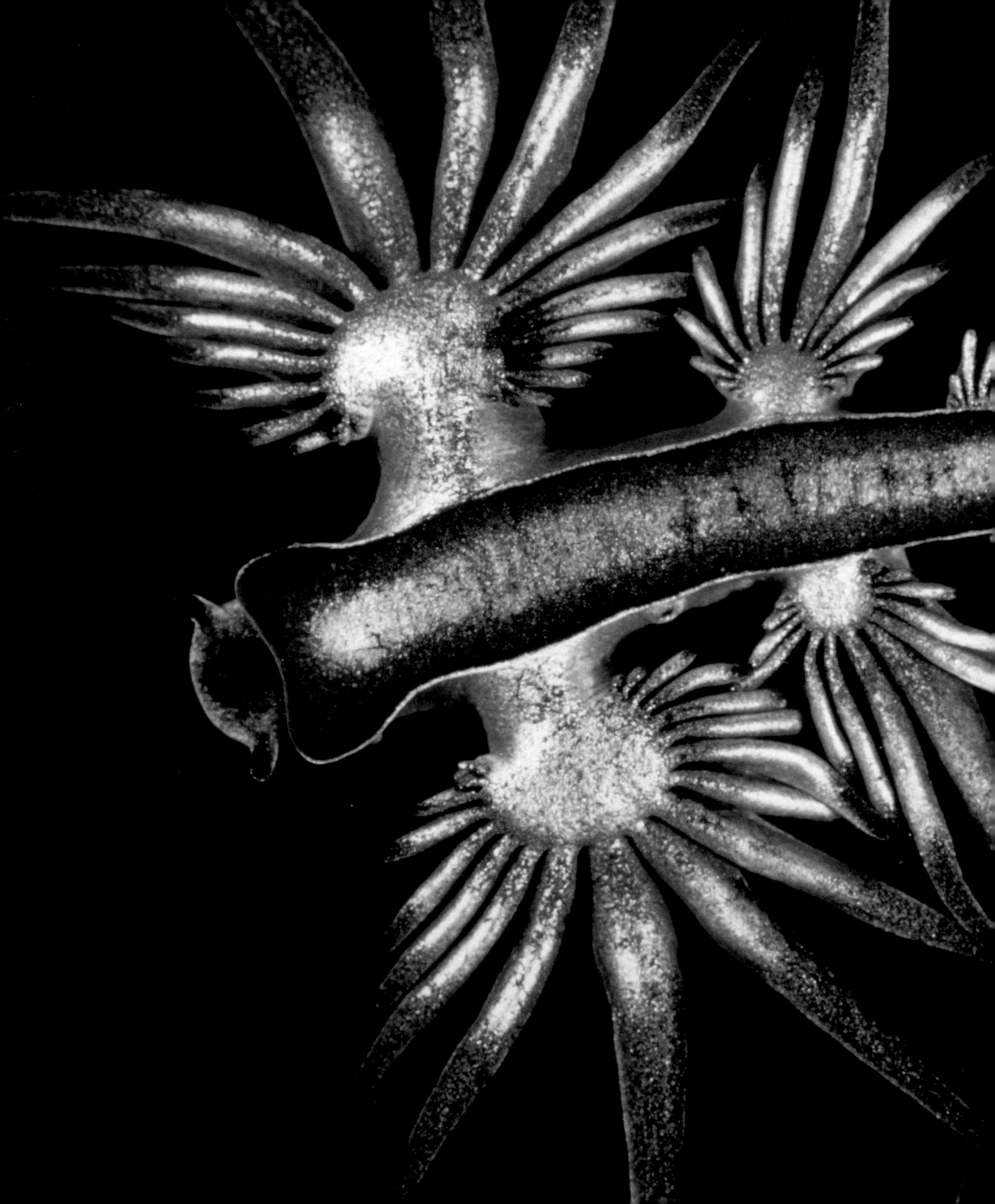

52

53

52 The file shell, *Lima pacifica*, a type of bivalve mollusc. This and some other file shells are able to swim using jet propulsion, augmented by a rowing motion of the numerous tentacles. The long tentacles are also a defence against predators as they're covered in sticky, distasteful mucus and they can be shed at will.

53 A squid with her mass of egg capsules (unidentified species).

Nemertea

(ribbon worms)
(Greek *Nemertes* = a sea nymph)

Diversity
c. 1200 species

Size range
0.5 mm to 55 m
(0.02 in. to 180 ft)

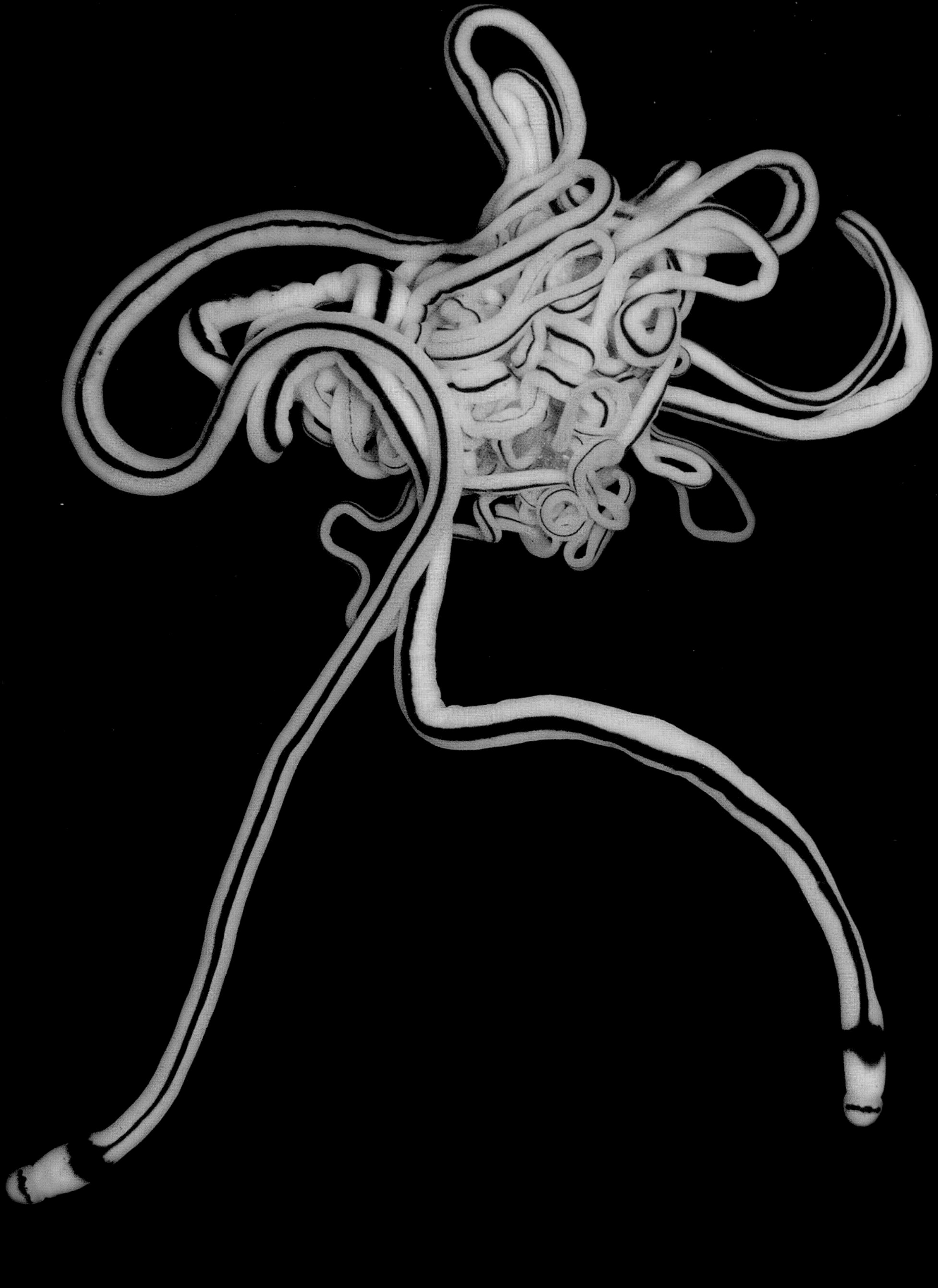

1 Ribbon worms are the longest of all animals, but just how long the largest specimens get is hard to define as they are very fragile. In this image, two individuals (*Baseodiscus hemprichii*) are tangled together.

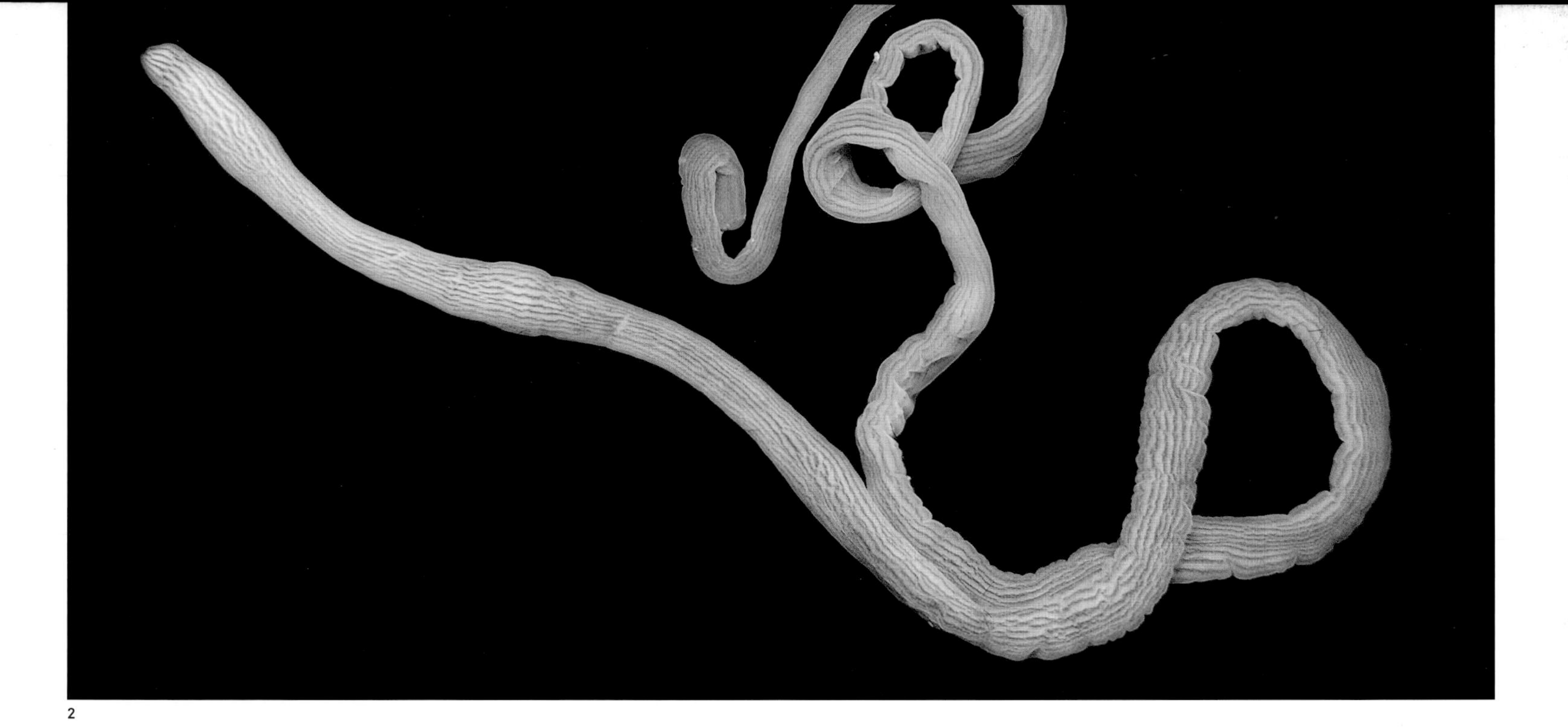

2

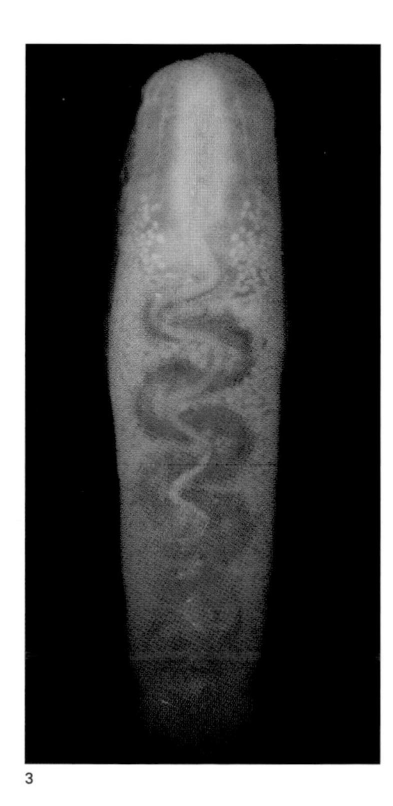

3

2 Ribbon worms are predominantly marine animals. Most species creep along the seabed in search of prey (*Baseodiscus delineatus*).

3 *Malacobdella arrokeana*, a commensal ribbon worm. This species lives inside the mantle cavity of the bivalve mollusc *Panopea abbreviata* (see p. 236).

Ribbon worms enjoy the accolade of being the longest animals on the planet. A specimen of the aptly named boot-lace worm washed ashore in Scotland in the 19th century was, reportedly, just shy of 55 m (180 ft) long. The typical ribbon worm is a long, thin predator or scavenger of the seabed, although a few species have managed to colonize fresh water as well as moist habitats on land. They are often to be found in abundance, but we know very little about them.

FORM AND FUNCTION

To the uninitiated, a typical ribbon worm might look a little like a long, thin flatworm (see Platyhelminthes, pp. 278–93) [1, 2, 6], but there are certain key differences, most fundamentally that the former have a well-developed body cavity and a through gut, whereas the latter do not have a body cavity or an anus. The tissues of ribbon worms are also served by circulatory system vessels, but their defining feature is a long, eversible proboscis housed in a fluid-filled cavity (running almost the entire length of the body in some species), which is completely separate from the gut and is believed to have evolved from an inward folding of the body wall. Muscles surrounding the structure evert the proboscis to catch and disable prey [4, 5]. The proboscis of some species, the so-called armed ribbon worms, brandishes a nasty looking stylet, used to repeatedly stab the prey [10, 11].

The nemerteans have nothing in the way of cuticle or an exoskeleton, but their exposed ciliated epidermis is protected by profuse secretions of sticky mucus often suffused with toxins to keep predators and settling organisms at bay (see below). Beneath the epidermis there is layer of connective tissue overlying a well-developed musculature. Gases diffuse in and out of the body across these layers of the body wall and in a small number of species bind to haemoglobin in circulatory system cells. The central nervous system consists of two pairs of connected ganglia (the brain) and a pair of nerve cords. There are often simple eye spots (sometimes as many as 100) [7], and some species have chemosensory cerebral organs that are important in detecting prey.

LIFESTYLE

All but around 40 species of ribbon worm are free-living, active predators or scavengers on the seabed, and very efficient they are too [14]. During the day they hole up in burrows, rock crevices or amongst algae, only venturing out as it gets dark to hunt their favourite prey, which depending on the ribbon worm in question can be bivalve molluscs, annelids or crustaceans. Ribbon worm eye spots can only detect the difference between light and dark, so these creatures rely on their cerebral organs to 'sniff out' their quarry. Some of the roving annelids leave an odour trail on the seabed and it is this the ribbon worm follows like a bloodhound. When it is very close to its prey or if it makes contact, the ribbon worm goes into feeding mode, explosively releasing its proboscis. In some species the proboscis simply grabs

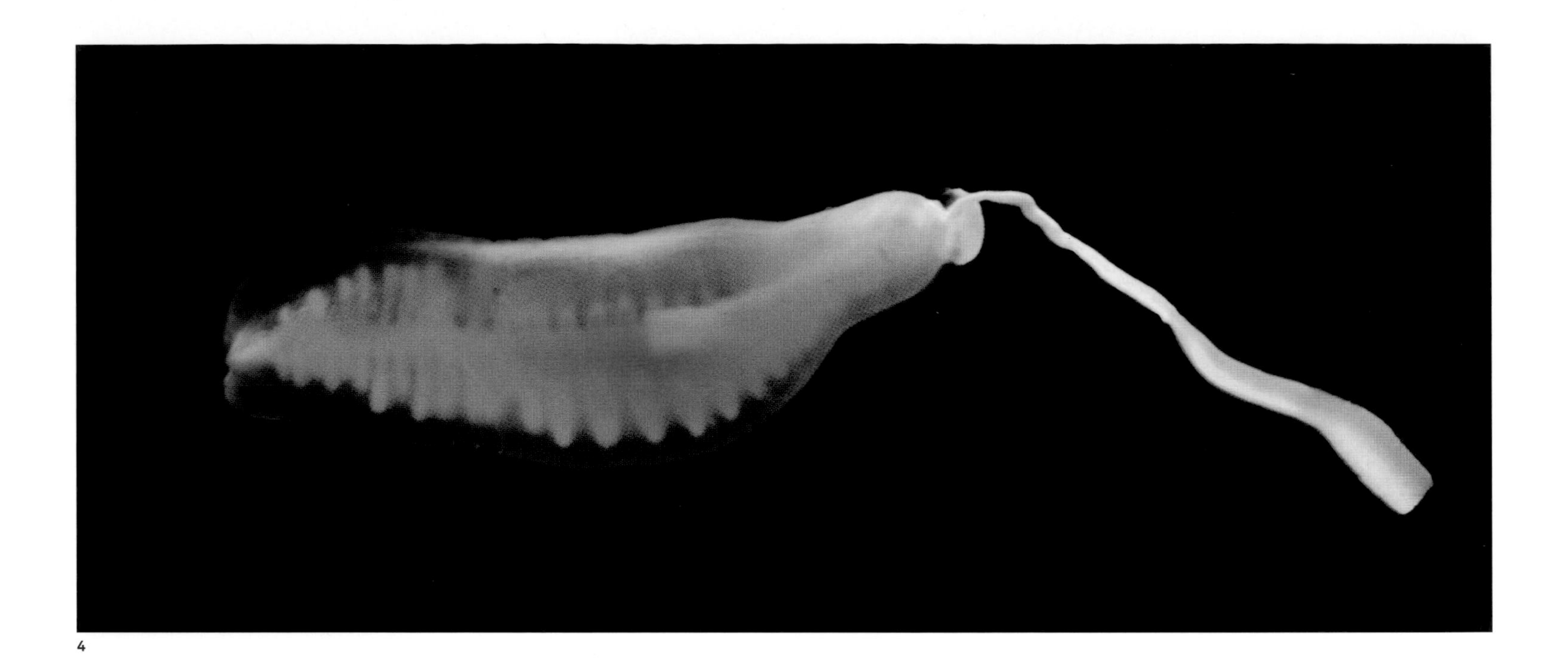

4

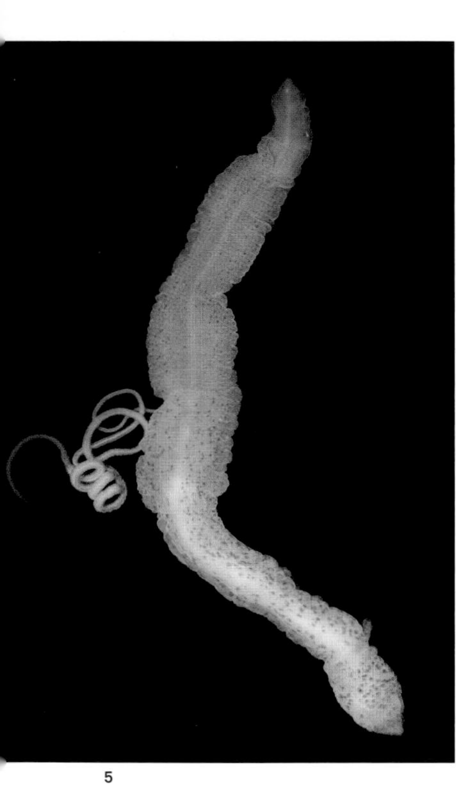

5

the prey and pulls it to the mouth, where it is engulfed. Others retract their proboscis leaving a tether of sticky mucus and then retreat to a safe distance until the victim is hopelessly tangled. In some species the mucus is thought to contain a cocktail of toxins including tetrodotoxin (see Chaetognatha, pp. 125), but it is unclear if these are secreted by the proboscis alone and also which toxins in particular are important in immobilizing the prey. The armed ribbon worms use their stylet-tipped proboscis to repeatedly stab the prey, permitting the entry of toxins [10, 11]. Once the target has been subdued, the ribbon worm moves in to feed. Some species just press their head to the hole made by their proboscis stylet, poke out their oesophagus and slurp up the bodily contents of the unfortunate prey. Others swallow the entire meal in one go, even if it is larger than themselves. Some ribbon worms are even specialist egg eaters, preying on the developing broods of crabs by piercing the egg shells with their stylets and sucking up the contents.

Around 40 ribbon worm species live on or in the bodies of other animals, but the exact nature of these interactions is not very well understood. Ribbon worms in the genus *Malacobdella* are superficially similar to leeches since they have a sucker and they get about in the same inch-worm fashion. Dwelling in the mantle cavity of several species of bivalve mollusc, they feed on some of the suspended food drawn in by their host and possibly also the mucus secreted by its gills, apparently without doing any damage – so they are considered to be commensal rather than parasitic [3, 15]. Depending on the species, a single bivalve might play host to anywhere between one and 190 of these commensals. Interestingly, it seems that when many individuals occupy the same host, only one of them is sexually mature, possibly due to this animal producing chemicals that temporarily neuter the others. This observation is merely a glimpse of the complexity of the lives of these animals and highlights how little we know about the more obscure corners of the natural world.

Ribbon worms have considerable regenerative powers. Many can shed their proboscis and grow a new one while others, especially some of the very elongate species, readily break into pieces if molested, each writhing fragment giving rise to a new individual; in fact, some species routinely reproduce by simply fragmenting. However, sexual reproduction with separate sexes is the norm. The nemerteans have lots of pairs of gonads, but these only develop a pore to the outside world when the gametes are mature. The males and females may simply shed their gametes at the same time in the hope that eggs and sperm will meet in the water, or a couple may take up residence in a mucus cocoon to increase the chances of successful fertilization.

Ribbon worms go through a larval stage before metamorphosing into an adult form. In a small number of species this larva is planktonic and similar to trochophore larvae (see Annelida,

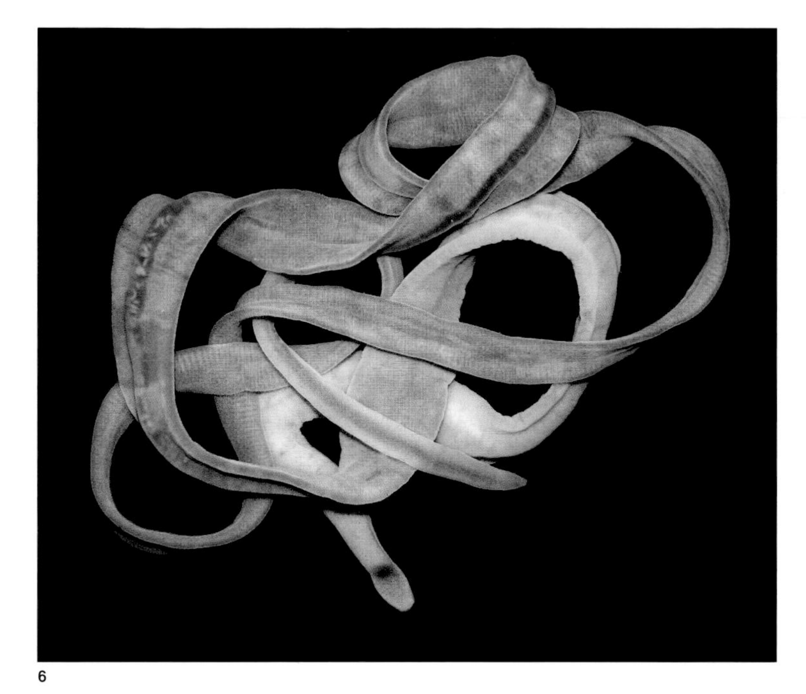

6

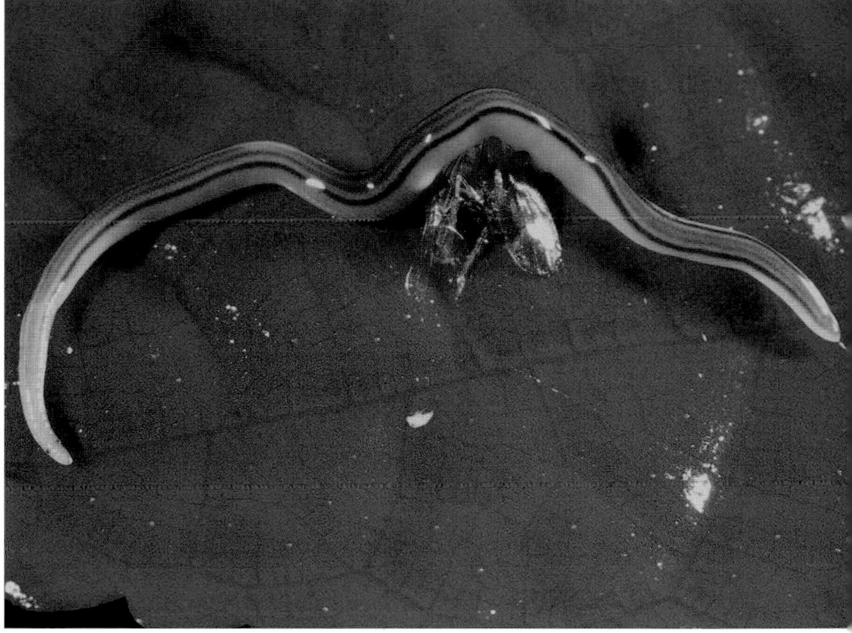

7

4 This species of ribbon worm (*Dinonemertes shinkaii*) was discovered in 2000 at a depth of 2343 m (7685 ft). The everted proboscis, which is about the same length as the body, is visible. The paired orange structures are branches of the gut that increase its surface area. This is a pelagic species that swims by beating the posterior part of its body.

5 Superficially, ribbon worms resemble flatworms, but there are a number of key differences. Most obvious of these is the unique ribbon worm proboscis, which is clearly visible here coiling to the left (unidentified species).

6 The long bodies of ribbon worms are typically very fragile and the longer species often fragment when handled. In many species this fragmentation is employed as a form of asexual reproduction, with each fragment growing into a new individual (*Baseodiscus* sp.).

7 A land-dwelling ribbon worm (*Geonemertes* sp.) scavenging the remains of a fly. Note the two clusters of eye spots (at left).

pp. 208–31, and Mollusca, pp. 232–59). In other species there is a unique larval stage known as the pilidium, which is similar in shape to a Sherlock Holmes' hat [9]. In these species metamorphosis is particularly odd because rather than the larva gradually taking on the young adult form, the juvenile actually grows inside the larva and eventually consumes it. In the few species where internal fertilization takes place, the young are brooded in the parent's gonads, but since the female lacks a dedicated birth canal the act of giving birth is brutal to say the least: the juveniles rupture their mother's gonads, penetrate the gut wall and squeeze out of her anus.

The ribbon worms are common animals around the globe, particularly in the oceans where they occupy important niches as bottom-dwelling predators. They have even branched out to a life in open water, and in freshwater and moist habitats on land [7, 13]. As we know so little about their ecology, we can only guess how their predations and commensal/parasitic interactions affect the populations of other animals. The toxins in the copious mucus they secrete, often advertised by their boldly coloured bodies, are effective in keeping many of their potential enemies at bay [8, 12], but there are lots of fish, birds and arthropods for whom the ribbon worms represent an important source of food.

ORIGINS AND AFFINITIES

Like most of the other soft-bodied animals, the fossil record of the ribbon worms is sparse to say the least. There is a long dead creature from the Carboniferous, dating to around 310 million years ago, that is widely accepted as being the earliest known fossil ribbon worm. The remains of Cambrian animals have also been touted as belonging to this lineage, but many experts remain sceptical.

Traditionally, the ribbon worms were considered to be very close relatives of the flatworms (see Platyhelminthes, pp. 278–93) because of a number of shared traits. Today, however, there is little support for a very close relationship between these lineages and the characteristics they share may be simply ancestral or a result of them adapting to similar niches, rather than a recent, common heritage. Not all the ribbon worms go through a planktonic phase as a larva, but in some that do, the larva is similar to the dispersal stage in the life of many annelids and molluscs; indeed this is one of the important characteristics uniting these animals. Today, most zoologists see the ribbon worms as close relatives of the molluscs and annelids based on evidence from morphology, developmental biology and DNA.

8

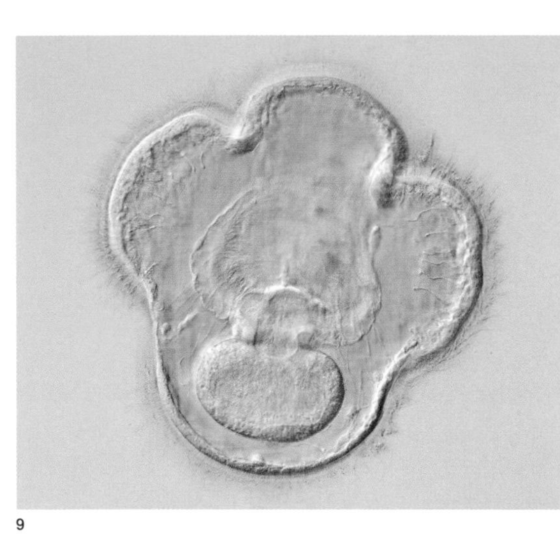

9

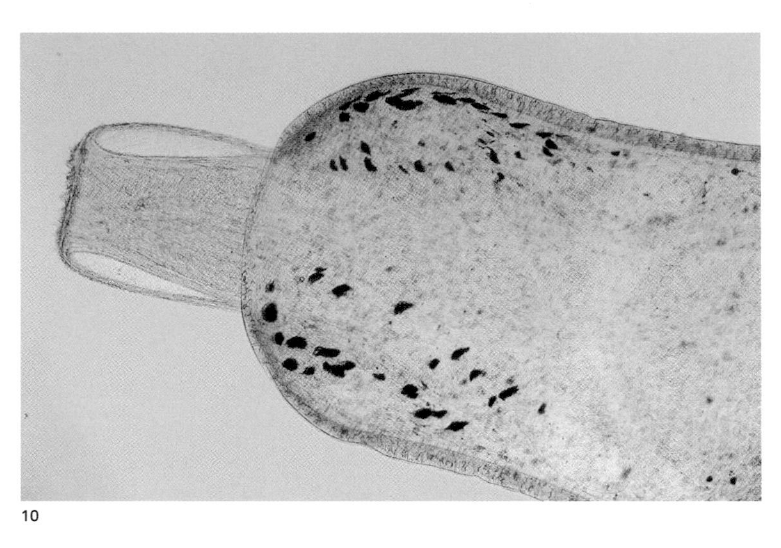

10

12

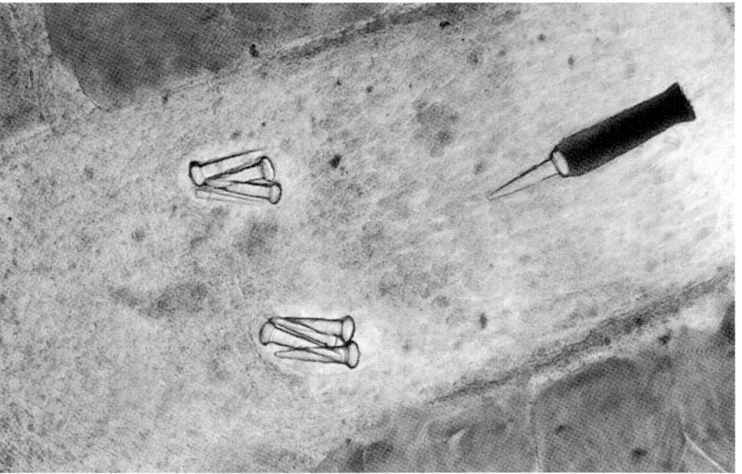

11

8 The so-called 'king ribbon worm'. The bright pattern on the head of the animal, in contrast to the dark body, may be to warn potential predators of its chemical defences.

9 Some ribbon worms have a distinctive larval stage – the pilidium. The juvenile ribbon worm grows inside the larva and eventually consumes it.

10 The partially everted proboscis of *Emplectonema echinoderma*, an armed ribbon worm, is equipped with a stylet, which is thrust into prey. Over time, the tip of the stylet wears out, but the tissues of the proboscis produce replacements.

11 In this close-up, the replacement stylet tips of an armed ribbon worm are clearly visible (*Emplectonema echinoderma*).

12 Many ribbon worms broadcast their chemical defences with aposematic colouring (*Monostilifera* sp.).

13 A few ribbon worm species have adapted to moist habitats on land (*Geonemertes* sp.).

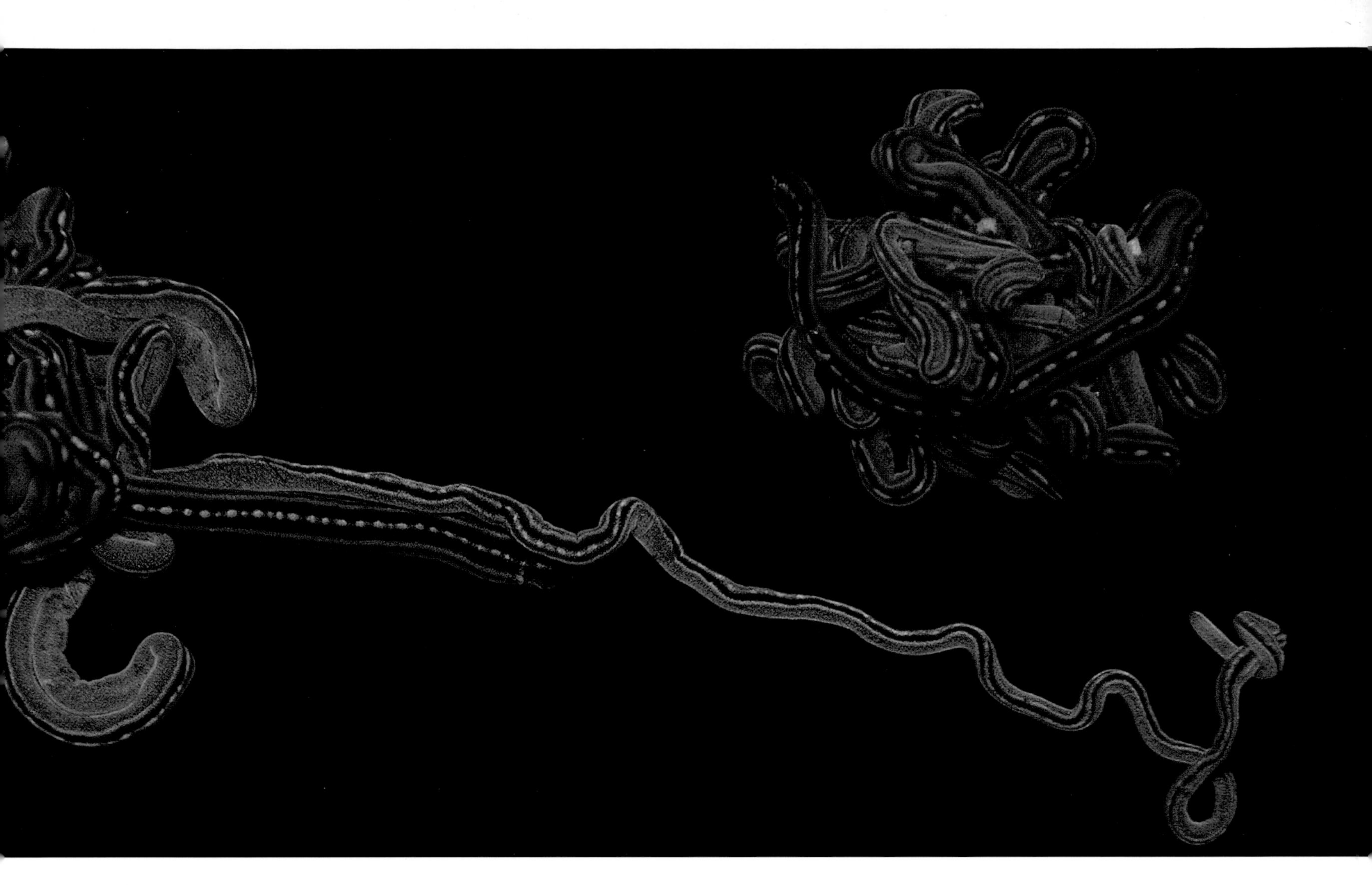

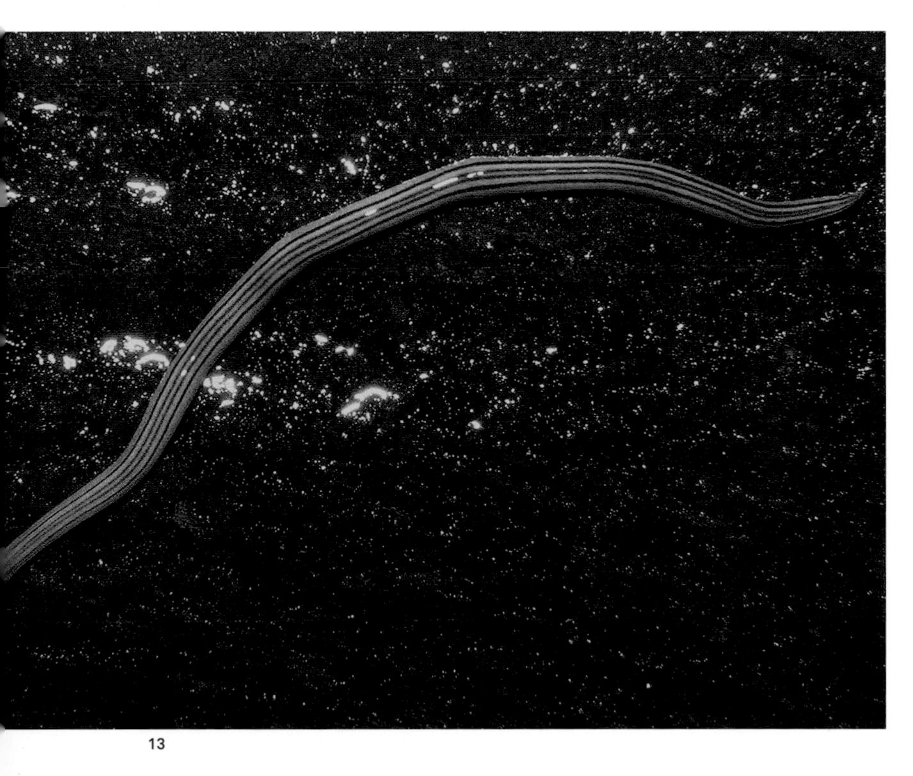

13

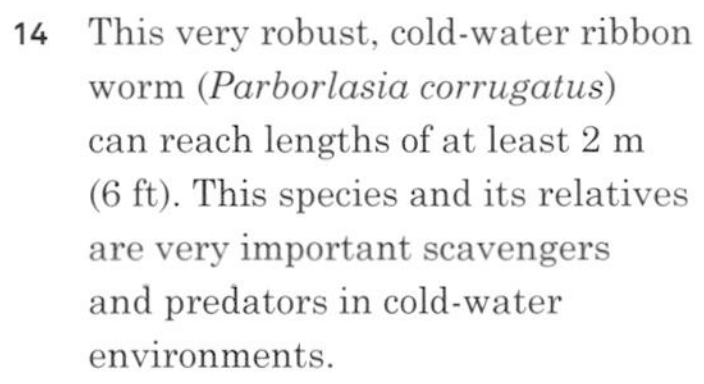

14 This very robust, cold-water ribbon worm (*Parborlasia corrugatus*) can reach lengths of at least 2 m (6 ft). This species and its relatives are very important scavengers and predators in cold-water environments.

15 The commensal ribbon worm, *Malacobdella arrokeana*, inside its host.

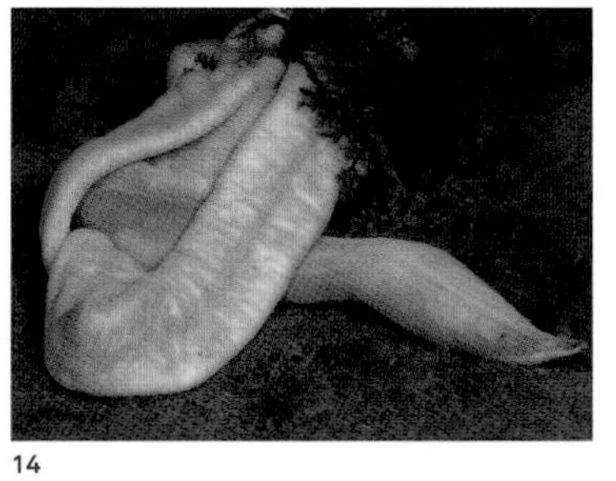

14

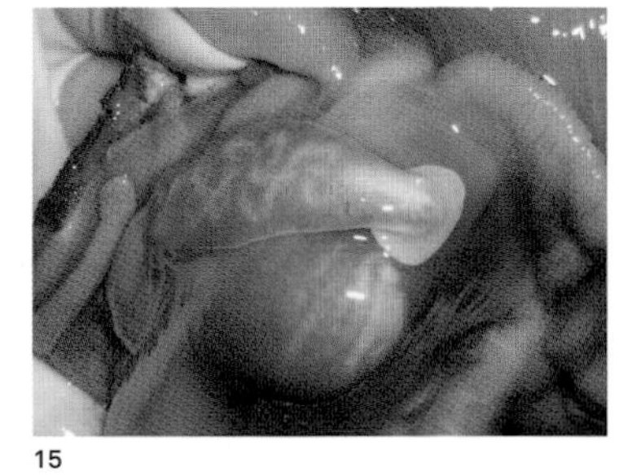

15

Brachiopoda

(lamp shells)
(Latin *brachium* = arm;
Greek *podos* = foot)

Diversity
c. 390 species

Size range
1 mm to 10 cm
(0.04 to 4 in.)

1 The pedicle of some brachiopods has evolved into a long stalk that anchors them in the sediment and allows them to withdraw to the deeper reaches of their burrows (*Lingula* sp.). These species can also move their shell valves in a slicing motion to burrow through the sediment. Also note the stiff bristles sticking out from the edge of the body.

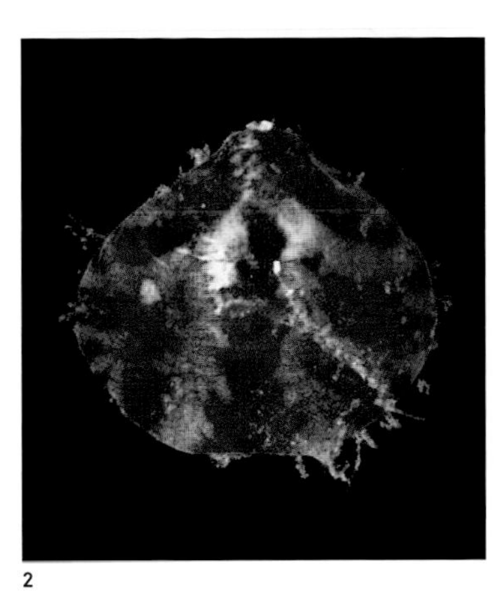

2

2 The body of a brachiopod is protected by a pair of shell valves (*Frenulina sanguinolenta*).

At first glance, you would be forgiven for thinking that a brachiopod was nothing more than a mollusc, perhaps just another a bivalve, but although these creatures do indeed bear a superficial resemblance to one another they are very different. The brachiopods are a distinct lineage of marine filter-feeders that have long since passed their heyday. They have perhaps the most detailed and far-reaching fossil record of all the animal lineages, but today their species diversity is a shadow of what it once was.

FORM AND FUNCTION

A brachiopod does look remarkably like a bivalve mollusc [2, 3], but they have lots of characteristics that are all their own. They have a pair of shells enclosing the body, but instead of being on the left and right of the animal – like in bivalves – they are ventral and dorsal, secreted from the underside and back, and they are typically different in size. Most species are firmly attached to the substrate via a short pedicle, sometimes cemented on. In adapting to a burrowing lifestyle the pedicle of some brachiopods (e.g., *Lingula*) has become very long, firmly anchoring them in the sediment and allowing them to withdraw into the deeper reaches of their retreat if needs be [1]. The shells are secreted by outgrowths of the body wall forming a central space that accommodates the organs of the animal and its substantial feeding apparatus – a crown of tentacles circling the mouth, the so-called lophophore [4]. These tentacles have been convoluted into what looks like a pair of arms, hence the scientific name of these creatures. Opening (at most a narrow gape) and closing of the shell is achieved by muscles or a combination of muscles and fluid pressure from inside the body, depending on the species. The burrowing species can move the halves of their shell in a slicing motion to cut their way through the sediment. Sticking out in the water from the edge of the body wall are stiff, mobile bristles that help to prevent larger particles from finding their way into the animal's filtering cavity when it is feeding. These bristles are particularly conspicuous in the burrowing species.

In adapting to a filter-feeding existence the brachiopods have lost many of the characteristics that we associate with more active animals. There is no head and the brain is no more than a nerve ring with one or two tiny clusters of nerve cells. The bristles and the edge of their body wall are sensitive to touch, while in some species receptors firmly close the shell in response to a passing shadow. Colourless blood, supplying the tissues with nutrients, is pumped through vessels and internal cavities by a heart, while gases are exchanged via the body wall and the considerable surface area of the feeding tentacles. The gut is normally a rather simple, generally U-shaped affair, although the majority of species have dispensed with their anus, so the digestive tract is a two-way passage.

LIFESTYLE

All brachiopods are filter-feeders. The cilia on the tentacles draw water and suspended particles into the mantle cavity where edible matter is separated out and conveyed to the mouth along ciliated grooves at the base of tentacles. In the sediment-dwelling species, filtering is achieved from the safety of their burrow entrance, their mantle bristles arranged to form two inhalant siphons sandwiching an exhalant siphon. Like all filter-feeders, all sorts of morsels are on the menu for the brachiopods, including phytoplankton and assorted particles of organic matter.

There are generally separate sexes and as the adults are not really in a position to seek each other out (since they are fixed to the spot), they shed their gametes into the water via their body cavity and the pores of their excretory system. Typically, the fertilized eggs develop into free-swimming, ciliated larvae that join the plankton for around one day before settling and metamorphosing into adults. Some of the brachiopods have what is known as direct development, where the fertilized egg develops into a juvenile (miniature adult) that spends quite some time amongst the plankton propelled by the cilia on its nascent feeding tentacles. Weighed down by its growing shell this footloose youngster eventually sinks to see out the rest of its days as a sessile animal. In a few species fertilization is internal and the female retains the developing young for varying lengths of time to improve their chances of survival.

Brachiopod diversity is just a shadow of what it once was and the ecological importance of these animals is assumed to have been on the wane for a very long time. However, although there are not very many brachiopods species living today, in many places around the world they are still an important component of the benthic fauna. They are to be found in relatively shallow coastal waters down to depths of at least 6000 m (20,000 ft). They often occur in profusion adhering to rock, coral and any other

3

substrates that afford a firm attachment. The sheer walls of fjords in British Columbia, Canada, are encrusted with dense aggregations of brachiopods, sometimes numbering as many as 945 per sq. m (785 per sq. yd). In some coastal areas of New Zealand there can be 20,000 individuals per sq. m (16,500 per sq. yd). It seems that far from being universally outcompeted by the bivalve molluscs, long thought to be more finely tuned to this way of life, there are lots of marine habitats, typically those we rarely visit, where the brachiopods still predominate, such as the upper continental slope areas where there is lots of sediment in the water but not much oxygen.

There is still a huge amount we don't know about the brachiopods and their interactions with other organisms. They do have their enemies, but many marine predators appear to find brachiopods distasteful.

ORIGINS AND AFFINITIES

Thanks to the durability of their shells, few other animal lineages have a fossil record as far reaching and as rich as the brachiopods. There are many places around the world where the fossil deposit is little more than the amassed remains of countless brachiopods, testament to success of these creatures in the ancient seas. Palaeontologists recognize some 20,000 species of extinct brachiopods, but identifying long-dead animals solely on the basis of external characteristics is likely to give us an inaccurate, probably exaggerated idea of their previous diversity. What we do know for sure is that they have been around for a very long time. Even before the Cambrian, perhaps 800–1000 million years ago, they were well established, but they reached the zenith of their diversity between 500 and 435 million years ago. Around 250 million years ago the brachiopods and all other life on earth was struck by an episode of great ecological upheaval (known as the Permian–Triassic extinction). Some of the brachiopods clung on during these extremely trying times, but their diversity never really recovered. To add insult to injury the bivalve molluscs, long-time rivals of the brachiopods, were undergoing an explosion in diversity before this event and they emerged the other side of it ready to exploit myriad vacant niches.

Unsurprisingly, the brachiopods were thought to be just another type of bivalve mollusc until the 19th century, when it was finally realized they were fundamentally different animals. Once this was recognized, the problem was working out where they belonged on the animal family tree, a quandary that has seen them shunted from one branch to another as new evidence came to light. Even today the evolutionary relationships of the brachiopods are among the most perplexing of all the animal lineages, but we are steadily edging closer to resolving them. They, along with their close relatives the phoronids (see pp. 270–73), are currently thought to be an offshoot of the branch that also includes the annelids, molluscs and nemerteans.

4

3 Two brachiopods (*Hemithiris psittacea*) on the seabed.

4 Brachiopods only open their shells to a narrow gape to filter feed. The white filter-feeding apparatus, the lophophore, is clearly visible here (*Hemithiris psittacea*).

Phoronida

(horseshoe worms)
(Greek *Phoronis* = another name for the Egyptian goddess, Isis)

Diversity
c. 10 species

Size range
1 to 50 cm
(0.4 to 20 in.)

1 The phoronids are exclusively benthic, marine animals. The vast majority of the long, thin body is concealed in a tube. They extend their crown of tentacles (lophophore) into the water to collect suspended, edible matter (*Phoronopsis californica*).

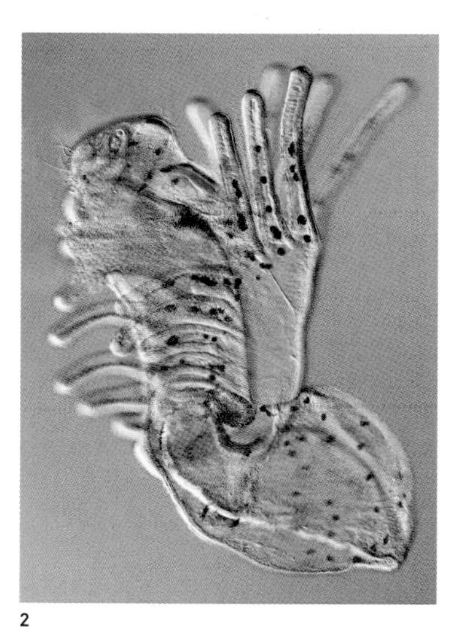

2

Slender and diaphanous, with a coronet of tentacles, the exclusively marine phoronids are rarely seen animals of which we know very little. The adults spend their entire life with much of their body hidden away in a tube embedded in soft sediment or encrusting hard substrate, only gingerly reaching up and extending their feeding tentacles into the water to filter edible matter from the water.

FORM AND FUNCTION

The body of a phoronid is a very slender stalk with a bulbous swelling at the posterior end, anchoring the animal in its tube, and a conspicuous crown of tentacles, the lophophore, at the head end [1, 7]. Even in the largest specimens this long stalk of a body is no more than 5 mm (0.2 in.) across, rendering them very fragile. The tentacles of the lophophore are distinctive since they are arranged in a horseshoe shape around the mouth, hence the common name of these creatures. The larger the species the more these feeding tentacles are folded and coiled to allow sufficient surface area for filter feeding and gas exchange [3, 4]. There is no exoskeleton or tough cuticle, but the bare, ciliated epidermis secretes a chitinous tube reinforced with sediment particles and it is this that lines the animal's vertical burrow, effectively functioning as a tough cuticle. Well-developed muscles running the length of the body enable the animal to withdraw any of its exposed soft bits into the safety of its burrow when its senses a threat.

Collected food makes its way into the mouth and through the long U-shaped gut that extends the length of the body. Blood with haemoglobin-loaded cells, a trait shared by very few lineages of animals, circulates in a network of vessels transporting gases throughout the body. The central nervous system consists of a nerve ring with one or two nerve cords and feeding information to this are various sensory receptors scattered over the body surface, particularly in and around the lophophore.

LIFESTYLE

With the exception of the Antarctic Ocean, the phoronids are found all over the world, albeit rarely and seemingly restricted to the relatively shallow waters of the continental shelf down to depths of around 400 m (1300 ft). The adult spends its whole life in its burrow, often having to survive in water where there is very little dissolved oxygen. However, the blood with its haemoglobin-packed cells is very good at grabbing whatever oxygen is available and getting it to where it is needed in the animal.

With the exception of a mantle and a shell into which water and suspended food are drawn, the way the phoronids feed is identical to the brachiopods: cilia on the feeding tentacles generate a current in the water and anything edible is trapped and conveyed to the mouth. Feeding in this way with their soft parts exposed, the phoronids are vulnerable little animals, but they are equipped with a few defences to help them survive the rigours of life on the seabed. Their retraction response to a shadow falling over them or being touched is very rapid – in the blink of an eye they shrink down into their burrow and out of the way of many predators. If they do not manage to retreat fast enough their last line of defence is to sacrifice the entire front portion of their body, including their feeding tentacles, mouth, nerve ring and excretory system, all of which are regenerated within three days. They even do this in order to release their eggs or if their environment becomes unfavourable, perhaps as a way of conserving energy until better conditions return. Some species also appear to be chemically defended too: as yet unknown chemicals concentrated in the front portion of the body are enough to put off a range of predators.

Sexual reproduction is the standard reproductive strategy and species are dioecious or hermaphrodite. The gonads surround the stomach in the bulbous posterior of the long thin body. The sperm are released into the body cavity, exit the body via the ducts of the excretory system and get packaged into little spermatophores by a pair of organs at the base of the tentacles. These packages are set adrift and if they are lucky they will be snagged by another individual of the same species. Sensing their goal, the sperm turn into creeping, amoeba-like cells capable of penetrating the body wall of the adult to reach the ripe eggs in its body cavity. Some species produce relatively few large and yolky eggs and once fertilized these are brooded until they hatch into larvae that spend around four days in the plankton. Other species produce as many as 500 small eggs that are released soon after fertilization. Larvae hatch from these eggs after a few days and spend the next three weeks or so drifting amongst the other organisms of the plankton. This planktonic larval stage, an bizarre-looking beast known as an actinotroch, is unique to the phoronids [2].

2 Phoronids have a distinctive larval stage known as the actinotroch. This is the dispersive stage of the life cycle, as the adult remains in the same burrow its entire life.

3 A pair of phoronids (*Phoronis australis*) on the seabed. Note the extreme coiling of the lophophore in this large species. They never leave the tube they construct in the sediment.

4 A small colony of phoronids with their lophophores extended. In these larger species the lophophore is coiled to provide sufficient surface area for filter feeding and gas exchange. (*Phoronopsis californica*).

5 One species of phoronid is known to live in close association with tube-dwelling anemones. Here, several *Phoronis australis* have made their home around the burrow of such an anemone.

3

4

5

6

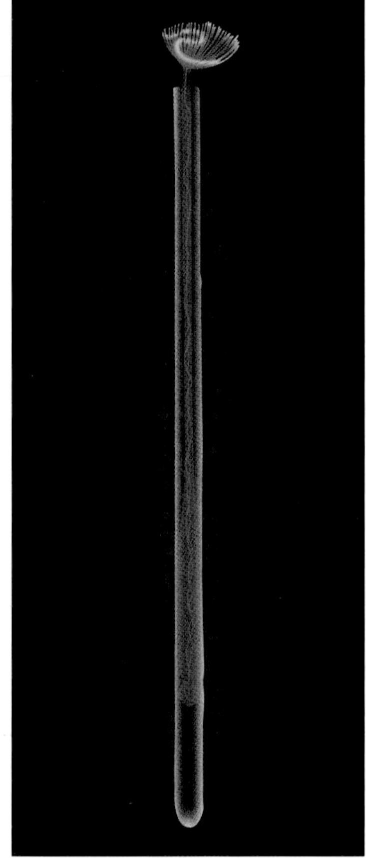

7

6 In some places, phoronids can reach very high population densities. Note the simpler, less coiled lophophores of this smaller species (*Phoronis vancouverensis*).

7 Illustration showing a phoronid in its tube. Most of a phoronid's long, thin body is hidden from view. Note the U-shaped gut that extends the length of the animal.

The drifting larvae eventually undergo a rapid metamorphosis where their bodies are reshaped and reconfigured into the adult form. The juvenile phoronids sink to the bottom and secrete a tube within the seabed where they will spend the rest of their lives.

Phoronids may be rarely seen, but in some places they are very successful animals, festooning the seabed in huge numbers [6]. Some species are known to occur in densities of more than 26,000 per sq. m (21,500 per sq. yd). At these high densities they are the dominant filter-feeding organisms by a long way and their burrows and secreted tubes shape the seabed, providing lots of nooks and crannies for many other small marine animals.

Our knowledge of how the phoronids interact with other sea creatures is limited to say the least. There are a range of animals that prey on them, including snails and fish, but chemical defences afford many of them protection from predators. It may be the noxious chemicals they produce that allow some species to reach very high population densities. Various parasites have been seen in the bodies of the adults, but the biology of these and how they affect the lives of their hosts are unknown. One phoronid species is known to live in very close association with tube-dwelling anemones using the tube of the cnidarian as a foundation for their own secreted home [5]. As well as perhaps taking advantage of any leftovers, these commensals also get an early warning of approaching danger as the anemone contracts its muscles to retract its tentacles, prompting them to do the same. Similarly, the phoronids may also alert the anemone to potential predators.

ORIGINS AND AFFINITIES

Fossilised tubes thought to be the work of ancient phoronids have been found in Cambrian deposits, some 540 million years old. However, the oldest fossils that have been confidently interpreted as long dead representatives of this lineage are 420 million years old.

In many respects the phoronids can be viewed as shell-less brachiopods (see pp. 266–69) and there are many zoologists who consider them as exactly that. For the time being, however, they are still recognized as a separate lineage, but very close relatives of the brachiopods and part of the same branch as the annelids, molluscs and nemerteans.

Gastrotricha

(gastrotrichs)
(Greek *gaster* = stomach;
thrix = hair)

Diversity
c. 790 species

Size range
0.05 to 4 mm
(0.002 to 0.16 in.)

1 The muscular pharynx and food-filled intestine of the gastrotrich *Polymerurus nodicaudus*.

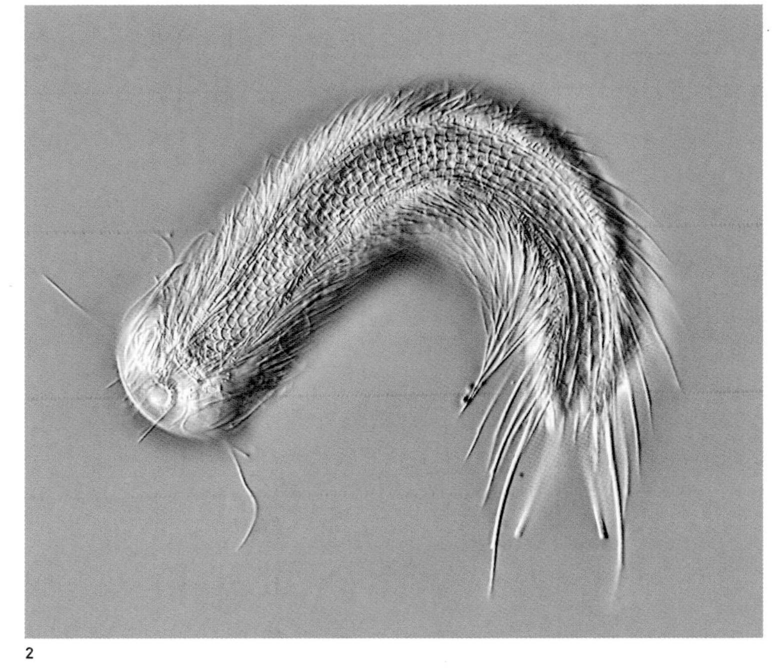

2

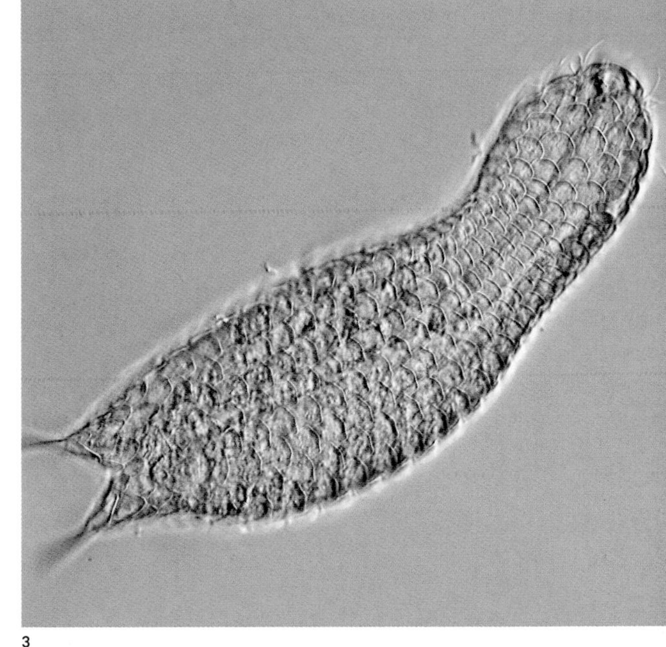

3

Few people will have heard of a gastrotrich, let alone seen one. But although these creatures might be small, they are far from rare. They are to be found in marine and freshwater environments where they glide around between sediment particles and on the surfaces of seaweeds and aquatic plants and animals. Any pond, especially those with a tangle of submerged plants, will be home to an array of species.

FORM AND FUNCTION

To the naked eye, most species are nothing more than tiny specks scooting around in a droplet of water, so it takes a microscope to fully appreciate these captivating little animals. They come in a variety of shapes and sizes, although they are typically elongate and many are distinctively bottle-shaped. The body is often adorned with bristles, spines and scales, the latter reminiscent of medieval armour [1, 2, 3]. Bands of cilia on the ventral surface provide propulsion. The ciliated underside of gastrotrichs inspired their scientific name, which roughly translates as 'hairy belly' [6]. At intervals on the underside and flanks the pores of duoglands (see Platyhelminthes, pp. 279) are visible as stubby little nozzles [6]. Their outer covering is unique since it is composed of numerous vanishingly thin layers, which completely enclose all of the cilia on their body.

Internally these are complex animals, regardless of their diminutive size. There is a fully developed gut, complete with robust pharynx, running the length of the body [1], well-developed muscle layers, a sizeable brain, kidneys and substantial gonads. The body is studded with various receptors, including eye spots, the gastrotrich equivalent of a nose and other sensory structures on the head [1]. There is no body cavity and being so small they have no need for specialized respiratory structures or an internal system of vessels or internal spaces to move dissolved gases and nutrients around the body. Oxygen and carbon dioxide simply diffuse in and out of the tissues across the body wall.

LIFESTYLE

The bands of cilia on their underside propel the gastrotrichs in a gliding fashion in much the same way as the flatworms (see Platyhelminthes, pp. 278–93). The muscles are also important in locomotion as they bend their body to change direction while the cilia are providing the propulsion. Some inch around their habitat by using their muscles to contract and extend the body, while muscles in the unusual free-swimming species move long bristles in a sculling motion, jerking the animal forward in very rapid bursts [2, 4].

The small size and fragility of the gastrotrichs makes it very difficult to study them, especially their lifestyle in natural or close-to-natural situations. They are known to eat a range of tiny organisms including diatoms, bacteria and protozoa, as well as particles of organic matter [1, 5, 7]. There are often long bristles on the head [2, 4] that may be important in 'tasting' whatever objects they touch and

2 Gastrotrichs are often clothed in scales and spines. The long spines on the head are sensory structures. The circular mouth of this specimen is visible (*Chaetonotus zelinkai*).

3 The scale armour of the gastrotrich *Lepidodermella squamata*.

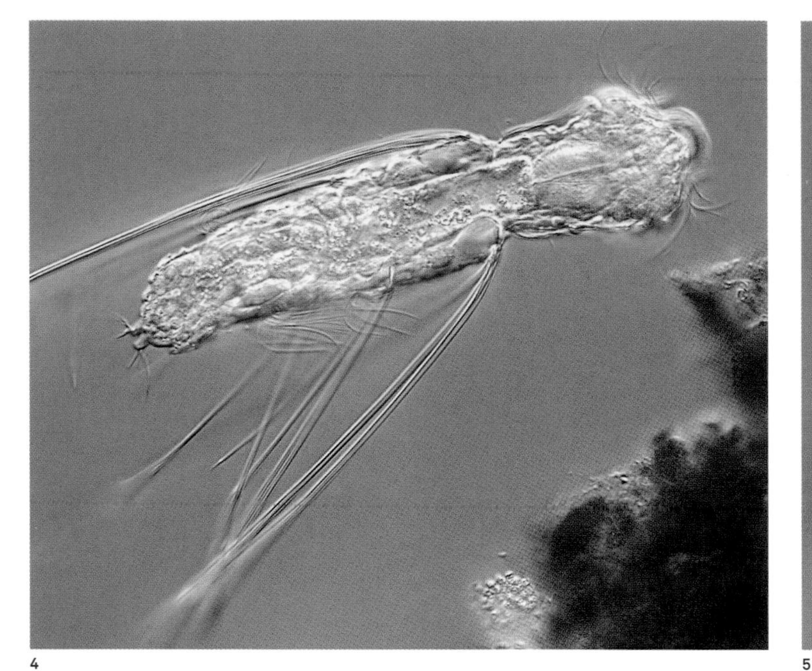

4

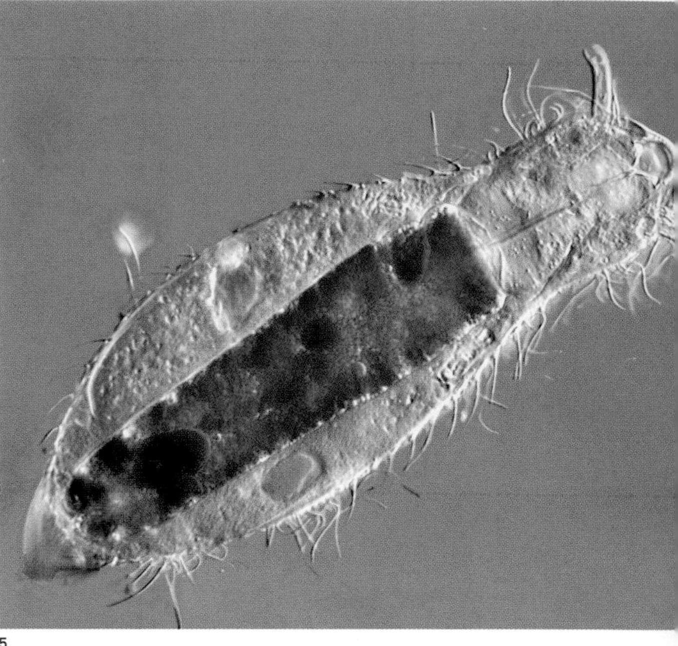

5

funnelling edible matter towards the mouth, where it will be inhaled by the sucking action of the muscular pharynx. The cilia also generate water currents to concentrate edible morsels before they are sucked into the digestive tract. Some marine gastrotrichs have pores branching from the pharynx and communicating with the body surface to channel excess water straight out of the body so it does not dilute the digestive enzymes in the gut. There is evidence they can supplement the food they eat by absorbing organic matter directly through their body wall.

The small, difficult to study animals seem to have a penchant for fiendishly complex and remarkable reproductive strategies and the gastrotrichs are no exception. We are only just scratching the surface of gastrotrich reproduction, but running themes are hermaphroditism, copulatory organs to facilitate the exchange of sperm between mating individuals, internal fertilization, massive eggs relative to the size of the adult, and no larval stage. As the pores for the male and female tract are only transient, the eggs can only be released by a similarly temporary hole that opens up in the body wall, the eggshell forming when the egg touches the water [8]. In the species investigated so far there is lots of variation on these themes, but all strategies can establish huge populations fast. Some species give birth to live young; in others the male gonads have degenerated to the point where only asexual reproduction occurs and the freshwater forms can produce two distinct types of egg. There is a thin-shelled egg that hatches in a matter of days and a thick-shelled one that goes into a state of dormancy for as long as two years to see out periods of extreme cold or drought. Young adults hatch from the eggs and go on to reproduce themselves after around three days. Species that have been observed in captivity live for around 40 days, the first ten days of which is spent producing four to five of their sizeable eggs.

The gastrotrichs are found all around the world. They inhabit the sediments of the deep ocean, the sand of the intertidal zone, the surface of the other marine animals, rivers, pools, lakes, bogs, ditches and just about every other type of freshwater habitat you can find. Some have even colonized the land to a degree, moving around in the thin film of water encapsulating and linking particles of soil. In some places where optimum conditions prevail they can reach staggering population densities, sometimes as many as 1000 individuals per sq. cm (6500 per sq. in.) of sediment, making them among the most dominant creatures of the meiofauna.

The densities and the species composition of their populations in the wild vary hugely even over a few centimetres of seemingly suitable habitat, suggesting they may be responding to minute changes in the oxygen concentrations or other factors we can only guess at. Like most of their biology there is still so much to find out about the gastrotrichs, including their actual diversity; wherever people take the time to look they find previously unknown species. There could be tens of thousands of species.

4 Most gastrotrichs use the bands of cilia on their underside to get around, but a few species move through open water jerkily, albeit very swiftly, by using long, muscle actuated bristles like oars. Also note the muscular pharynx (large ovoid structure at the anterior end) and the sensory bristles adorning the head (*Stylochaeta fusiformis*).

5 This gastrotrich has been busy feeding and the intestine is packed with food (*Neogossea* sp.).

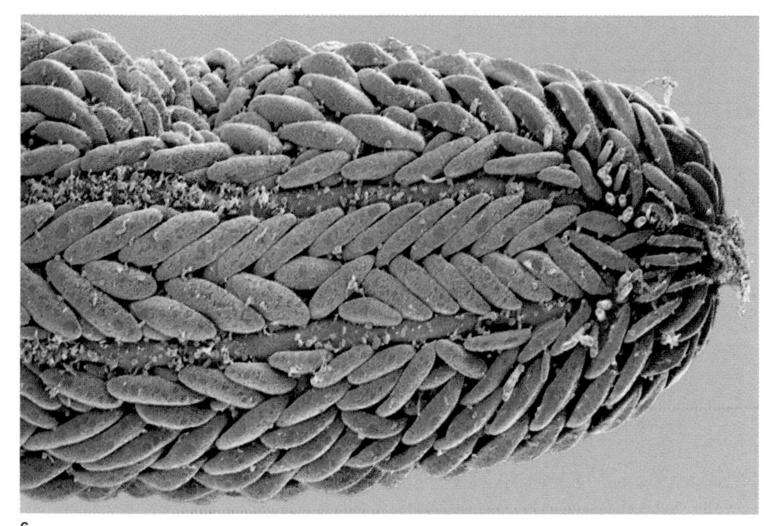

6

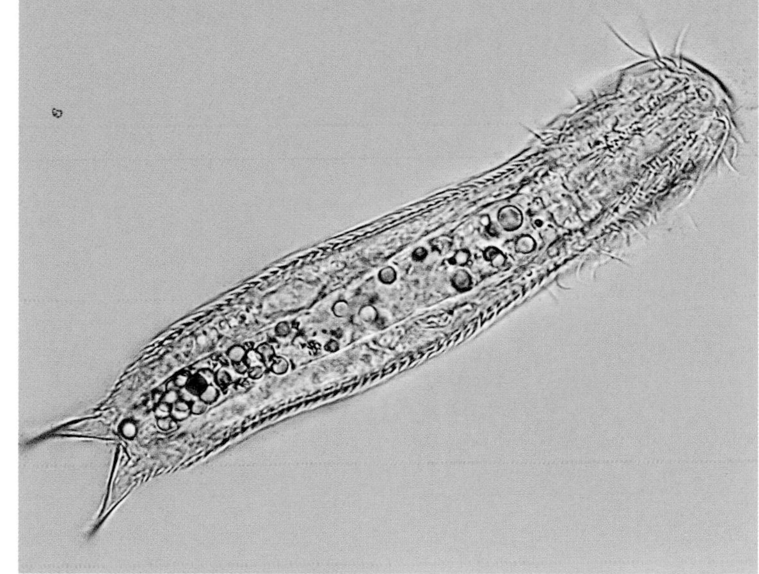

7

8

6 Coloured SEM of two tracks of cilia running along the scale-covered body a gastrotrich (*Lepidodasys* sp.). Cilia enable the gastrotrich to move swiftly over and in the substrate and through the water, while adhesive glands (small tubes at right) allow it to cling tightly to grains of sediment.
7 The gut of this gastrotrich contains numerous food particles (*Aspidophorus paradoxus*).
8 Gastrotrichs release their eggs via a temporary hole that opens up in the body wall (*Acanthodasys* sp.; coloured SEM).

Being so numerous and ubiquitous the gastrotrichs make a significant contribution to the normal functioning of marine and freshwater habitats. In hoovering up microorganisms, both alive and dead, and getting eaten themselves by a huge range of other animals they are an indispensable link in the recycling of nutrients and energy. With our fixation on larger animals we completely overlook such tiny animals and their role in regulating the natural world that is our home.

ORIGINS AND AFFINITIES

To date, no gastrotrich fossils have come to light – a result of their small, soft bodies being inconspicuous and not fossilizing all that well. We therefore have little to go on in working out how long they have been around for. Before the advent of molecular techniques allowed us to compare the DNA of different lineages, the gastrotrichs were thought to be closely related to the nematodes (see pp. 126–33) by virtue of various shared characteristics including the structure of the pharynx. However, today the consensus is that they belong to the branch of animal life that includes the platyhelminths and a posse of other interesting, rarely seen animals, some of which are also surprisingly common.

As more and more evidence is amassed we will be able to work out the evolutionary relationships of the gastrotrichs with greater confidence. This, in turn, will contribute to our understanding of the key events in animal evolution that underpin the incredible diversity of species we can see around us.

Platyhelminthes

(flatworms, flukes, tapeworms)
(Greek *platus* = flat;
helmins = parasitic worm)

Diversity
c. 29,300 species

Size range
<1 mm to 30 m
(<0.04 in. to 100 ft)

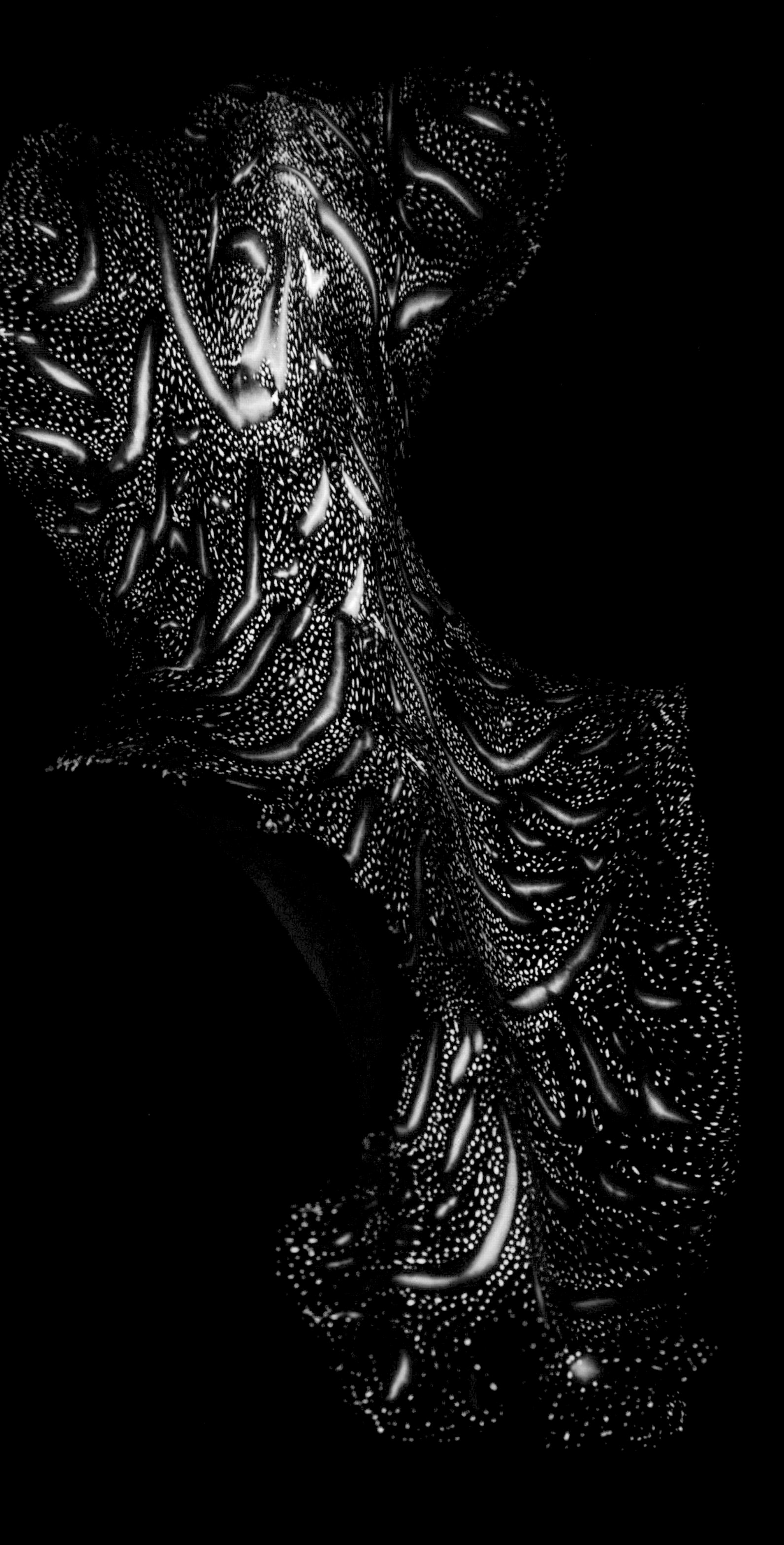

1 The aptly named Persian carpet flatworm swimming using a typical undulating motion (*Pseudobiceros bedfordi*). This species is hermaphroditic and has two penises. A pair of mating individuals 'fence' with their penises, attempting to stab and inject their sperm into the body of the mate.

Flatworms, flukes and tapeworms are relatively simple animals, yet they exhibit a bewildering diversity of forms and lifestyles. Collectively these creatures are the platyhelminths, ubiquitous and extremely numerous animals – but ones that few people will have taken notice of or even ever seen.

Typically when we speak of 'flatworms', we mean the free-living members of this lineage. Most of those are aquatic, though some live in moist environments such as damp soil. Flukes and tapeworms, in contrast, are parasites and almost never leave their host except in the form of newly released eggs. Nevertheless, even though there is hardly a fish, amphibian, reptile, bird or mammal without at least one species of fluke or tapeworm of its own to parasitize it (and there are usually several of both), the parasitic platyhelminths still are animals of the water in essence, since the inside of a host's body can be likened to an aquatic environment.

Their gruesome lifestyles do nothing to promote a positive public image, but it is hard not to marvel at the incredible range of platyhelminth adaptations.

FORM AND FUNCTION

Platyhelminths have a relatively simple body plan, omitting both respiratory and circulatory systems, although some species do have a lymphatic system. Many of the free-living species have a pharynx that can be extended out of the mouth to engulf prey or else poked through small gaps to reach food that would otherwise be inaccessible.

Most platyhelminths do not have a gut, but in those species that do it is a sac-like structure with a mouth but no anus, so solid waste has to be regurgitated. In some species this sac-like gut has become very complex with numerous branches [12]. To maintain optimum internal conditions, platyhelminths also eliminate excess water and waste products of metabolism via specialized cells known as flame cells. The tapeworms and some of the flukes have completely dispensed with their gut – instead, they have outer coverings that, at a microscopic level, look like the inside surface of the typical intestines of other animals. The reason for this is that these creatures feed by simply absorbing food, their outer covering becoming their 'gut'.

The flukes and tapeworms have also developed awesome capacities for reproduction; in fact, most of their bodies are devoted to this endeavour. Many tapeworm species have hugely elongated, ribbon-like forms divided into segments known as proglottids, each one an egg-making factory [22]. The bodies of some whale tapeworms (*Hexagonoporus* spp.) comprise as many as 45,000 proglottids, making them among the longest of all animals – sometimes reaching more than 30 m (100 ft). These little reproductive parcels sprout from the tapeworm's 'neck' region throughout its adult life. Each in turn reaches the end of the proglottid queue, detaches, and reaches the outside world, generally via the anus of the host. They assist in their escape from the host's body, partly by irritating the host's gut and partly by squirming and wriggling by applying their own nerves and muscles. Some prefer to regard each proglottid as an individual in its own right and the tapeworm as a colony; each proglottid certainly has its own complete, albeit minimal, set of organs.

Free-living flatworms have a number of adaptations for getting about: cilia on their skins propel them over surfaces and through water, their musculature is well developed and it supplements and steers the ciliary propulsion, and they can contort their bodies at will, extending, contracting, twisting and somersaulting. Species that glide about on the bottom of seas, lakes and streams have various structures with which they resist being dislodged and carried off in water currents. Many have powerful suckers and also adhesive organs aptly named duoglands. The latter secrete adhesives that literally glue the animal down when it wants to cling; they then produce compounds that dissolve the adhesive when the animal wants to relinquish its grip. Some of the free-living flatworms are even able to swim effectively with undulations of their body [1, 13, 27].

In contrast, adult flukes and tapeworms have so little need to move that they have become veritable weaklings. They also do without many organs that are taken for granted in most free-living animals. Nonetheless, depending on their lifestyles, parasitic platyhelminths still may be at risk of being dislodged from their non-stop banquet, so they are equipped with impressive adaptations to help them stay put (in the same way as many other parasites). Tapeworms, for example, are pin-headed creatures, but those heads are equipped with elaborate suckers and hooks for keeping a grip on the slippery wall of their host's digestive tract [9]. The diverse

BOX 1 ***Neodermis***

Adult parasitic flatworms do not have an epidermis in the conventional sense. During their early development the epidermis is shed to be replaced from below by the neodermis, created by deep cells growing and fusing to form a single membrane with lots of nuclei (a syncytium – see Acoelomorpha, p. 119). Although the fused membranes of these cells envelope the animal, their vulnerable nuclei are buried further down in the parasite's tissues.

The neodermis is thought to be key to the success of the flukes and tapeworms; it has no chinks through which a host's acids and proteins can pass unregulated. Even if it is damaged, the important bits of the cell, safely buried out of harm's way, can quickly get to work to make repairs. The neodermis may also protect these parasites from the osmotic challenges they face during their typically complex life cycles, such as when they move from seawater to the inside of a host.

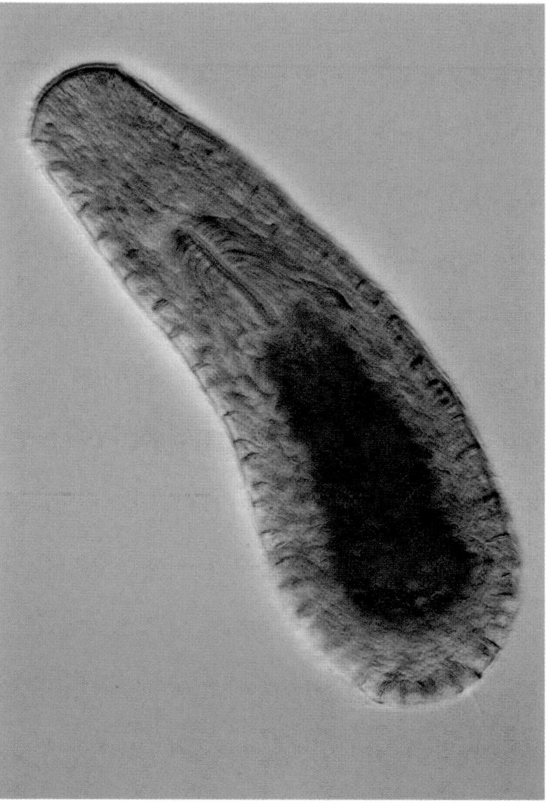

2 This small, free-living flatworm is able to tackle large prey by engulfing them in its capacious gut (brownish structure). The slit-like mouth is visible just off centre (*Macrostomum* sp.).

ectoparasitic flukes live on the gills and skin of vertebrates, principally fishes [18–20], and they too employ similar means of latching on to a host.

Flukes and tapeworms may be surrounded by food, but life inside another animal remains challenging. The host's gut produces flesh-digesting enzymes and acids, and elsewhere in the body visitors are by no means welcome either – so parasites have to withstand the constant, often very sophisticated, onslaught of the host's immune system.

Parasitic platyhelminths also have special characteristics that are less obvious than separate organs like suckers and guts, but just as important; for many of them the first line of defence in their ongoing battle for survival within a host organism is their unique outer covering – the so called neodermis [BOX 1].

There are many forms of nervous system among the various species of platyhelminths. Some are equipped with extremely complex sensory organs, such as eye spots (to detect light [26]) and chemosensory pits (to 'taste' the world around them); consequently they have evolved tiny, but well-defined, brains to process the sensory signals. Large nerve fibres connect the brain to all parts of the body. The organization of the fibres of such species gives the nervous system a ladder-like appearance. This suggests an impressive sophistication of cross-connection of the controls of the two sides of the body. In other flatworms the nervous system is more diffuse, having a net-like appearance.

LIFESTYLE

Flatworms are aquatic animals – the few species found in terrestrial habitats are nocturnal or subterranean, and limited to very humid places [14–17]. They scavenge the remains of dead animals, and may hunt living food (including small worms, arthropods and molluscs) [2, 25]. Having no limbs whatsoever they have evolved some remarkable ways of subduing and killing their prey. Many flatworms simply wrap themselves around their quarry, while others use their adhesive organs and copious amounts of slime to pin it down. Some secrete toxic mucus and there even a few flatworms that stab their prey to death with a pointy penis that they poke out of their mouth. Still other species have an eversible proboscis that either is very sticky or has fearsome hooks to secure a victim. Once it has been subdued, digestive enzymes eat a hole in its body. The flatworm then can poke its eversible pharynx into the hole, secreting further enzymes to digest the creature from within, producing a mush that is easy to suck up. Thin

BOX 2 ***Flukey life cycles***

Adult parasitic platyhelminths are little more than bags of reproductive tissue, producing huge numbers of eggs in order to offset the low odds of any one offspring negotiating the pitfalls of their often extremely elaborate life cycles.

The liver fluke, *Dicrocoelium dendriticum*, requires a snail, an ant and a grazing mammal to complete its development. A snail inadvertently ingests the eggs deposited in the faeces of grazing animals; they hatch and the resultant juveniles eventually end up in the snail's lung, where they cause irritation and are enveloped with thick mucus to form slime-balls that are expelled, sometimes forcibly, from the snail's breathing hole. Ants are very partial to these slime-balls and in eating them they get infected by the juvenile flukes, one or two of which encyst near their brains to take control of the host.

During the night the ants cling to blades of grass using their mandibles while their nest-mates have retreated underground. By making the ant do this, the parasite is increasing its chances of being eaten by a grazing mammal, the definitive host, during usual feeding time in the evening and early hours of the morning. Amazingly, if the ant survives the night the parasite relaxes its control and the host resumes normal activities until darkness falls again. This prevents the ant from being baked in the midday sun whilst clinging to a blade of grass.

The bird fluke, *Leucochloridium paradoxum*, increases the chance of its intermediate snail host falling prey to the definitive bird host by altering the snail's behaviour and appearance. When this parasite is ready to seek its definitive host the parts of its body in the snail's tentacles swell, become brightly coloured and pulsate rapidly.

Normally, snails are retiring creatures that prefer to hide during the day; however, the parasite manipulates its behaviour so that it wanders into the open. Standing out like a beacon, its hugely distended tentacles looking like succulent caterpillars, the snail soon catches the beady eye of a bird. But in eating this molluscan morsel, the bird inadvertently becomes the fluke's new home. The fluke finds a comfortable spot in the bird's gut, reaches maturity and starts producing eggs. These eggs are passed out in the bird's faeces and a small number of them will be lucky enough to be ingested by a snail, thus completing the life cycle.

These are just two examples of the myriad and extraordinary interactions flukes have with other animals.

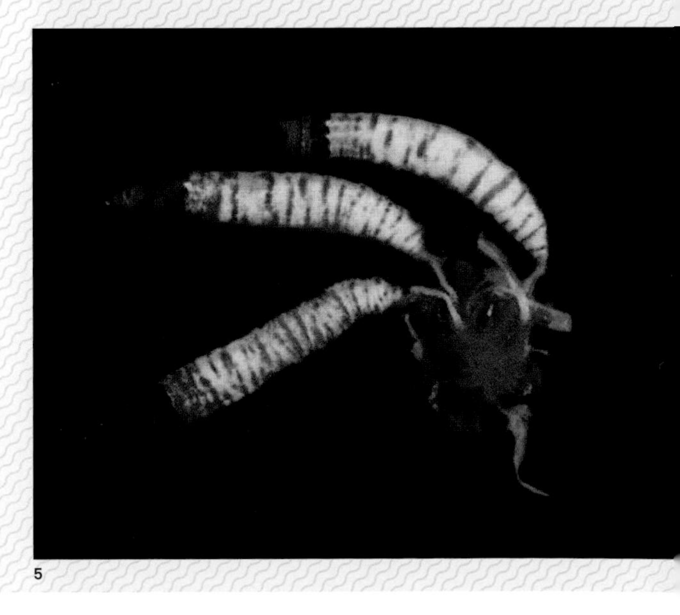

3 An adult of the liver fluke, *Dicrocoelium dendriticum*. The red/orange branches are the gonads, while the black branching structures produce yolk cells for the eggs. The circular structures at the head end are suckers that keep them attached to the intestinal lining of the host.

4 This stage in the life cycle of the bird fluke (*Leucochloridium paradoxum*), known as the sporocyst, turns the intermediate host – an unfortunate snail – into a beacon to attract a feeding bird.

5 Sporocysts of the bird fluke dissected out of the snail host. The brood sacs (white, grub-like structures) are packed with the next stage of the life cycle that will infect a bird host.

pickings often subject flatworms to long bouts of starvation, but these adaptable animals can digest their own insides to survive until prey is more numerous.

In contrast, and as already mentioned, flukes and tapeworms are surrounded by food; they lack many of the adaptations for feeding seen in their free-living flatworm relatives. Primarily, the flukes rely on their muscular mouth and pharynx to suck up cells, cell fragments, mucus, tissue fluids and blood, while the highly specialized tapeworms and blood flukes (such as the schistosomes [23]) absorb all their nutrients directly through the neodermis.

Many of the free-living flatworms, especially freshwater species, can reproduce asexually by wrenching themselves in two and growing their missing halves to form two new individuals. Other species grow into a chain of identical individuals that eventually split apart and go their own way. The powers of regeneration of some species of flatworm are such that a suitable sliver, a mere one three-hundredth of the animal's original size, can completely rebuild a new individual.

Most platyhelminths also reproduce sexually and the majority of species with a sexual reproductive capability have both male and female reproductive organs [11, 23]. Cross-fertilization is the rule, although individuals of many species, especially among the flukes and tapeworms, are known to fertilize their own eggs with their own sperm. In some flatworms, the exchange of sperm is a rather brutal affair, with the penis (known as a cirrus) used to puncture the body wall of the copulating partner [1]. In the bizarre fish fluke, *Diplozoon paradoxum,* a life-long partner for the exchange of sperm is guaranteed when two juveniles seamlessly fuse their bodies together [8]. In the tapeworms each proglottid is a tiny, self-contained egg-making unit equipped with male and female reproductive organs, allowing self-fertilization within and between proglottids. However, when there's more than one tapeworm of the same species in the same host, release of sperm permits cross-fertilization.

Tapeworms produce millions of eggs in their lifetime. Flukes are similarly fecund, but instead of producing eggs directly, they have evolved an incredible adaptation whereby each egg develops into an embryo that goes through a number of generations, each of which spawns further embryos. The end result is a huge number of embryos from a single egg. The fecundity of these animals says a lot about the exceptionally low chance of a fluke or tapeworm egg successfully reaching adulthood.

In the forms of the flukes and tapeworms, the platyhelminths include some of the most elaborate and specialized parasitic life cycles known. The flukes in particular go through a complex sequence of life stages, from a free-swimming existence to their first ('intermediate') host, where they do a lot of their growing. From this stepping stone they must infect the animal in which they will mature and reproduce (the 'definitive' host). Some flukes have more than one intermediate host, further complicating the journey to adulthood. All manner of animals are used as intermediate hosts, including molluscs, arthropods, annelids and fish. The definitive hosts are normally larger predators such as mammals, birds and bigger fish. These complex life cycles are difficult to study, and we have only scratched the surface of the remarkable ways in which these parasites make use of other animals to complete their development.

Fluke life cycles are so varied that generalizing is very difficult, but they begin with fertilized eggs from an adult finding their way into the environment in the faeces, urine or mucus of the definitive host. Tiny, ciliated larvae hatch from these eggs and these find their way into an intermediate host by being inadvertently eaten or by digesting a hole in its skin. In the intermediate host the eggs go through a number of developmental stages, ultimately giving rise to a huge number of juveniles whose job it is to infect a definitive host [7]. This is no mean feat and many flukes have evolved some incredible tricks to reach their goal [BOX 2].

Free-living flatworms are important predators in aquatic ecosystems and they in turn provide food for other animals higher up the food chain [6, 10, 24–27]. However, their ecological importance pales into insignificance next to their parasitic relatives. All large animals (see Craniata, pp. 84–97) play host to at least one fluke or tapeworm species; consequently their ecological significance must be considerable, but this is something which is still very poorly understood. For example, by subtly manipulating host behaviour to get from one animal to another, parasitic platyhelminths must play a role in driving food chains as well influencing the way in which animal species interact with one another.

Even more poorly understood is the role these parasites play in the ecosystem that

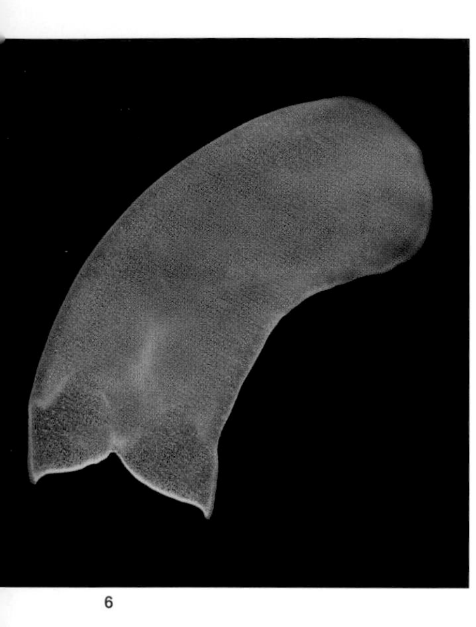

6

6 Most species of free-living platyhelminths are small, rarely exceeding 1 cm (0.4 in.) in length. They glide around on the seabed or in amongst the sediment searching for prey (unidentified flatworm).

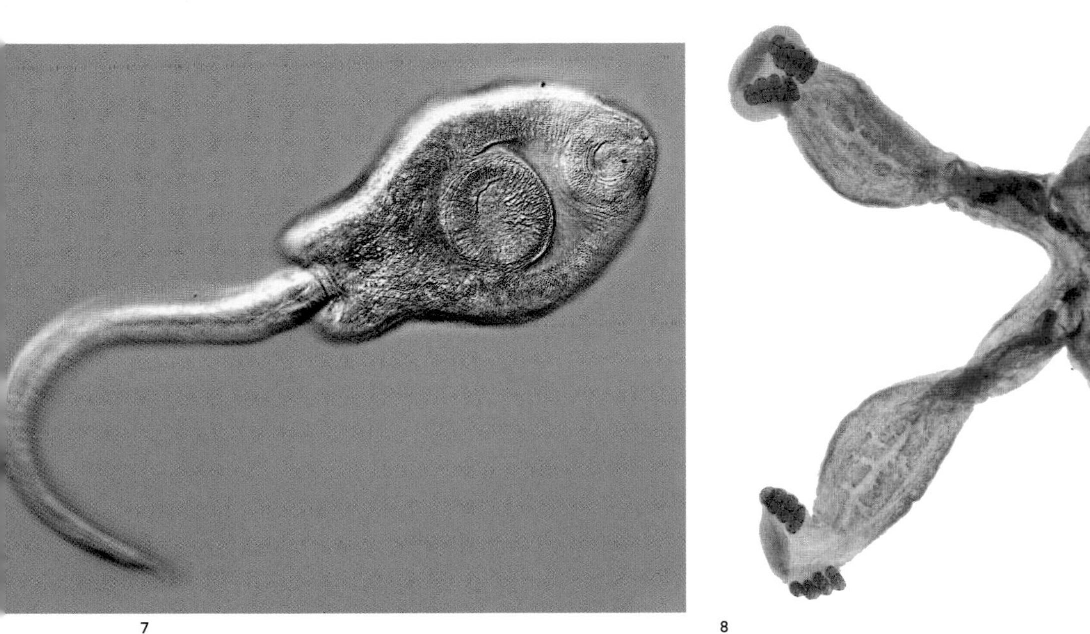

7

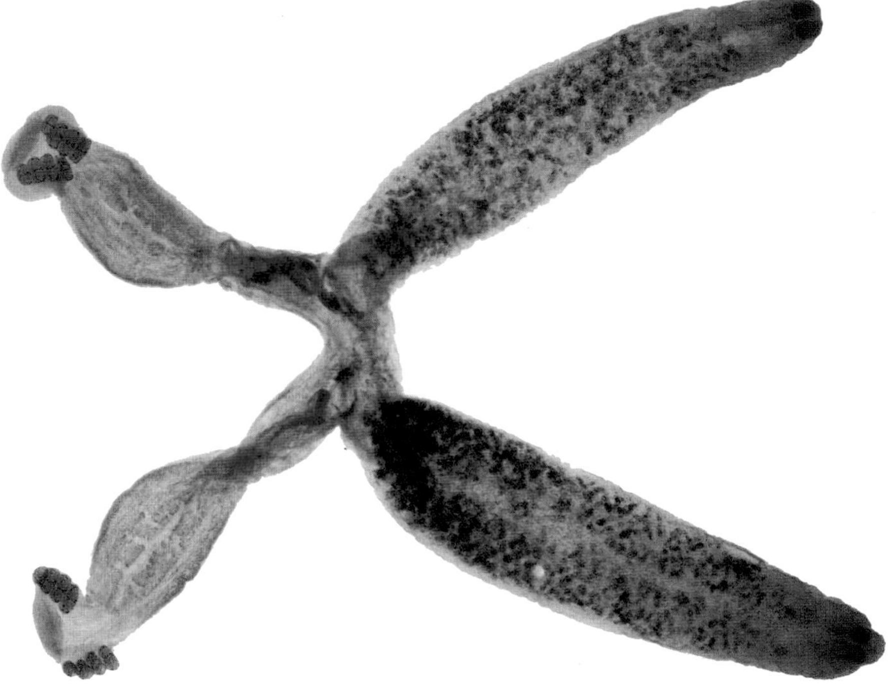

8

includes the host and all the other organisms that dwell in or on its body. In providing niches for other organisms over hundreds of millions of years, an animal is far more than simply the sum of its own DNA; rather it is a community of intimately connected, co-evolved organisms. Flukes and tapeworms are integral parts of these communities and we are only beginning to grasp the importance of these creatures in the proper functioning of their hosts.

It is true that some parasites can cause serious disease in their hosts, and for this reason we humans have tried our level best to eradicate them from ourselves and from the animals we have domesticated. But this may have been a little too hasty, since it seems that the elimination of certain key species has a destabilizing effect, precipitating dysfunctions in the immune system, the intestinal lining and perhaps even other organs of humans and domesticated animals. In terms of human and domesticated animal health, we still don't know what the implications are of living without various parasitic 'worms' (both platyhelminths and nematodes – see pp. 126–33). But there is growing evidence that some refractory immune and autoimmune conditions seem to respond well to carefully managed infections with suitable parasites.

ORIGINS AND AFFINITIES

Platyhelminths are soft, thin and typically small, so it's not much of a surprise that the fossil record reveals little of their evolutionary history. For conclusive evidence we must await increasingly elaborate DNA studies. It seems likely that flukes and tapeworms evolved from flatworms. This is not too hard to imagine, since many flatworms are small, flattened creatures that spend their time sliding between grains of sediment in aquatic habitats or on the surfaces of larger animals, and because larger creatures also frequently swallow them. These are essentially preadaptations for a parasitic existence.

As platyhelminths lack a through gut and body cavity (coelom), they were historically considered to represent the common ancestor of all the bilaterally symmetrical animals (see Introduction, pp. 10–25), but recent evidence suggests that they may have actually evolved from a more complex ancestor, secondarily losing the through gut and the coelom.

The platyhelminths have certainly been around for a very long time. The ubiquity and host specificity of the flukes and tapeworms, for example, suggests the forebears of these animals were well established as parasites long before their hosts underwent the diversification that saw them colonize all the habitats on earth.

7 The life cycles of parasitic platyhelminths are enormously complex and consist of several different stages. This stage in the life cycle of a fluke, the cercaria, infects the definitive host (normally a vertebrate) and develops into the adult parasite. Equipped with suckers, it gets into the host either by being inadvertently eaten, by penetrating the skin of the host, or turning itself into a kind of living syringe (unidentified species).

8 Two *Diplozoon paradoxum* individuals fused together.

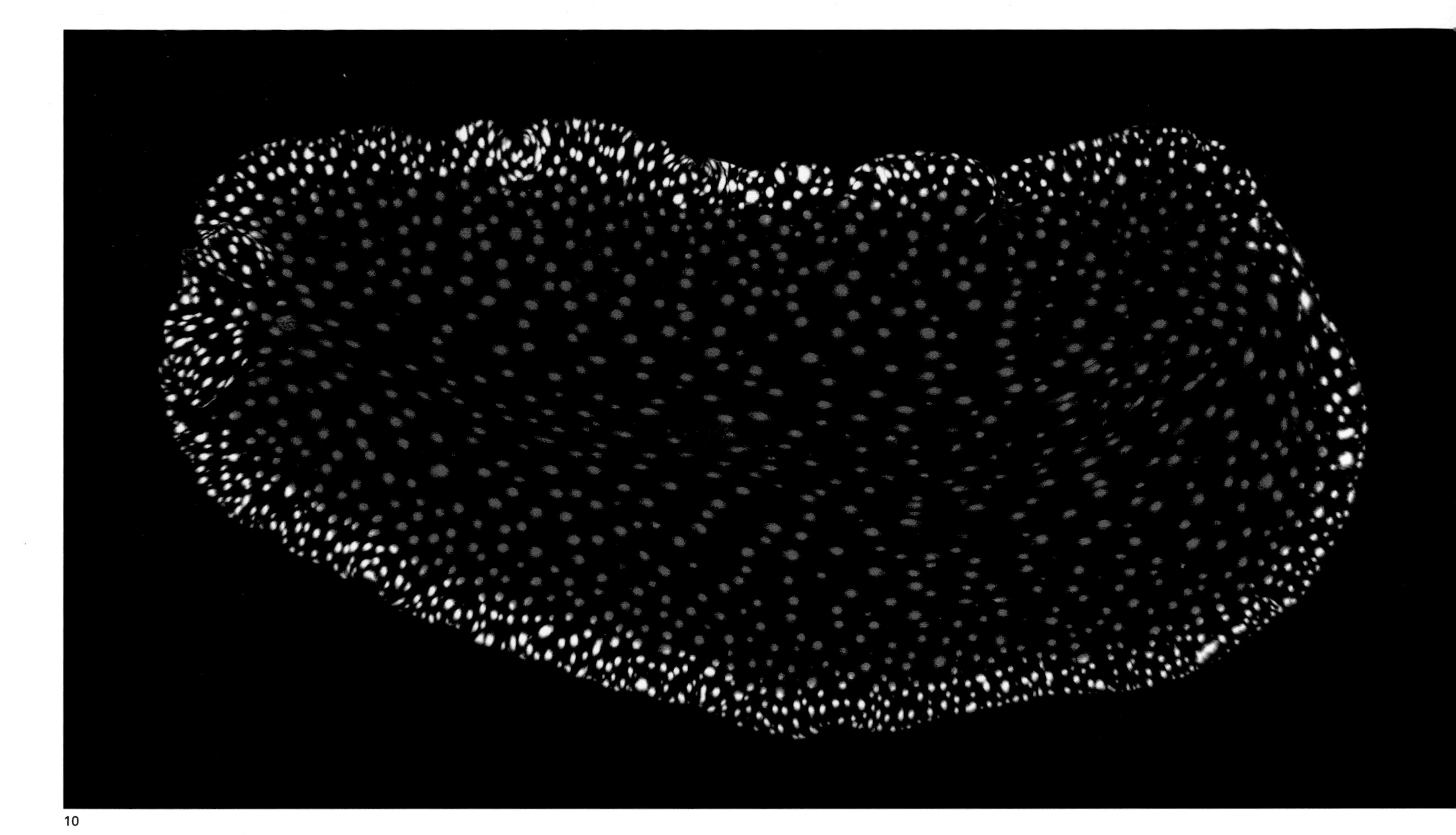

10

‹9 Tapeworms have an elaborate structure (the scolex) at their anterior end that anchors them to the intestinal lining of their host. The top two individuals (both unidentified species) were found in the intestines of sharks; bottom left is the scolex of *Platybothrium auriculatum*, found in the intestine of a blue shark (*Prionace glauca*); the bottom right example came from the intestine of a bony fish (unidentified species).

10 Some of the marine flatworms are very boldly coloured, either to advertise chemical defences or to mimic other well-defended marine animals, such as sea slugs (*Pseudoceros* sp.).

11 A small, free-living flatworm (*Gyratrix* sp.). Like many of its kind it is hermaphrodite, and in this specimen you can make out the copulatory stylet (dark, narrow structure at posterior), used to introduce sperm into a partner and the egg capsule (dark brown ovoid). Much of the body is taken up by yolk glands, running alongside the food-packed gut (green). At the anterior end are two eye spots and a sticky proboscis that can be everted to snag prey.

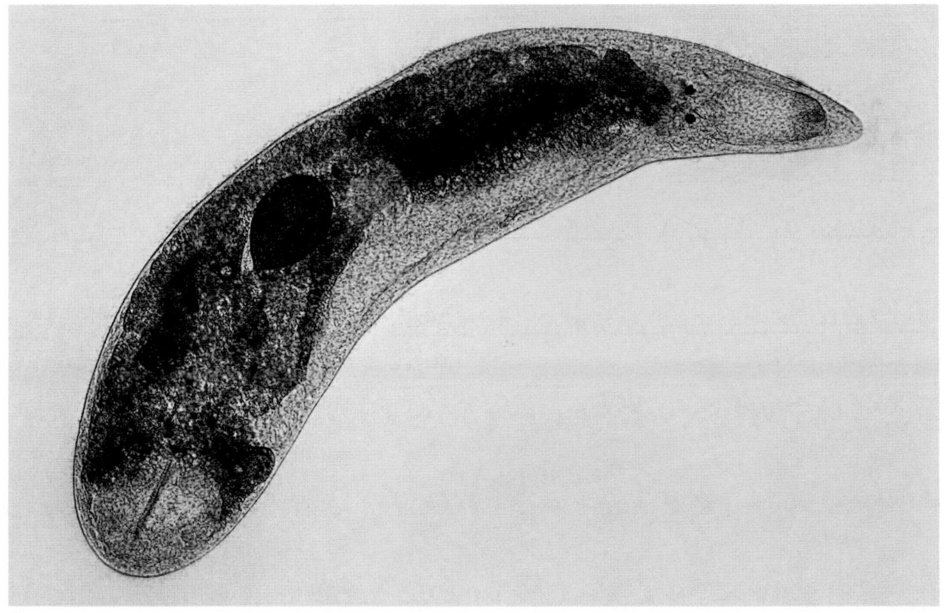

11

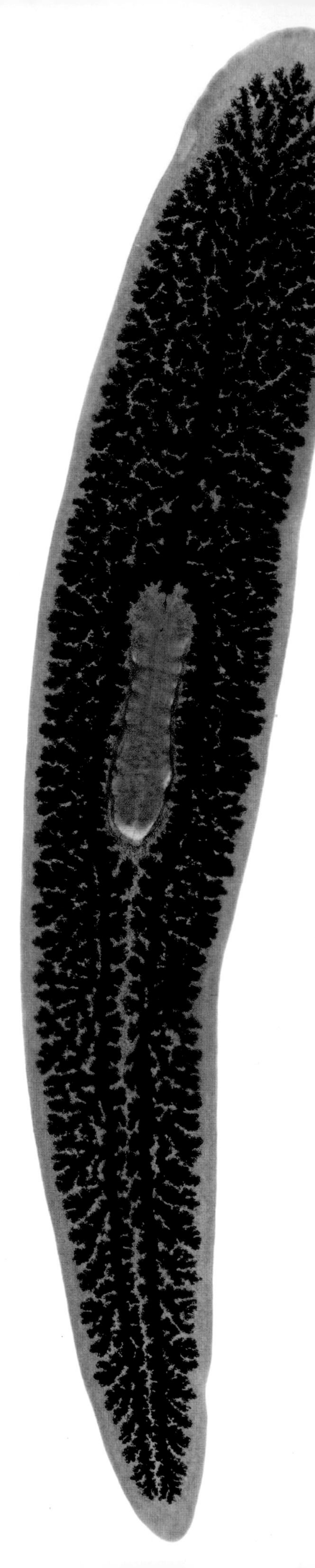

12 Around a quarter of platyhelminth species are free-living flatworms that glide around on the substrate using cilia and mucus. In many of these the gut is an elaborate branching structure (black in this image) (*Dugesia* sp.).

13 A marine, free-living flatworm swimming (unidentified species).

14 Some of the platyhelminths are found on land, but they are restricted to moist habitats (*Bipalium* sp.).

15 Brightly coloured terrestrial flatworm (unidentified geoplanid).

16 The land-dwelling flatworms, like their aquatic relatives, are active predators and scavengers. Using their cilia they patrol the leaf litter and the vegetation in search of prey (unidentified geoplanid).

17 A terrestrial flatworm (unidentified species).

13

14

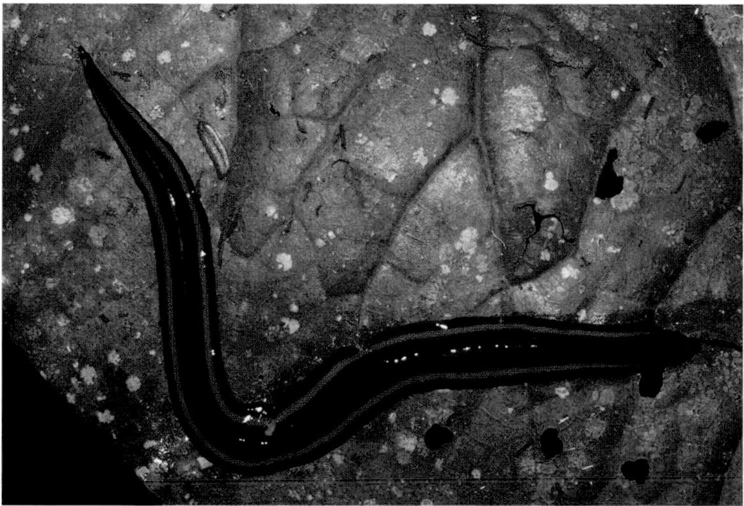

15

16

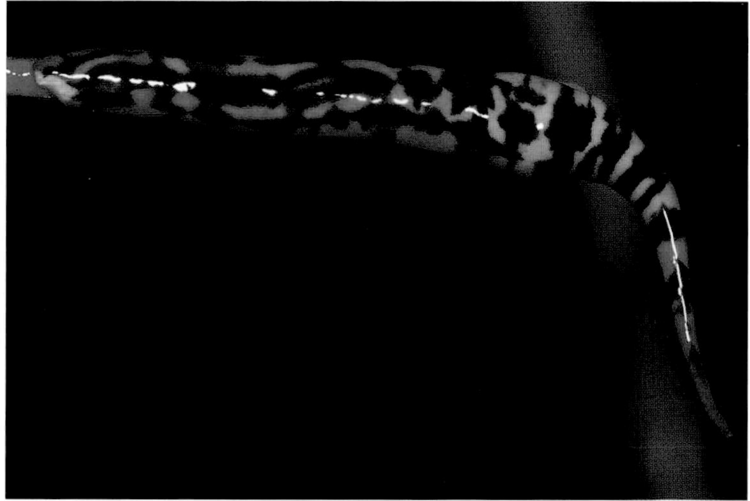

17

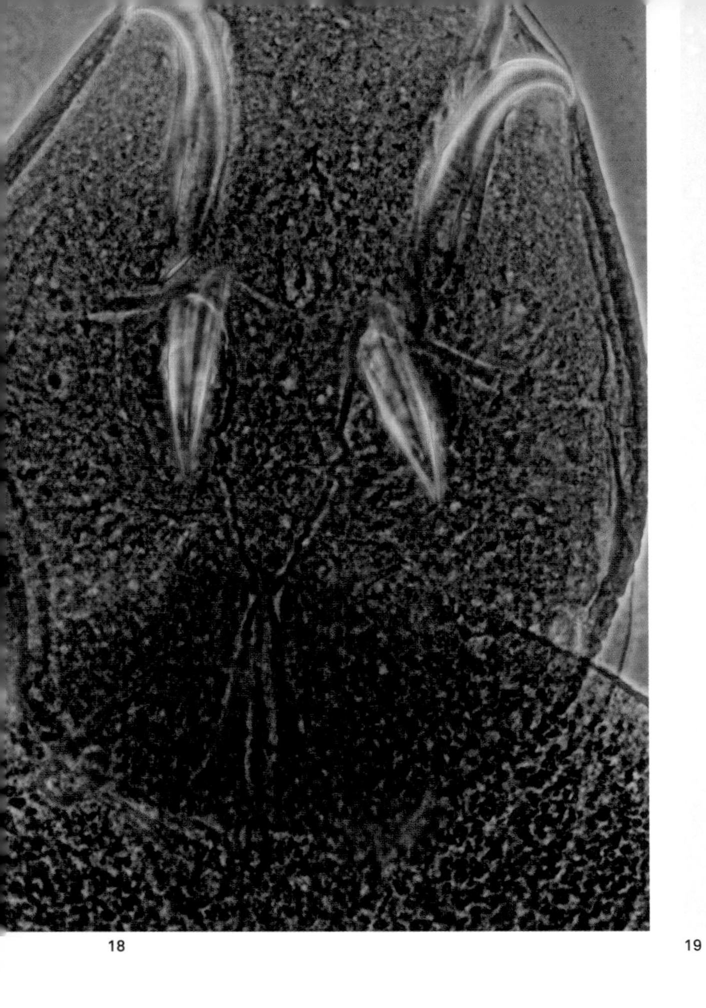

18

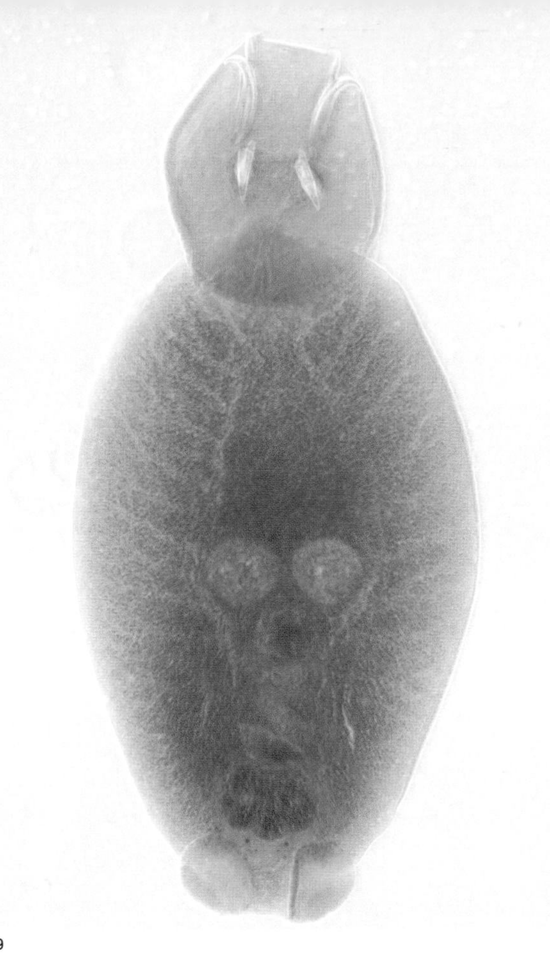

19

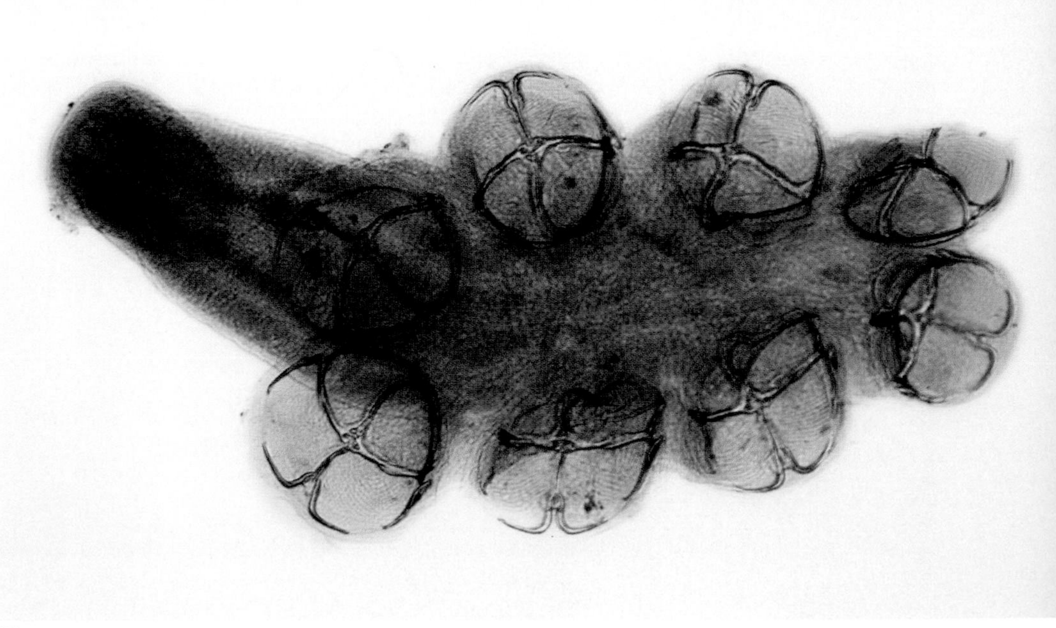

20

18 Many species of fluke live on the outside of their hosts, typically fish, but also amphibians and reptiles. This species, *Benedenia seriolae*, clings on to the skin of bony fish using suckers and a brutal, hook-bearing attachment organ (haptor) at its posterior end.

19 Close up of the attachment organ of *Benedenia seriolae*.

20 This fluke adheres to the gills and mouth cavity of marine fish using its eight large suckers. Related species have clamps to cling onto the host's gill filaments (*Eurysorchis* sp.).

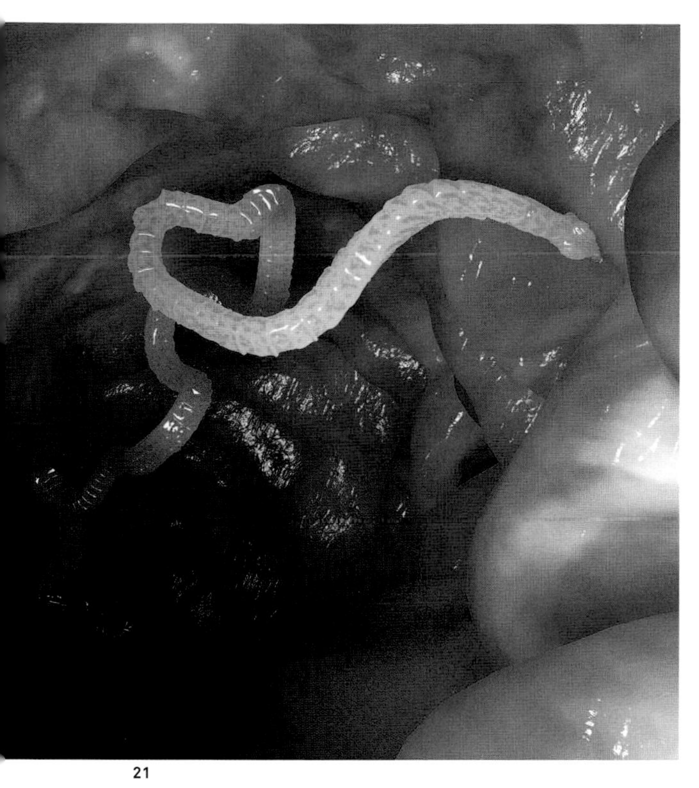

21

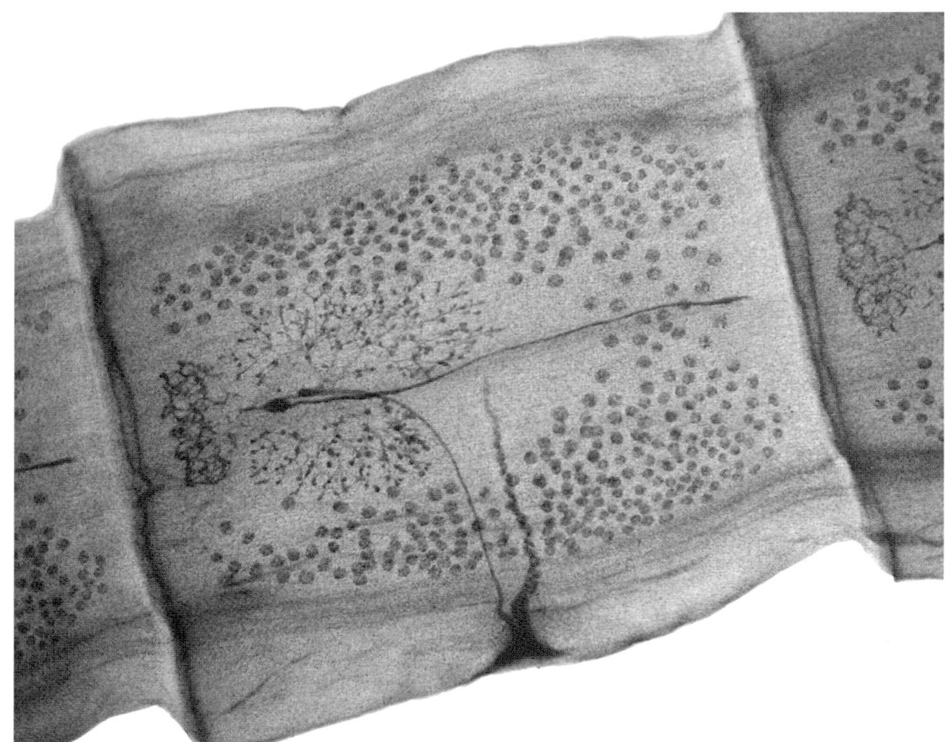

22

21 An adult tapeworm inside a human intestine. Note the scolex at the head end and the long chain of proglottids (illustration based on the pork tapeworm, *Taenia solium*).

22 Much of a tapeworm's body consists of a chain of egg-making factories, known as proglottids. Each unit usually contains a complete male and female reproductive system. In this *Taenia serrata* proglottid, you can see the ovary (branching structure); testes (dark dots); uterus (central band running almost the length of the proglottid); vagina (curving dark line) and the sperm duct (dark, tapering structure diverging from the vagina).

23 Like so many other flukes, schistosomes have a very elaborate life cycle, involving fresh-water snails and ultimately larger animals, such as birds and mammals, including humans. These blood flukes are unusual because there are separate sexes. Here, a male and female are mating: the relatively thin female is 'embraced' in a ventral groove on the broader male (*Schistosoma* sp.).

23

24

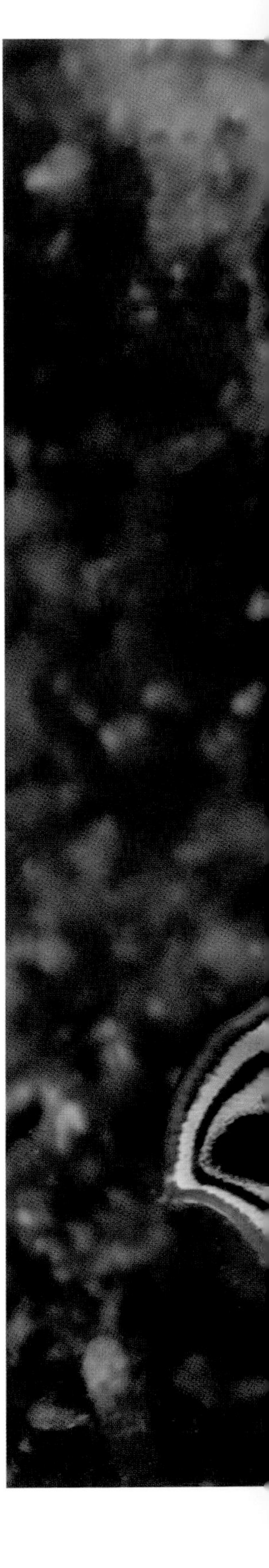

24 A marine, free-living flatworm (unidentified prosthiostomid).
25 The free-living flatworms are important predators and scavengers of marine and freshwater benthic habitats (*Prostheceraeus bellostriatus*).

25

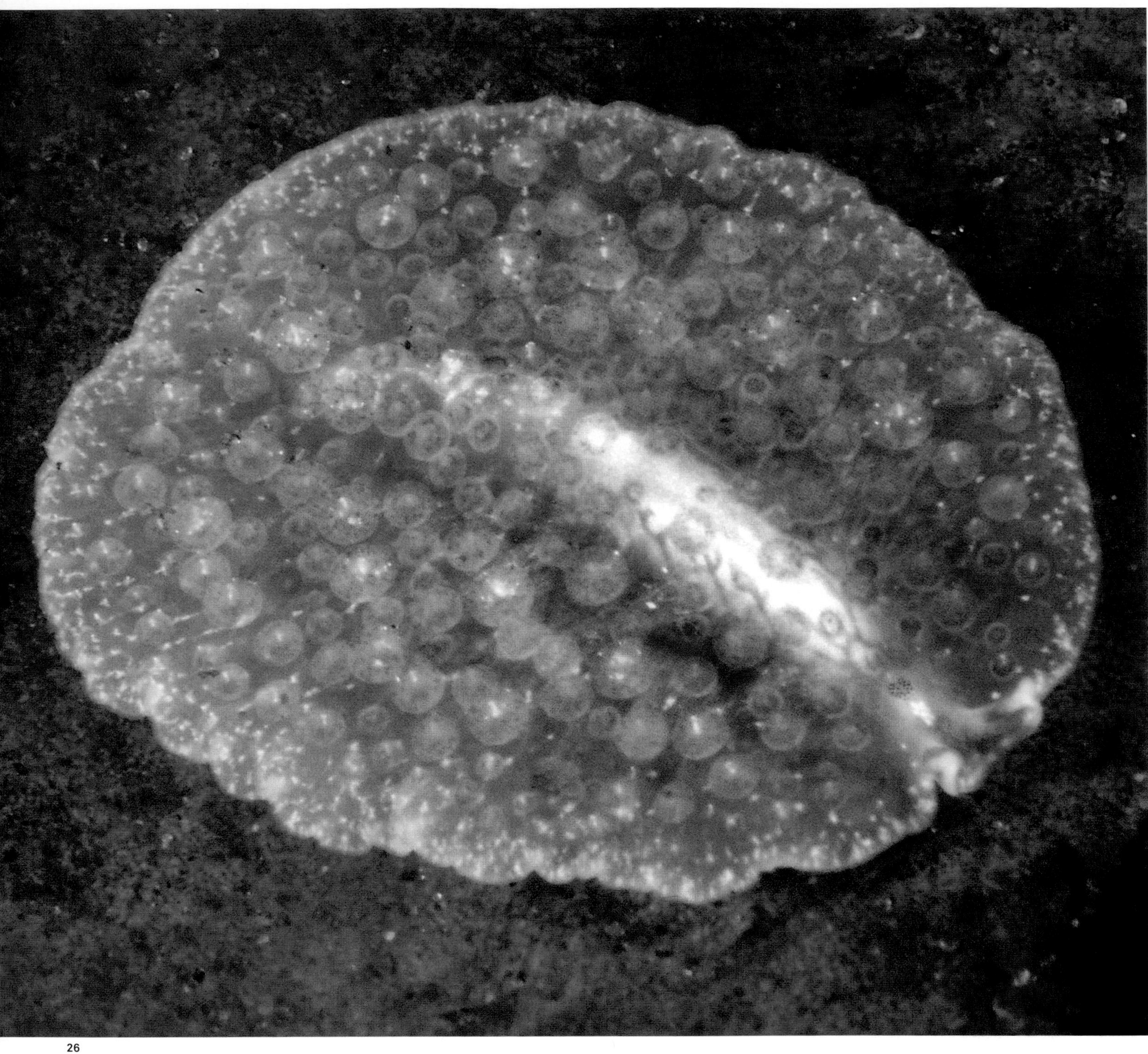

26

26 A cluster of black eye spots is visible at the anterior end of this flatworm (unidentified species).

› **27** Some of the free-living flatworms can swim through the water by undulating their extremely thin body (unidentified species).

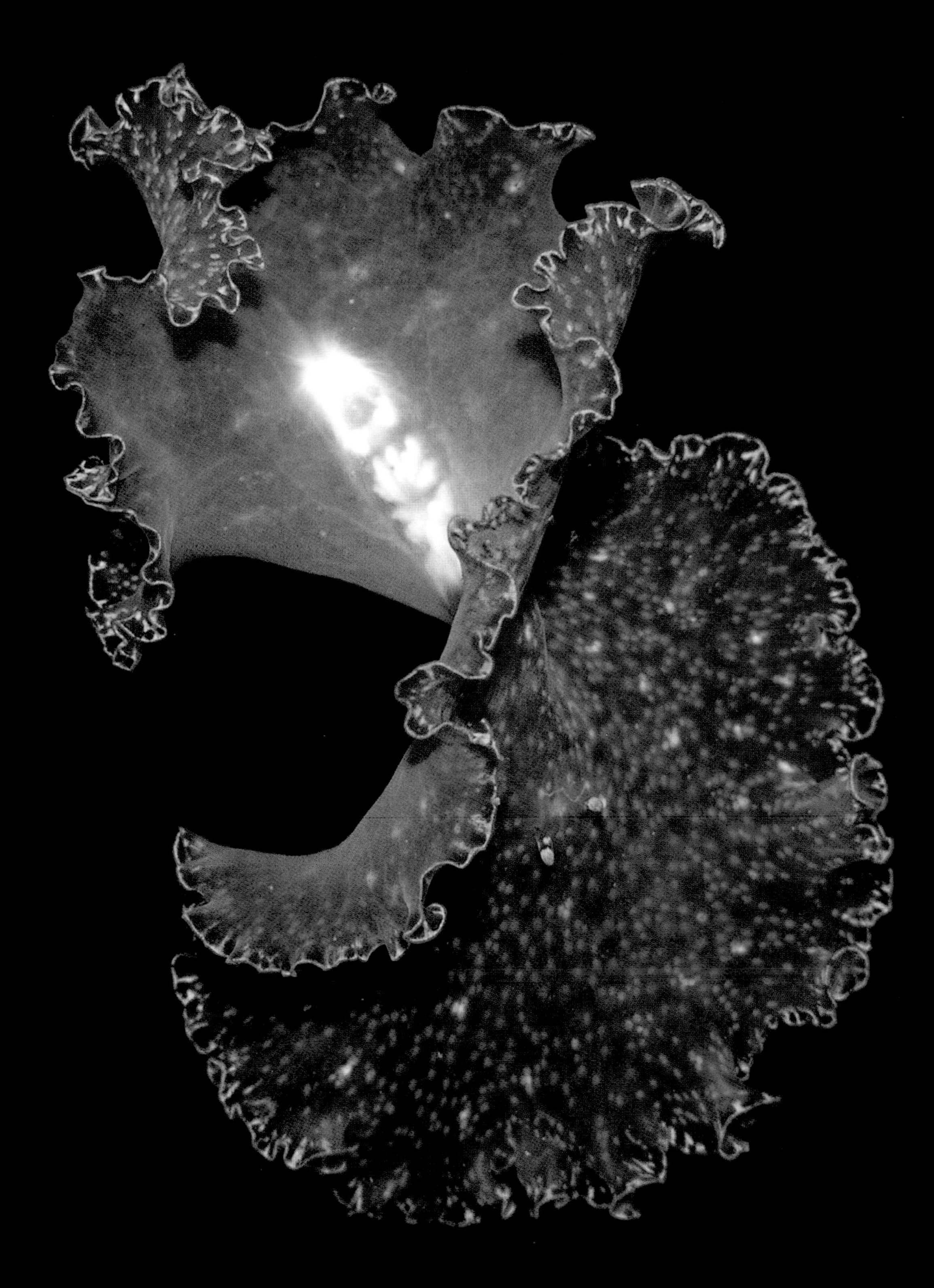

Gnathostomulida

(jaw worms)
(Greek *gnathos* = jaw;
stoma = mouth)

Diversity
c. 110 species

Size range
0.5 to 4 mm
(0.02 to 0.16 in.)

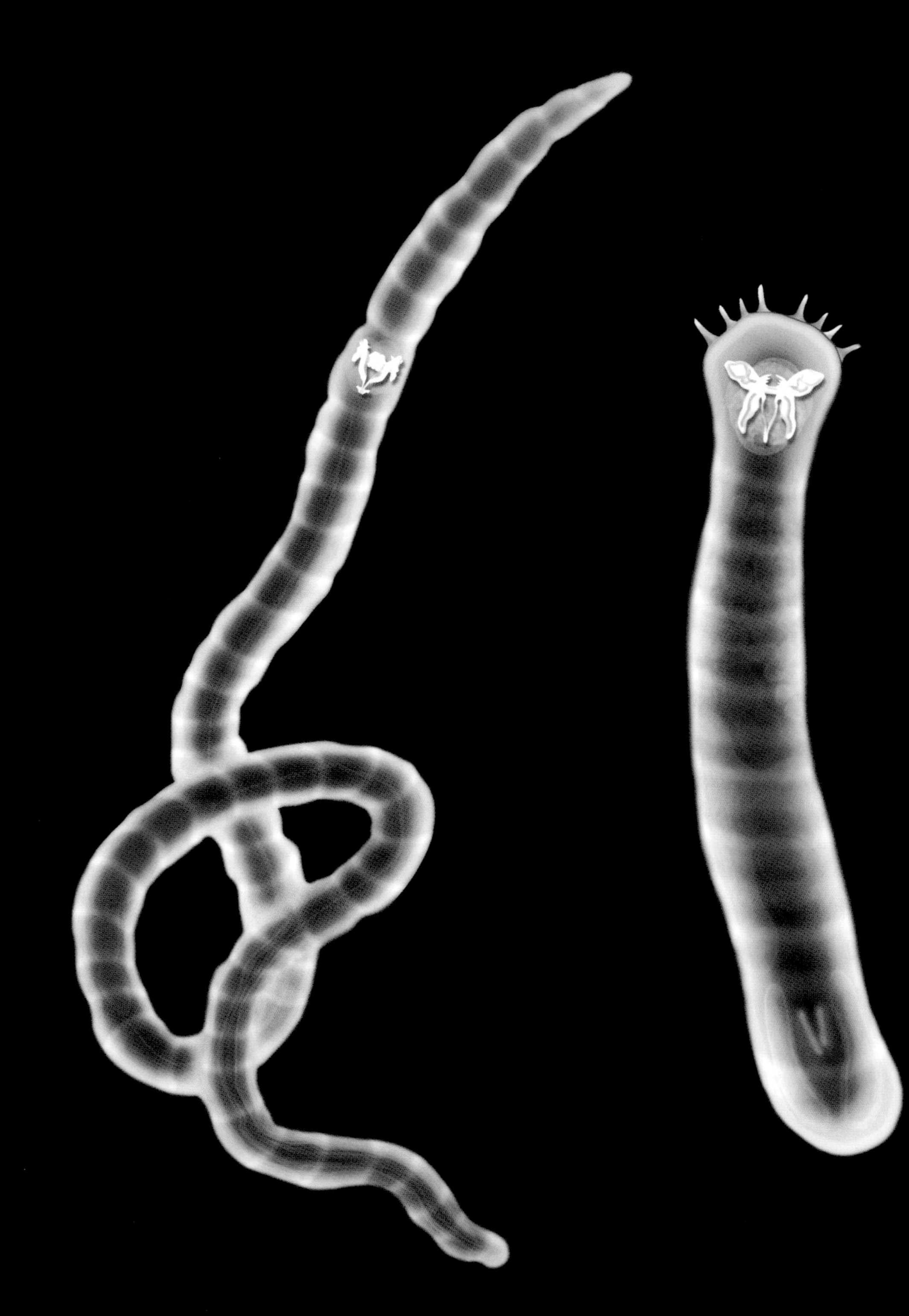

1 The majority of gnathostomulids are stout, worm-like creatures, while a smaller number of species are longer and thinner. They are ubiquitous and sometimes very abundant in marine sediments.

Like the gastrotrichs, the gnathostomulids or jaw worms are surprisingly common animals living in aquatic sediments, but they are exclusively marine. They have been found all over the world, sometimes in incredible densities, but it will probably come as no surprise that these animals have received little attention and are very poorly understood.

FORM AND FUNCTION

Jaw worms are elongate animals and with an average thickness of 0.045 to 0.065 mm (0.0018 to 0.0026 in.), they are pretty thin [1,4,7]; many, in fact, are decidedly thread-like [1,5]. A slight constriction of the body delineates the head and the trunk. There is no cuticle, but some of the epidermal cells secrete mucus affording a degree of protection from the outside world. Each of the epidermal cells also bears a single cilium. Beneath this outer covering and its supporting layer there is a sheath of weak muscles. Further inside the body there is a gut with a mouth but no anus (some species have an anal pore), kidneys, a brain and all the apparatus associated with reproduction including large male and female gonads, a vagina on their back and a penis on their underside. There is no sign of a central body cavity and nothing in the way of a specialized respiratory or circulatory system – these animals are so small they have no need of such extravagances. On the head there are a range of sensory pits and conspicuous, whisker-like cilia known as sensoria, which are probably very sensitive to touch.

A characteristic feature of these tiny beasts is their elaborate pharynx, which includes a set of very complex jaws [2,3,6]. The synchronized movements of the muscles, membranes and jaws that make up this pharynx are controlled by a dedicated nerve centre, a little cluster of nerve cells (a ganglion) just beneath the front portion of the gut.

LIFESTYLE

The jaw worms are animals of marine sediments, their elongate form allowing them to move freely around. Propelled by the cilia sprouting from their epidermal cells they slowly glide, swim and twist their way through the labyrinthine spaces between the sediment particles, nodding their head from side to side as they go. The relatively weak musculature does not really feature in locomotion apart from shortening the body to recoil away from potential danger. The various sensory structures on the head are important in detecting the chemicals that betray the presence of food, such as bacteria, single-celled fungi and particles of non-descript organic matter, all of which are scraped from the surfaces of sediment particles by the tough jaws. Following digestion, solid waste is regurgitated out of the mouth or else it may be disposed of through a temporary conduit between the gut and the body surface – a transient anus.

All the known gnathostomulid species are hermaphrodite, having male and female gonads at the same time. The sperm come in a variety of shapes and sizes, including thread-like, mushroom-shaped and exceedingly tiny spherical forms. The act of jaw worm mating has never been seen, but in many species one of the copulating individuals is believed to use its stylet stiffened penis to pierce the body of its mate in order to deposit a mucus ball of sperm. Following their brutal entry, the sperm make a bee-line for a storage area near the female gonads and the most mature egg in the ovary is fertilized. This single egg grows until it is enormous in relation to the size of the adult, eventually being released to the outside world in the same way as in the gastrotrichs (see pp. 274–77) – i.e., a temporary hole opens up in the side of the adult and out pops the egg. There is no larval stage, so what hatches from the egg is a juvenile jaw worm ready for a life amid the sediment.

Jaw worms have been found in a wide range of marine sediments and there are undoubtedly many, many more species yet to be discovered and described. It is possible that this lineage contains several thousand species, but we just do not know enough about them to estimate their global diversity. One species appears to have a preference for fine sands, where there is little or no oxygen and high concentrations of the very malodorous hydrogen sulphide. Exactly how it survives in such extreme environments is unknown, but some of the brush-heads (see Loricifera, pp. 182–85) are known to thrive in similar settings by possessing organelles that produce cell fuel in the absence of oxygen. It remains to be seen if the jaw worms do something similar.

In suitable habitats gnathostomulids can be found in staggering population densities, making them easily the dominant representatives of the meiofauna in such locations. Some 6000 individuals per litre (or around 22,000 per gal.) of sediment have been reported and in each cubic centimetre (or 0.06 cu. in.) of sediment from around polychaete

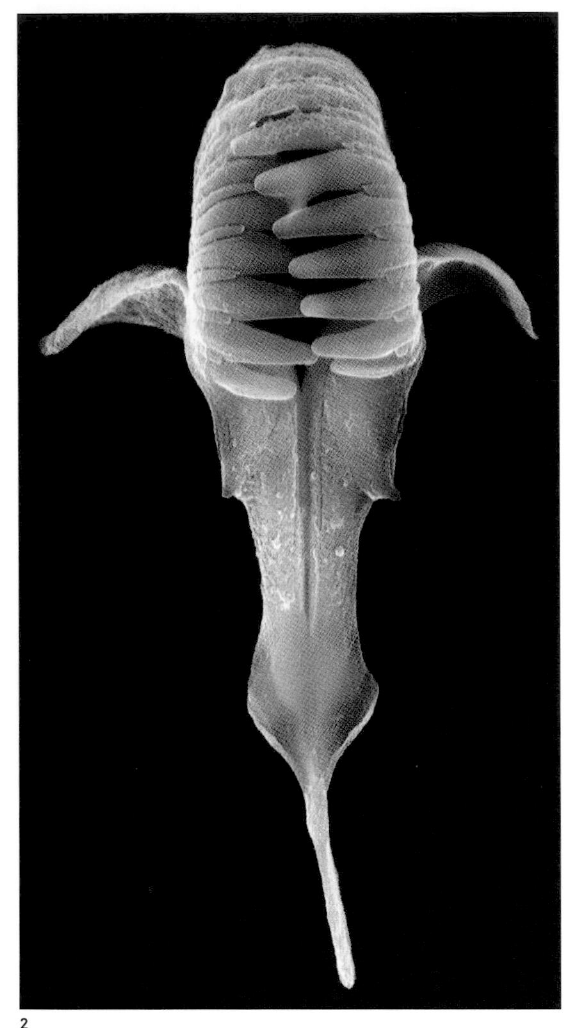

2

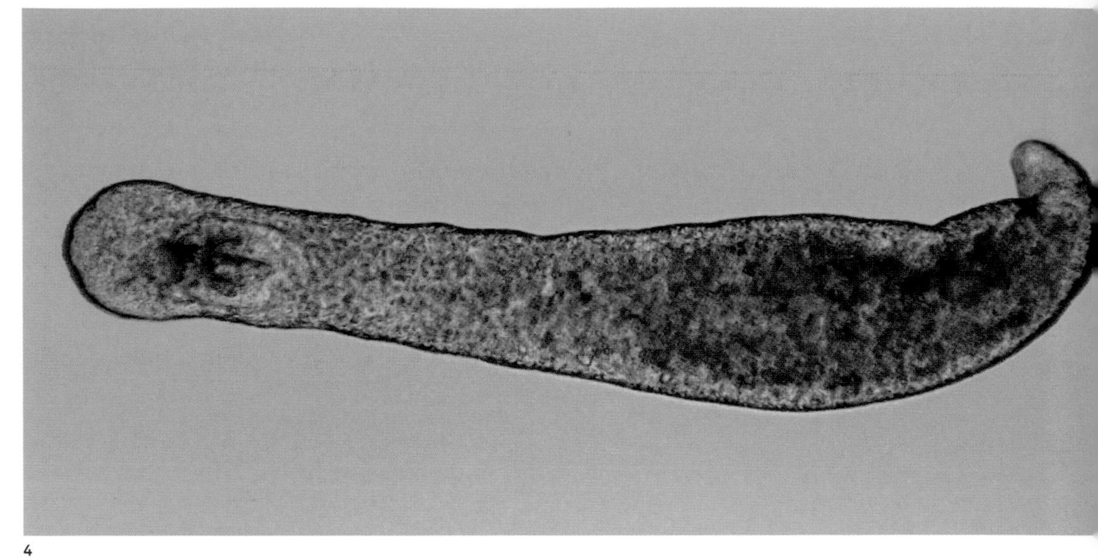

4

2 The soft tissues have been dissolved away from the jaw apparatus of a gnathostomulid leaving the very elaborate hard parts that grind and rend the animal's food (*Rastrognathia macrostoma*).
3 The complex jaw apparatus of *Gnathostomula armata*.

4 Most gnathostomulids are stout, worm-like animals of marine sediments (unidentified species).
5 Some jaw worms have a very slender, thread-like form (*Haplognathia simplex*).

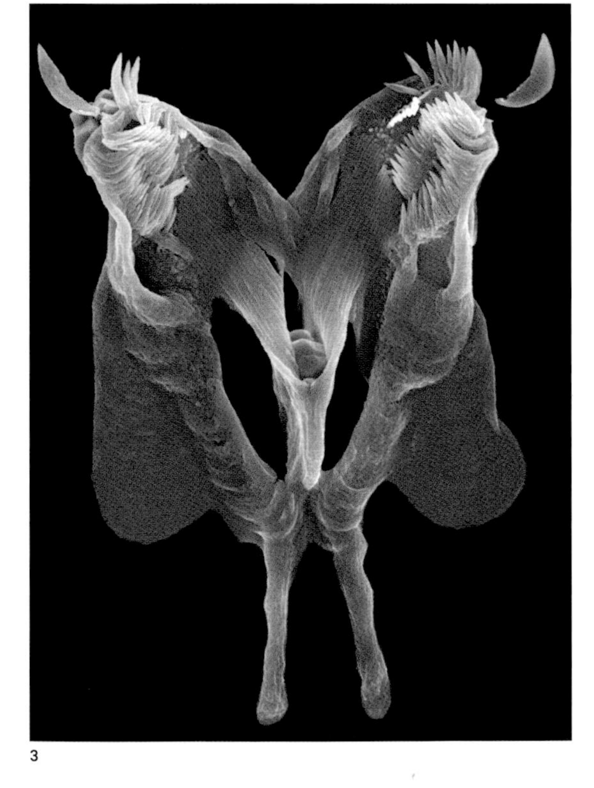

3

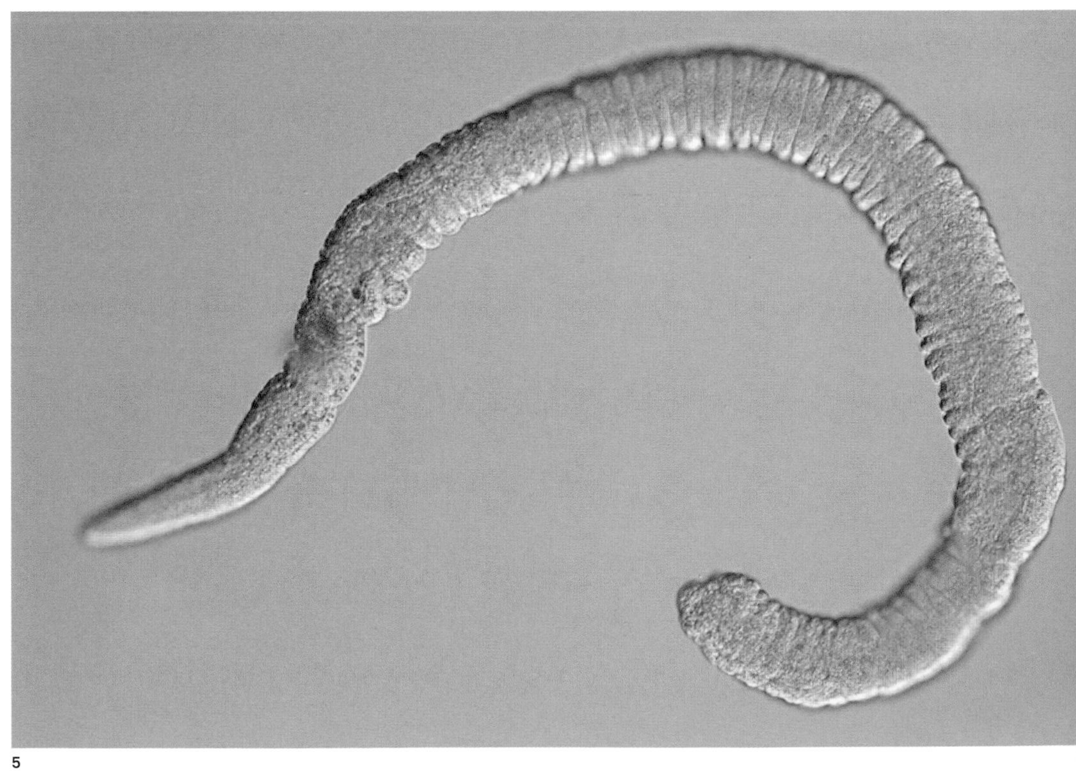

5

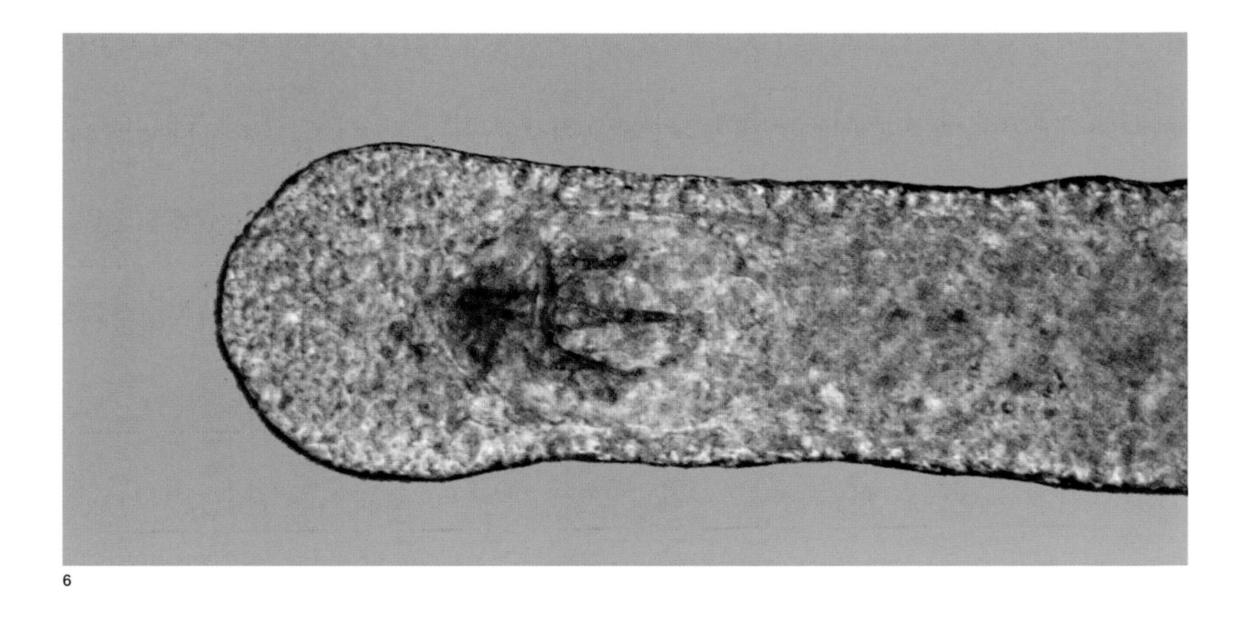

6

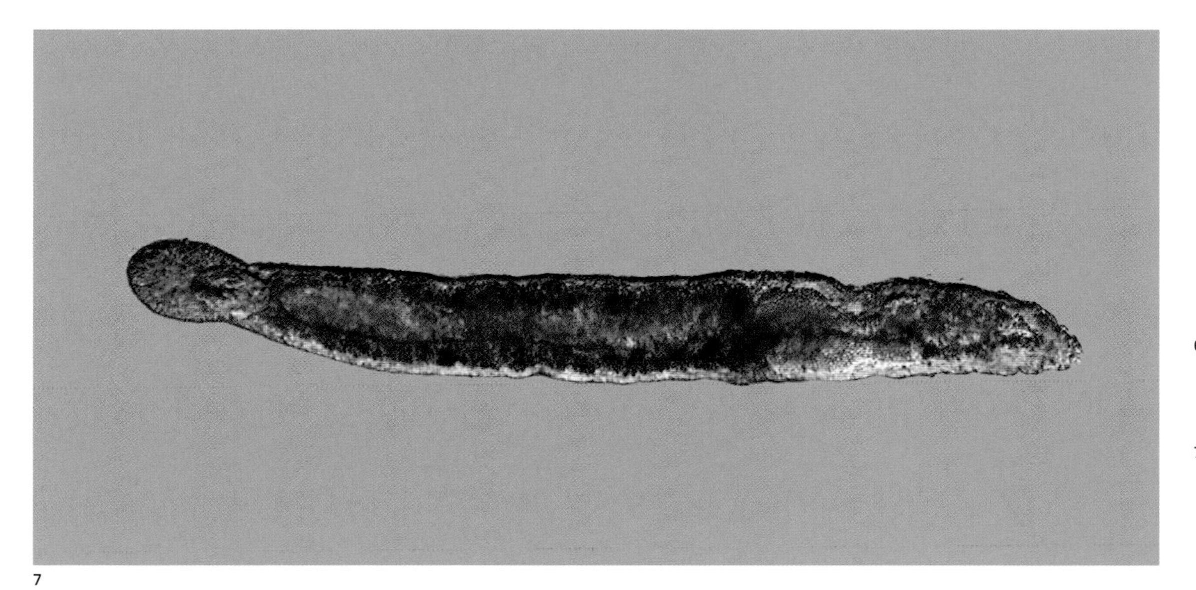

7

6 The complex gnathostomulid jaw apparatus is typically visible through the transparent body wall of the animal (unidentified species).
7 Very little is known about the biology and diversity of jaw worms even though they very easy to find in marine sediments (*Onychognathia rhombocephala*).

burrows, a micro-habitat believed to be very desirable for a range of sediment-dwelling animals, as many as 29 jaw worms, represented by five species, have been observed. Like the other members of the meiofauna, the jaw worms play a significant, but overlooked role in the healthy functioning of marine ecosystems. One only has to imagine how many of them there are in the sediments of the oceans, how much food they collectively consume and how many other organisms are sustained by predating them.

ORIGINS AND AFFINITIES

The jaw worms have nothing in the way of a fossil record. At various times, a handful of specimens have been interpreted as ancient members of the lineage, but these have failed to convince many experts. Working out their affinities to the other animal lineages is also proving to be something of a challenge. Following their discovery in 1956, the jaw worms were considered to be platyhelminths, and it was not until 1969 that they were recognized as a distinct lineage of animals due to their unique characteristics. Since that time they have also been touted as very specialized annelids, based purely on the structure of their complex jaws and their musculature. In the light of emerging evidence, it now seems that the jaw worms are very closely related to the rotifers and micrognathozoans, the latter a very obscure lineage of animals recently discovered in a freshwater spring in Greenland (see pp. 308–10). This little grouping is thought to be an offshoot of the branch that also includes the gastrotrichs and the platyhelminthes.

Rotifera

(wheel animals,
thorny-headed worms)
(Latin *rota* = wheel;
ferre = to bear)

Diversity
c. 2800 species

Size range
~0.04 mm to ~80 cm
(~0.0016 to ~32 in.)

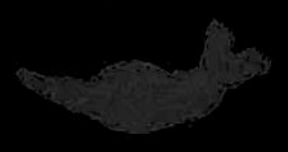

1 Rotifers are tiny, diverse animals. One of their characteristic features is a crown of cilia called a corona. This sessile species secretes a tube in which it lives, and it has perhaps the most elaborate corona of any member of the lineage (*Stephanoceros* sp.).

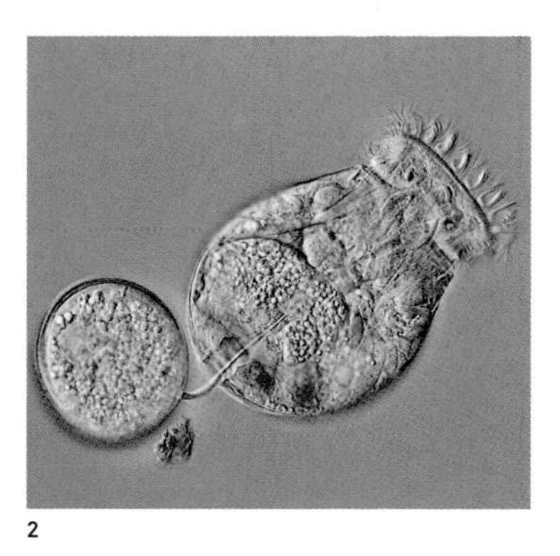

2

Rotifers are incredible little animals: they come in an amazing variety of forms, their sex lives are bewilderingly complex and they can travel through time in a state of suspended animation. They are also extremely common, usually living in fresh water and in the temporary water films enveloping objects on land. If you have a bucket in your garden with some murky looking rainwater in it, it will be a hotbed of rotifer action – all you need to see them is a microscope. If they were larger, they would no doubt generate huge amounts of interest, such are the intricacies of their lives. As it is, they are tiny and for the most part overlooked.

FORM AND FUNCTION

The vast majority of the rotifers are minuscule animals, around 0.1 to 1 mm (0.004 to 0.04 in.) long, but although small they are complex, their bodies made up of around a thousand cells – which is what they are born with and what they die with. There is no cell division, only cell growth, so for this reason they cannot regenerate lost body parts nor can they repair any damage that occurs during their normally short lives. Of their characteristic features perhaps the most conspicuous is a crest of cilia, known as the corona, which is important in feeding and locomotion [2, 4]. An optical illusion makes the corona with its beating cilia look like a rotating wheel or wheels, hence the rotifers' common names of 'wheel bearers' and 'wheel animals'. The body tapers to the rear into what is known as a 'foot', equipped with a number of 'toes' [13]. The latter are normally rather stubby projections through which adhesive substances are dispensed to anchor the animal to the substrate. Some species have adapted to a parasitic way of life, and thus have lost many of the superficial features that characterize this group [BOX 1].

The rotifers have an unusual arrangement of the body wall because the layer that effectively functions like a cuticle by giving the animal some rigidity and providing a surface against which the muscles can act is actually *inside* the epidermal cells. In most other animals it is the epidermis that secretes the cuticle, exoskeleton and so on. In some species this functioning cuticle is very thick and adorned with a variety of spines and surface sculpting [4, 7].

If we descend further into the animal, we find a well-developed musculature and beyond this a fluid-filled body cavity accommodating more muscles, the gut, nervous system, reproductive system and kidneys. Like the jaw worms (see pp. 294–97) and jaw animals (see pp. 308–11), the rotifer pharynx contains an intricate jaw apparatus known as the mastax, which is made up of several separate elements [2, 8]. Rotifers that feed on small particles suspended in the water have plate-like elements in their mastax for grinding food, while predatory species have toothed pincers for grasping and puncturing prey. A few species do not have an anus, so they cannot defecate, but they have evolved an ingenious, albeit unsavoury solution: forceps-like elements of the mastax pluck bits of faeces from the gut, so they can be discarded out of the mouth.

The central nervous system is well developed, with a brain in the head and a couple of mini-brains (ganglia), one serving the pharynx and its complicated mastax and another dedicated to the control of the foot and toes. Depending on the species, a rotifer brain is made up of around 200 neurones. It receives information from the rotifer's senses, including touch sensitive bristles, small pits that detect chemicals and simple eyes [2]. Rotifers also have antennae – small patches of sensory cells and cilia on the head and trunk. Also in the head is a unique retrocerebral organ that may secrete mucus to lubricate the coronal cilia.

LIFESTYLE

Rotifers have evolved a number of ways of getting around. Many of them creep, almost leech style, across the substrate, telescoping their body and temporarily anchoring their rear with adhesive secretions from the toes. Some species can also swim, normally propelled by the beating cilia of their corona. There are even rotifers that propel themselves at velocities of around 175 body lengths per second by quickly flicking what look like small feathers [12]. There are also lots of species that spend most of their life amongst the plankton, using a range of adaptations to stay afloat, including thin, flexible bodies, long spines and oil droplets in the body cavity [11].

Still others have adapted to a completely sessile existence, spending their entire adult life in an elegant, secreted tube, which is often reinforced with particles of sediment and other bits of detritus [1, 9, 10, 13]. In some of these tube-dwellers the anus has migrated to the anterior part of the body, so they can defecate freely without sullying the inside of their lair.

When it comes to securing food, the trend in the rotifers is suspension feeding – currents of water generated by cilia on the corona and around the mouth bring particles of food in the

2 Red eye spots and the mastax (jaw structure beneath eyes) are visible in this rotifer, which is also dragging one of its thick-shelled eggs (*Pompholyx sulcata*).

BOX 1 ***Thorny-headed worms***

The rotifers, thorny-heady worms (acanthocephalans) and the animals known as seisonids were once widely thought to represent distinct lineages, a view that is still held by some zoologists, but there is now mounting evidence that all of them are simply rotifers. The seisonids and thorny-headed worms, in adapting to living on in the bodies of other animals, have lost, to varying degrees, the typical outward rotifer features, but they are still rotifers, just very specialized ones.

Adult thorny-headed worms live inside the intestines of large animals, most commonly bony fish and birds, but also cartilaginous fish, amphibians, reptiles and mammals. Although not as common as tapeworms and some nematodes (animals that they have come to resemble as a consequence of adapting to the same niche), they are still consummate parasites with complex life cycles always requiring at least two hosts. Most are around 1–2 cm (0.4–0.8 in.) long, but some, such as the giant thorny-headed worm that lives in pigs, can be 80 cm (32 in.) long.

In adapting to a parasitic way of life their gut has all but disappeared, since they absorb all the nutrients they need from the fluids of their host, which are then distributed around the body by a network of vessels known as the lacunar system. They grip on to the intestinal lining using a short, retractable proboscis bristling with recurved spines, a unique feature (and the inspiration for their name).

Thorny-headed worms can be male or female. To inseminate the female the male squeezes out part of his reproductive tract to form a cup-like structure that engulfs his mate's posterior end, allowing the penis to do its work. With copulation complete, the male's exposed reproductive structures retract and leave a little mucus cap over the entrance to the female's reproductive tract in a bid to slow down other amorous males. Once fertilized, the eggs are shed, eventually reaching the outside world in the faeces of the host. If an egg is ingested by a crustacean or an insect, it will hatch and the larva will use its hooked proboscis to tear its way into the body cavity of this intermediate host. Here, the larva transforms into a cyst-like form in the hope that the hapless arthropod will be eaten by a vertebrate, completing the life cycle. These immature stages are known to change the behaviour of an intermediate host in order to increase the chances of it being eaten by the definitive host. For example, some species use a bottom-dwelling crustacean as their intermediate host and to get into the definitive host, a duck, they force the crustacean to swim near the surface or to grip onto a rock where it is very obvious to the bird.

Thorny-headed worms can often be present in very large numbers in their host, with a thousand individuals observed in the intestine of an individual duck. At such high population densities, these creatures can severely damage the host's intestine with their rasping proboscises, eventually causing disease and death.

3 A thorny-headed worm. These animals are extremely specialized parasitic rotifers. The brutal-looking spiny proboscis allows them to latch on to the slippery lining of their host's digestive tract. This species (*Polymorphus* sp.) can alter the behaviour of its intermediate host (an amphipod crustacean) so that it is more likely to be consumed by the definitive host (e.g., a duck).

water towards the animal and into a food groove lined with yet more cilia. The food particles are passed to the mouth, where the muscular pharynx sucks them down towards the gnashing mastax. To prevent unwanted bits of material finding their way into the mouth, large cilia can screen it off and those drawing the particles in can reverse their beating motion.

Predation is also common among the rotifers and they are surrounded by suitably sized prey, including other rotifers (interestingly, the jaws of a rotifer that falls prey to another rotifer are not digested, so by looking at the predator's gut contents you can see which members of its lineage it has been eating). Some predatory rotifers simply grab their prey with the pincer-like elements of their mastax, while others have adapted a more passive technique relying on a hood-like corona or a wreath of long bristles around their mouth forming what looks like a tiny basket [6, 10, 13]. When an organism swims into this basket the lobes bearing the bristles bend inwards forming a cage to trap the prey, which is then sucked into the mouth by the muscular pharynx.

A considerable number of rotifers have turned to a parasitic way of life and of these the thorny-headed worms are the most specialized by quite some margin [BOX 1]. There are other, more typical, rotifers that also live on or in the bodies of other organisms. The elegant seisonids grip on to the exoskeleton of certain crustaceans by their toe, feeding on assorted organic debris and the host's eggs. Other parasitic/commensal rotifers attach to the gills of polychaete worms and the tube feet of echinoderms. Yet more live inside the intestine and body cavity of earthworms and there are even some that drain snail's eggs of their contents before laying their own eggs in the empty shell. Freshwater algae are also exploited by a variety of parasitic rotifers. Some of these invade filamentous algae stimulating the host to produce a gall in which they lay their eggs, while others run amok in the green, disco ball-like colonies of *Volvox* algae.

Reproduction is delightfully complex in the rotifers. Depending on the species there can be normal males and normal females, dwarf males and normal females or just females who are able to produce young without ever needing the help of a male. A female rotifer has anywhere between eight and 20 immature egg nuclei, depending on the species. Because these animals are born with a fixed number of cells this is all that she will ever have.

Generally, male rotifers, when they occur, are rather unfortunate creatures. In a cruel quirk of evolution they have been reduced to little more than a mobile gonad, their sole purpose to stab a female with their oversized penis in order to transmit sperm. Like the females, males are born with a set number of around 30 sperm, although this varies between species. There are two forms: normal-shaped sperm, but with a flagella at the front rather than the rear, which is unusual in itself, and simple rod-shaped sperm, which are believed to assist the flagellated sperm in some way. To stretch out his finite supply of gametes the male only uses two to three sperm each time he mates.

In some species males are completely unknown and the females simply give birth to clones of themselves. In a few species these clones stick together to form floating colonies [5]. More normal is a complex life cycle that includes both asexual and sexual phases. During the asexual phase females produce so-called amictic eggs, thin-shelled, unable to be fertilized and which quickly hatch into clones of the mother. Without all the hassle of males and fertilization, this asexual phase can rapidly generate large populations. However, at some point during this asexual phase, perhaps triggered by changing day length, temperature or lack of food, the asexual females somehow become sexual and start laying mictic eggs. If left unfertilized these eggs give rise to dwarf males that go off and inseminate other sexual females brooding mictic eggs with their massive penis [BOX 2]. A mictic egg, once fertilized, develops into a thick-shelled, dormant egg that can remain viable for at least 20 years if unfavourable conditions prevail [2]. Females hatch from these dormant eggs and they will carry on the asexual phase of the life cycle, perhaps switching to sexual reproduction if they live long enough and conditions dictate.

Typically, the thin-shelled eggs are produced during the summer, while the thick-shelled eggs are laid in the winter, but this is not always true and both types may be produced throughout the year. Why should the rotifers have evolved such an elaborate life cycle? It seems to be a response to the pressures of living in places that can quickly become unsuitable, such as a small pool in a woodland clearing, an ephemeral stream or the film of water around a low-growing plant. The asexual phase allows rotifer numbers to grow quickly, while the production of tough, dormant eggs via a sexual phase is an insurance policy against things turning bad and allows

BOX 2 ***Haplodiploidy***

This strange phenomenon is common in animals, especially the arthropods and the rotifers. In a haplodiploid species a fertilized egg develops into a female while an unfertilized egg develops into a male; therefore, a female has a full complement of chromosomes, while a male has only half. This creates the unusual situation where a male has a grandfather, but no father. What is the significance of this peculiar breeding system? Haplodiploidy, in rotifers at least, may be a response to the pressures of inbreeding or perhaps a way of dealing with alternating periods of suitable and unsuitable environmental conditions.

a species to disperse to new habitats (not to mention the manifold benefits of being able to swap genetic material with other individuals, which is essentially what sexual reproduction is).

Regardless of whether the eggs are produced asexually or sexually, the rotifers also display varying degrees of parental care. In some species the eggs dangle like baubles from the body of the female [2, 12], while others attach their eggs to the bodies of other rotifers. Some even brood the eggs inside their body until the young are ready to hatch. With the exception of the thorny-headed worms, the rotifers have no larval stage, so it is young adults that hatch from the eggs. Males, when they occur, are sexually mature when they hatch; females, on the other-hand, do grow even though their cell number remains constant, with some species increasing in weight almost 30-fold.

The rotifers are predominantly animals of fresh water, with only about 5 per cent of known species being found in the oceans. Even though they are technically aquatic, their small size has enabled them to colonize the land to a degree by making use of the water film enveloping soil particles and the surfaces of plants and detritus. Capable of inhabiting the tiniest of habitats and able to see out long periods of unfavourable conditions as dormant eggs or in the death-like state known as cryptobiosis (see Tardigrada, pp. 138–43), the rotifers are very successful animals. It is only the most arid terrestrial environments where they do not really thrive.

Like so many other small and easily overlooked animals, the rotifers can exist in huge population densities. In a single litre (or about 0.25 gal.) of water from a suitable habitat there may be as many as a thousand of them and a single square metre (or about 11 sq. ft) of soil may support as many as two million individuals. They are even found in tap water. Their diversity and abundance makes them very important animals from an ecological point of view since they are crucial in the cycling of nutrients and energy through freshwater, terrestrial and marine ecosystems. Collectively, they consume huge quantities of smaller organisms and organic matter, converting this material into biomass, which is utilized by a large number of other creatures. As dormant eggs or in a state of cryptobiosis they can survive conditions that can obliterate the populations of other animals, establishing pioneer populations when more favourable conditions return, the first steps in restoring the habitat.

ORIGINS AND AFFINITIES

Apart from some fossils of extant forms dating to around 45 million years ago, the rotifers do not really have a fossil record, so our understanding of their heritage is based on what we can infer from their relationships with other animal lineages. In earlier times they were thought to be part of a rag-tag assortment of obscure animals known as aschelminthes, in reality little more than a taxonomic dustbin, but as more and more evidence became available this grouping was thankfully abandoned. The rotifers have since been grouped with the nematodes and platyhelminths at various times, but today we are getting close to their real affinities.

The rotifers (including the seisonids and the thorny-headed worms) are currently thought to be closely related to the gnathostomulids and micrognathozoans, creatures they share a number of characteristics with, most notably the complex pharynx with its distinctive jaws (see p. 310). This clade is, in turn, an offshoot of the branch that also includes the platyhelminths and the gastrotrichs.

4 The well-developed corona of *Keratella quadrata*. Many rotifers also have a tough, spiny cuticle.
5 A colonial rotifer. A founding individual produces offspring via parthenogenesis, some of which will remain attached to their parent while others leave to start a new colony (*Conochilus unicornis*).
6 This rotifer (*Cupelopagis* sp.) uses its hood-like corona to engulf prey.

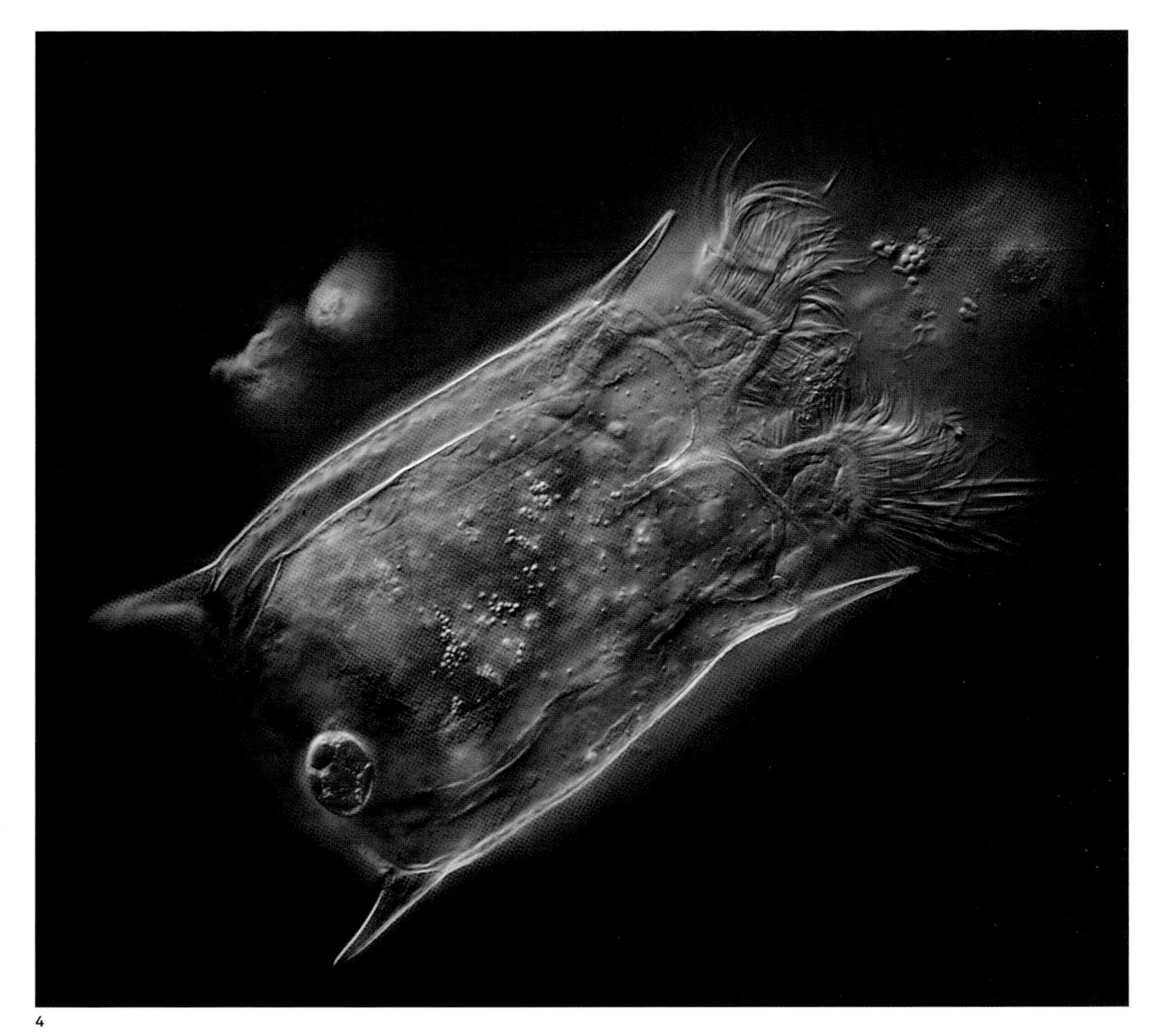

4

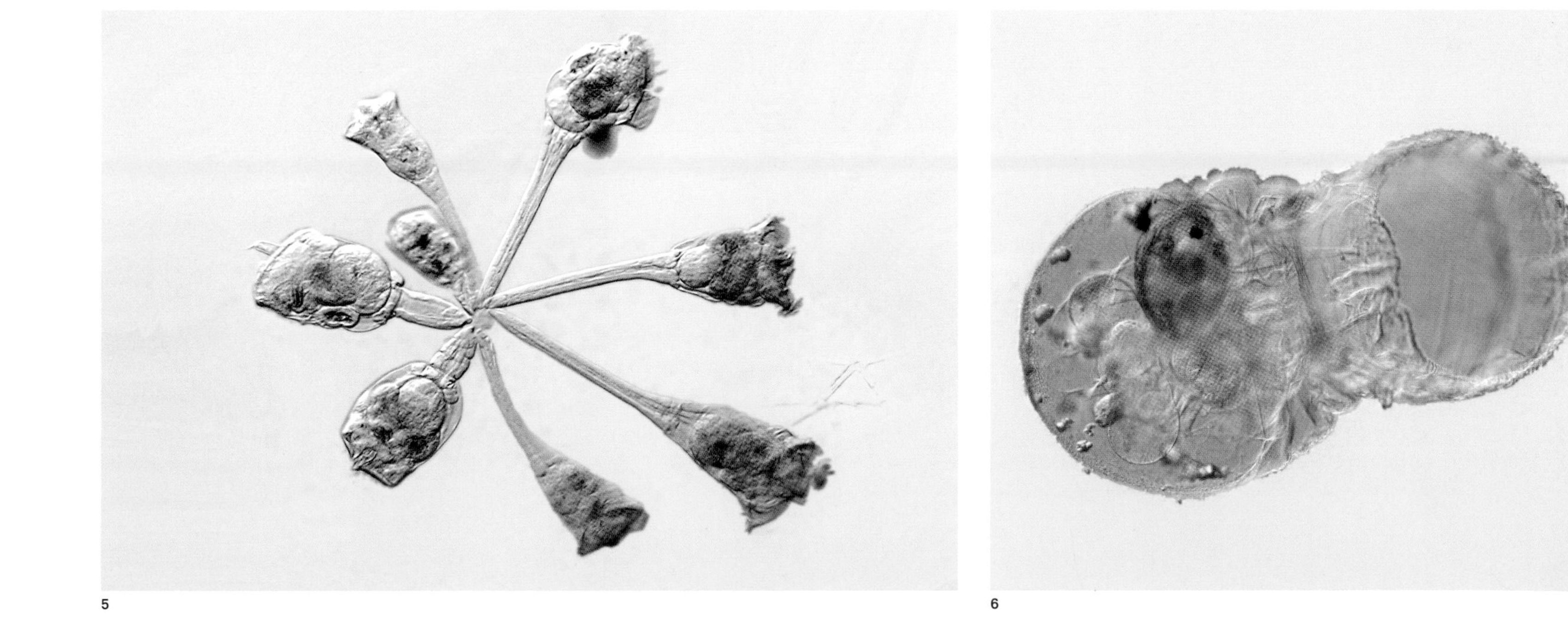

5

6

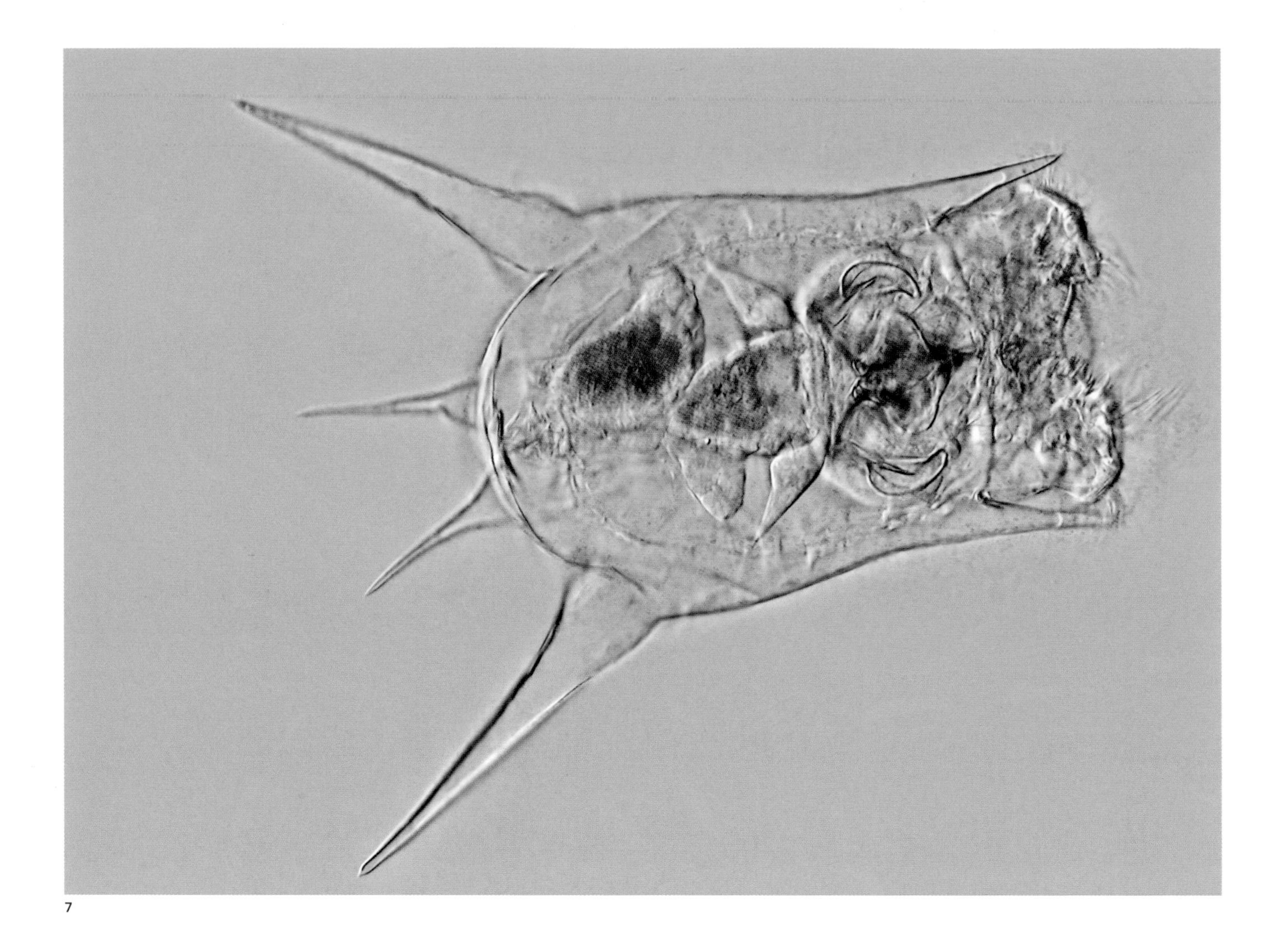

7

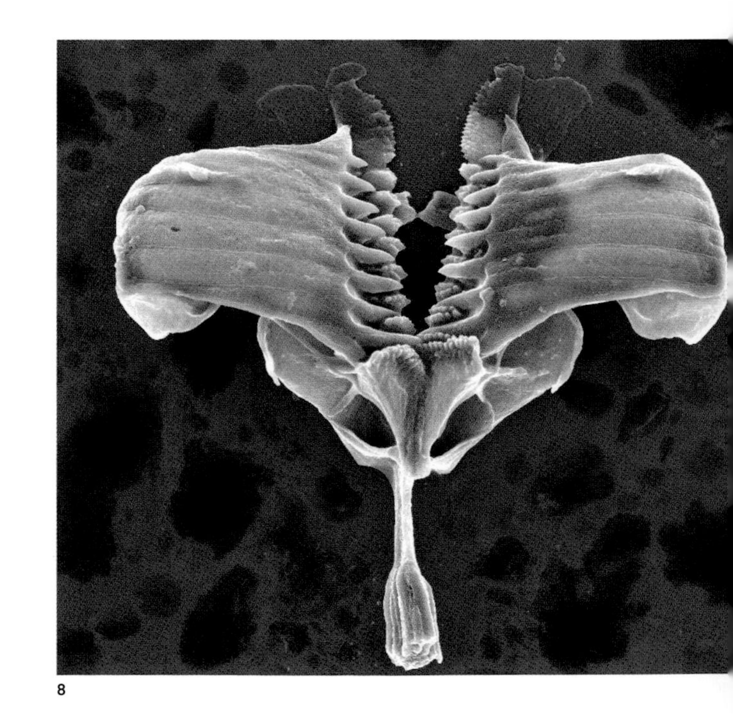

8

7 The elaborate spines of this rotifer (*Brachionus calyciflorus*) are a defence against its predators, typically other rotifers (such as *Asplanchna* spp.).

8 The mastax is a tiny, yet extremely complex jaw apparatus within the pharynx of rotifers. It consists of hard parts (trophi) to break prey up, muscles, ligaments and nerves. The trophi shown here, from the mastax of *Microcodides chlaena*, are only 0.028 mm (0.0011 in.) across.

9 Some rotifers live a sessile existence in a secreted tube, which is often reinforced with particles of sediment and detritus. Note the elaborate corona of this tube-dwelling species (*Limnias* sp.).

10 The long bristles around the mouth of the tube-dwelling Collotheca rotifers function like a basket to trap prey (*Collotheca ornata*).

11 Rotifers can have long spines as well as fat droplets in their bodies to help them stay afloat in open water (*Kellicottia* sp.).

9

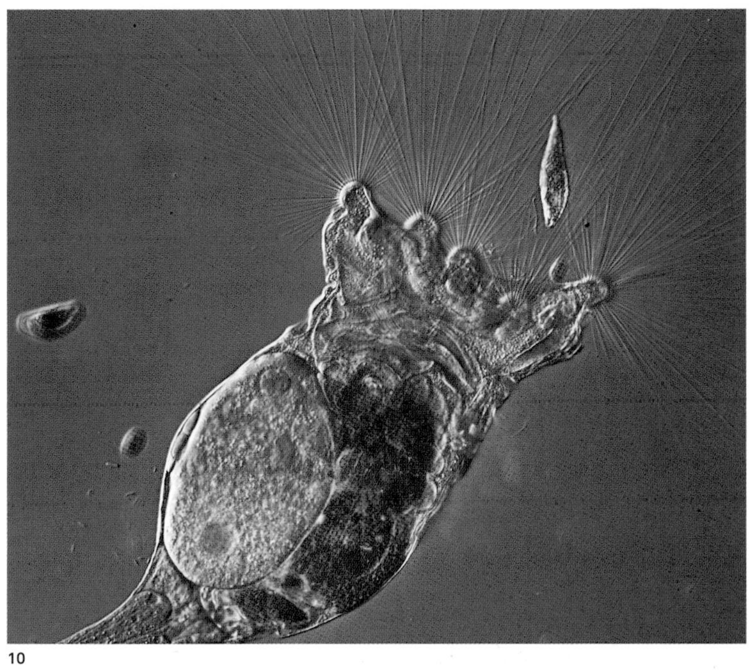
10

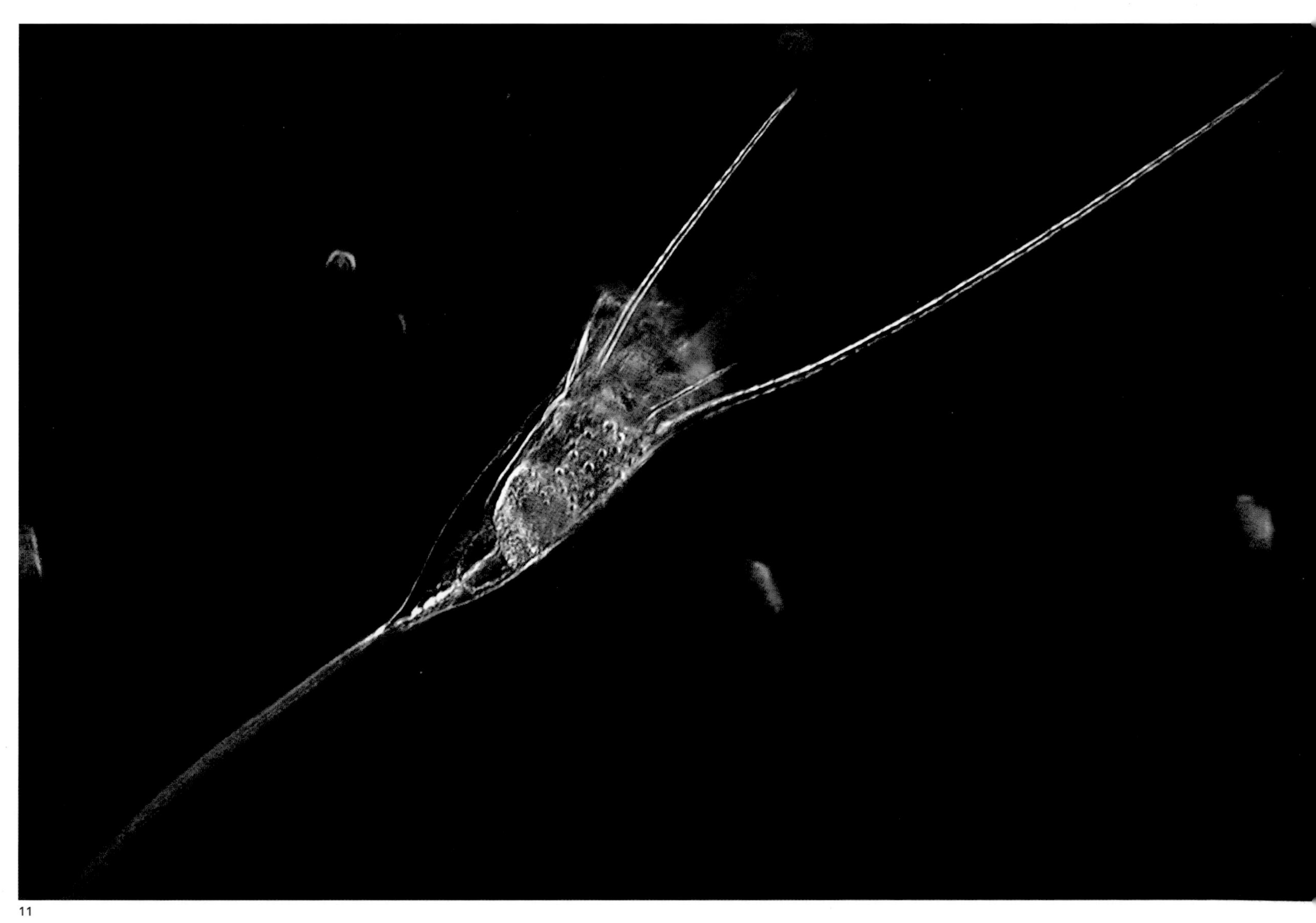
11

‹ **12** Using specialized, feather-like appendages this rotifer (*Polyarthra vulgaris*) 'skips' through the water at around 35 mm (1.4 in.) per second, which is extremely swift for an animal that is only around 0.2 mm (0.008 in.) long. Note the animal's eggs attached to its rear.

13 *Collotheca* spp. are sessile and secrete a mucus tube. Their corona of cilia is reduced or absent, but the extensible bristles at their anterior end are very obvious (*Collotheca mutabilis*).

13

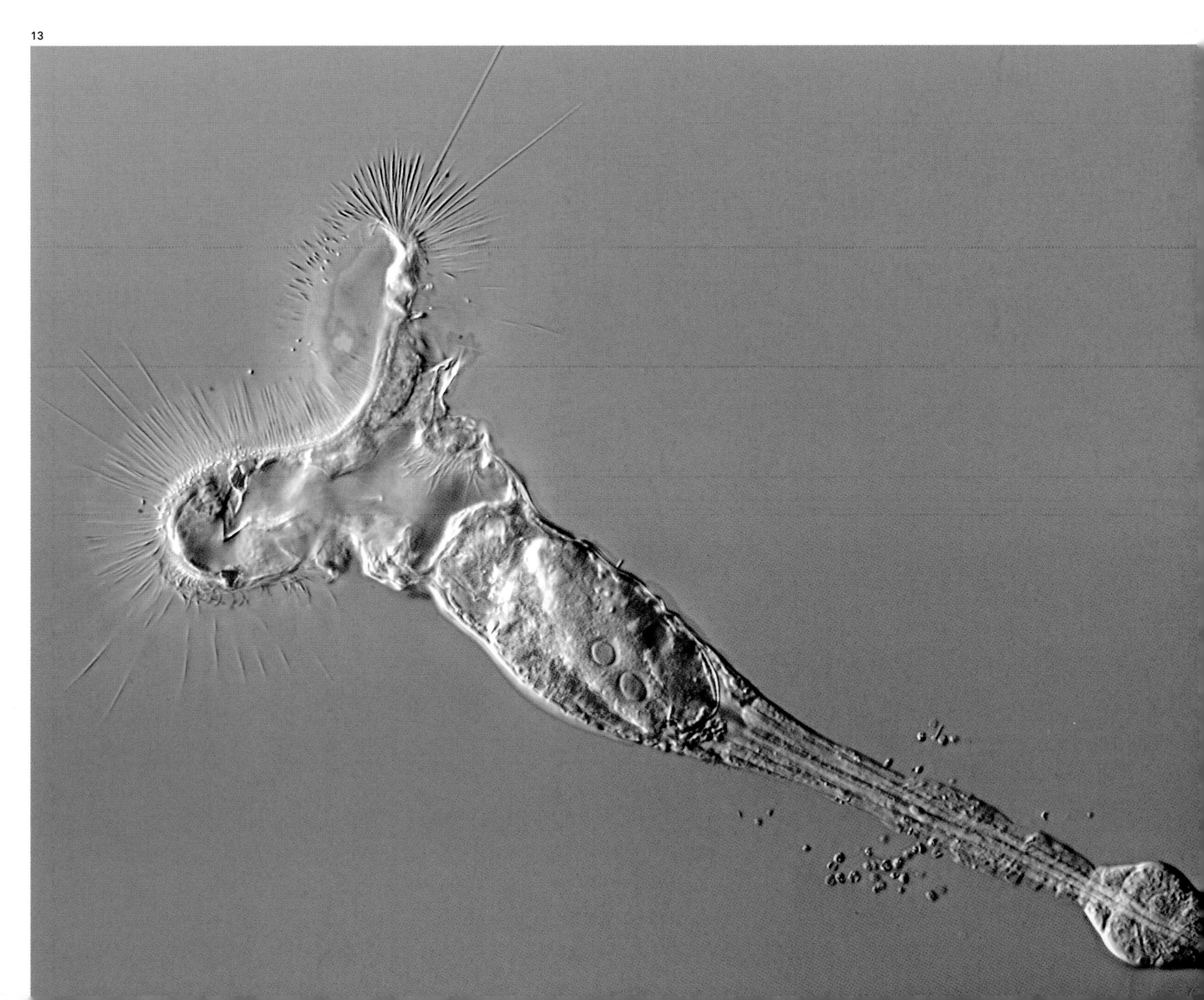

Micrognathozoa

(jaw animals)
(Greek *micro* = small;
gnathos = jaw;
zoion = animal)

Diversity
1 species

Size range
~0.15 mm
(~0.006 in.)

1 Illustration of a micrognathozoan. The complex jaw apparatus is a distinctive feature of these animals, as are the pads of cilia (ciliophores) on the ventral surface (*Limnognathia maerski*).

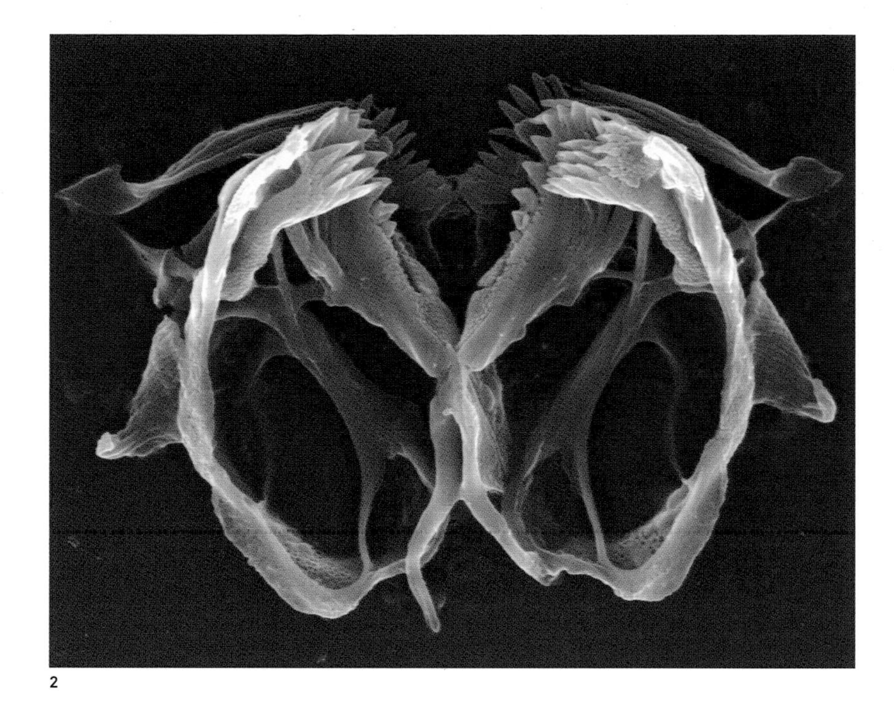

2

2 SEM of the tough jaws (trophi) of the pharynx. All the soft parts (muscles, ligaments and nerves) have been dissolved away to reveal these elaborate, hard structures (*Limnognathia maerski*).

Discovered in 1994 and formally described in 2000, this fascinating lineage is the most recent addition to the animal family tree. To date only one species is known and it was found living amongst mosses growing in a freshwater spring on Disko Island, Greenland. Now, more than ten years after it was first described, we have learnt a fair amount about the body plan of this animal, but much of its biology is still a tantalizing mystery.

FORM AND FUNCTION

These are extremely compact animals, smaller than many single-celled organisms, but they have considerable internal complexity. The body can be divided into a head, an accordion-like thorax and an abdomen [1]. Apart from a portion of the back that is covered in plates, the rest of the epidermis is naked and mostly non-ciliated. On the underside of the animal some of the epidermal cells bear lots of cilia forming little bands known as ciliophores [1]. Ten cells at the rear of the animal form a ciliated, adhesive pad.

Inside, there is a brain, two nerve cords, kidneys, a relatively simple female reproductive system, a few muscle cells and a blind-ending gut, which opens temporarily to the outside via a pore. But without doubt the most conspicuous organ is the pharynx and its complicated arrangement of tough jaws [2, 3]. The latter comprise no less than 15 separate units ranging in size from 0.004–0.014 mm (0.00016–0.00055 in.), all roughly the same size as the diameter of a human red blood cell, and all of which are linked together by a system of vanishingly tiny muscles and ligaments. To coordinate the movements of the complex pharynx there are sensory cells and a dedicated mini-brain.

Sprouting from various epidermal cells are sensory bristles, each of which is composed of one to three cilia. These are most abundant on the micrognathozoan head and are presumed to be sensitive to touch. In addition there are a number of other receptors that probably detect various chemicals in the water.

LIFESTYLE

The micrognathozoans use the pads of cilia on their underside to trundle around on and to swim amongst the leaves of mosses, moving their head from side to side as they go to detect the presence of food. The mouth, also located on the underside, is at the open end of a horseshoe arrangement of ciliary bands. The currents generated by these wafting cilia bring food particles within range of the muscular pharynx and complex jaws. By poking certain elements of the jaw apparatus out of the mouth, the animal uses them like tiny hands to grasp individual, single-celled organisms, such as diatoms, and to scrape at patches of bacteria and cyanobacteria. These are pulled inside by the retracting jaws, crushed and passed into the gut for digestion.

All the fully-grown individuals so far observed of the only known species, *Limnognathia maerski*, have been female, so

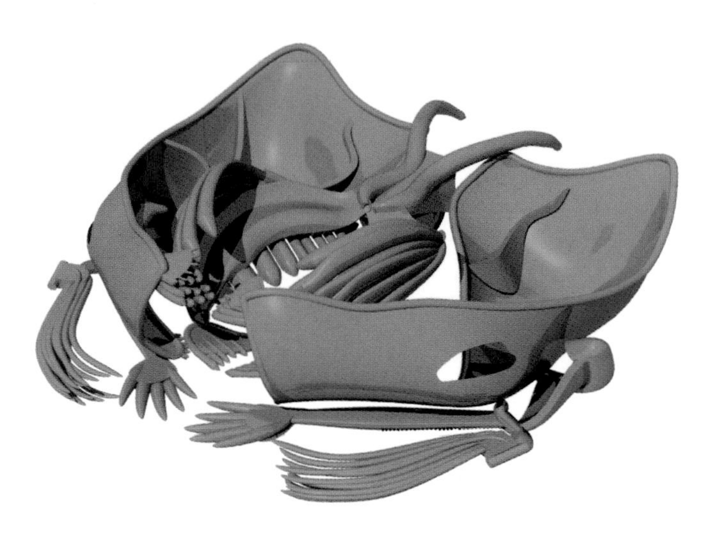

it is assumed that these animals reproduce asexually. Having said that, it has been suggested that micrognathozoans hatch as males and subsequently transform into females (but this still needs to be confirmed). It is not known how these youngsters go about swapping sperm, if at all. Regardless of how the eggs are fertilized, one egg in the adult grows until it takes up most of the abdominal space. Again, there are no ducts for it to be laid through, so it can only be released via a temporary rupture that opens up in the underside of the animal. Two types of egg can be produced: a smooth, thin-shelled one and a thick-shelled one with a sculpted surface [4]. The former hatches quickly, in a matter of days, while the latter can remain dormant to see out the harsh Greenland winter, the only way populations can persist from one short Arctic summer to the next, since the adults cannot survive the harsh winter conditions. In the rotifers (see pp. 298–307), thick-shelled eggs are only produced after being fertilized, which adds further weight to the theory that these animals start off life as males and then rapidly mature into females.

As the micrognathozoans are an only recently discovered lineage, as yet we have no idea of how these little animals fit into the freshwater ecosystems in which they are found. Following their initial discovery they have been found in Wales and Antarctica, but, generally speaking, the habitats they are associated with are harsh to say the least. For much of the year these animals are entombed in ice and snow, the onset of the fleeting summer a signal for them to burst into frantic activity for a couple of months. There are lots of other places around the world where other representatives of this lineage may be found, but since they are so small and delicate they will be difficult to find.

3 The jaw apparatus is made up of no less than 15 separate units, as shown in this 3D reconstruction (*Limnognathia maerski*).

4 A thick-shelled winter egg (*Limnognathia maerski*).

ORIGINS AND AFFINITIES

The discovery of the micrognathozoans came at a time when big developments in our understanding of how the animal lineages are related were happening, mostly thanks to the penetrating insights provided by comparing DNA sequences. Their characteristics, particularly the microscopic structure of their complex jaw apparatus, have helped to answer some important questions about how some of other lineages of animals are related, uniting some groups that have previously proved difficult to place on the animal family tree. The gnathostomulids, micrognathozoans and rotifers all have a very complex jaw apparatus and under extremely high magnifications it can be seen that these are composed of densely packed, rod-like structures, a feature that appears to be unique to these three groups of animals. This shared trait suggests that these animals share a common ancestor, thus uniting them as a distinct branch on the animal family tree. However, analysis of DNA has been inconclusive, so it will be some time yet before we have a better understanding of where these groups fit in the wider story of animal evolution.

Summary Table
Glossary
Further Reading
Acknowledgments
Index

Summary Table

LINEAGE	KNOWN SPECIES	SIZE RANGE	HABITAT	DEFINING CHARACTERISTICS
Ctenophora *(comb jellies)*	*c.* 240	~5 mm to 1.5 m (with tentacles) (~0.2 in. to 5 ft)	Marine; typically pelagic, but some benthic forms	Biradial symmetry; adhesive colloblast cells; eight rows of ciliary plates (ctenes); cydippid larvae
Porifera *(sponges)*	*c.* 8350	2 mm to 2 m (0.08 in. to 6.5 ft)	Marine and freshwater; benthic	No true gut, tissues or organs; choanocytes; inner and outer cells divided by mesohyl; supportive scaffold composed of spongin and calcareous or siliceous spicules
Placozoa *(placozoans)*	*c.* 8	1 to 3 mm (0.04 to 0.12 in.) across; 0.025 mm (0.001 in.) thick	Marine; benthic	Two cell layers sandwiching a space filled with fibre cells; external digestion via the formation of a temporary digestive chamber
Cnidaria *(jellyfish, etc.)*	*c.* 12,500	0.01 mm (myxozoa spores) to ~40 m (colonial forms) (0.0004 in. to ~130 ft	Marine and freshwater; benthic, pelagic and other organisms	Radial symmetry (polyp); cnidocysts; musculature originates from ectoderm and endoderm; typically, reproduction alternates between polyp and medusa stages; planula larva
Bilateria	*c.* 1,479,400			Bilateral symmetry; anterior–posterior axis reflects their direction of travel and the movement of food through their body; senses and nerve tissue concentrated to form an anterior head; three body layers; mesoderm gives rise to body cavity
Deuterostomia	*c.* 75,750			Mouth develops from the second pore that opens in the early embryo; body cavity forms in a distinctive way; dorsal nervous system; endoskeletons
Cephalochordata *(lancelets)*	*c.* 30	4 to 8 cm (1.6 to 3.2 in.)	Marine; benthic	Notochord; hollow nerve cord; pharyngeal slits; wheel organ; endostyle
Tunicata *(sea squirts, etc.)*	*c.* 2860	~1 mm to >20 m (colonial forms) (0.04 in. to >65 ft)	Marine; benthic and pelagic	Body enclosed in tunic; two siphons corresponding to mouth and anus; large and complex pharynx; tadpole-like larvae have notochord, dorsal nerve cord and tail
Craniata *(vertebrates, etc.)*	*c.* 64,830	~10 mm to ~30 m (~0.4 in. to ~100 ft)	Marine, freshwater, terrestrial and aerial	Cranium; dorsal nerve cord, notochord (replaced in vertebrates by the vertebral column); post-anal tail
Hemichordata *(acorn worms)*	*c.* 120	~1 mm to ~2.5 m (~0.04 in. to 8 ft)	Marine; benthic (some species may drift just above the seabed)	Acorn-shaped proboscis; collar; trunk; gill slits

LINEAGE	KNOWN SPECIES	SIZE RANGE	HABITAT	DEFINING CHARACTERISTICS
Echinodermata *(sea stars, etc.)*	c. 7500	~10 mm to ~2 m (~0.4 in. to ~6 ft)	Marine; benthic and a few pelagic forms	Pentaradial (five-part) symmetry; internal calcium carbonate skeleton; coelom with discrete compartments, e.g. water vascular system
Xenoturbellida *(strange worms)*	1	3 to 4 cm (1.2 to 1.6 in.)	Marine; benthic	No through gut, organized gonads, coelom or excretory structures; epidermal nervous system; superficial furrows corresponding to thickenings in nerve net
Acoelomorpha *(acoelomorphs)*	c. 400	<0.5 to 15 mm (<0.02 to 0.6 in.)	Marine (typically), freshwater; benthic and a few pelagic forms	Unique ciliary structure; diffuse nervous system; most have no gut
Protostomia	c. 1,403,650			Mouth develops from the first pore that opens in the early embryo; body cavity forms in distinctive way; ventral nervous systems; exoskeletons
Chaetognatha *(arrow worms)*	c. 180	~1 to ~12 cm (0.4 to 4.7 in.)	Marine; pelagic (some benthic forms)	Chitinous grasping spines; hood, ciliary fences
Nematoda *(nematodes)*	c. 24,800	<1 mm to ~9 m (<0.04 in. to ~29.5 ft)	Marine, freshwater, terrestrial (restricted to 'aquatic' habitats on land) and other organisms	Non-cellular cuticle, longitudinal muscles only; no motile cilia
Nematomorpha *(horsehair worms)*	c. 350	~5 cm to ~1 m (~2 in. to ~3.3 ft)	Freshwater and marine in other organisms; pelagic/benthic short-lived adults	Vestigial gut; sculpted, non-cellular cuticle; no excretory system; larvae with distinctive hooks and proboscis; endoparasitic
Tardigrada *(water bears)*	c. 1160	~0.08 to 2 mm (~0.003 to ~0.08 in.)	Marine, freshwater and terrestrial (restricted to 'aquatic' habitats on land)	Four pairs of claw-bearing walking legs; sensory cirri; distinctive structure of eye spots; mouth stylets; brain linked to first ganglion by additional nerve connection
Onychophora *(velvet worms)*	c. 180	~5 mm to ~15 cm (~0.2 to ~6 in.)	Terrestrial (restricted to moist habitats)	Slime and slime glands; unique trachea structure; no continuous layer of smooth muscle; subcutaneous circulatory system
Arthropoda *(arthropods)*	c. 1.2 million	~0.1 mm to ~3.5 m (~0.004 in. to ~11.5 ft)	Marine, freshwater, terrestrial, aerial, on and in other organisms	Segmented; chitinous exoskeleton; jointed appendages; compound eyes, silk production; no locomotory cilia (most sense organs incorporate modified cilia); flagella are present in the sperm of the some groups

LINEAGE	KNOWN SPECIES	SIZE RANGE	HABITAT	DEFINING CHARACTERISTICS
Priapulida *(penis worms)*	*c.* 20	~0.5 mm to ~40 cm (~0.02 to ~16 in.)	Marine; benthic	Retractable introvert; large, fluid-filled body cavity containing two types of cell
Loricifera *(brush-heads)*	*c.* 30	0.25 to 0.85 mm (0.01 to 0.03 in.)	Marine; benthic	Retractable introvert; scalids; lorica composed of chitinous plates
Kinorhyncha *(mud-dragons)*	*c.* 180	~0.1 to ~1 mm (~0.004 to ~0.04 in.)	Marine; benthic	Retractable introvert; scalids; mouth cone bearing nine stylets; two to three chitinous plates cover each of the 11 trunk segments
Ectoprocta *(bryozoans)*	*c.* 5500	~0.5 mm to ~1 m (colonies) (~0.02 in. to ~3.3 ft)	Marine, freshwater; benthic, typically colonial and sessile	Retractable lophophore; anus outside of lophophore; hollow tentacles; secondary loss of circulatory, respiratory and excretory systems
Entoprocta *(goblet animals)*	*c.* 170	0.1 to 10 mm (zooids) (0.004 to 0.4 in.)	Marine, freshwater; benthic, typically colonial and sessile	Non-retractable lophophore; mouth and anus inside ring of the lophophore; solid tentacles; lime-twig glands
Cycliophora *(cycliophorans)*	3	0.04 to 0.6 mm (0.0015 to 0.02 in.)	Marine; commensal	Distinctive feeding funnel; dwarf males; asexual and sexual life stages each with distinctive larvae; complex life cycle
Dicyemida *(dicyemids)*	*c.* 125	0.1 to 5 mm (0.004 to 0.2 in.)	Marine; inside cephalopods (possibly symbiotic)	Extreme simplification; long, thin central cell surrounded by 8–30 ciliated peripheral cells; calotte (attachment structure)
Orthonectida *(orthonectids)*	*c.* 45	0.05 to 0.8 mm (0.002 to 0.03 in.)	Marine; inside other organisms	Extreme simplification; rings of ciliated and non-ciliated jacket cells surrounding a mass of gametes; two distinctive types of female
Annelida *(annelids)*	*c.* 18,950	~0.1 mm to 3 m (~0.004 in. to 10 ft)	Marine, freshwater, terrestrial (restricted to moist habitats)	Segmentation (superficially lost in some groups); closed circulatory system; chitinous epidermal structures (chaetae)

LINEAGE	KNOWN SPECIES	SIZE RANGE	HABITAT	DEFINING CHARACTERISTICS
Mollusca *(molluscs)*	*c.* 117,350	<1 mm to 20 m (<0.04 in. to 65 ft)	Marine, freshwater, terrestrial (restricted to moist habitats)	Muscular foot (lost in some species); mantle; mantle cavity; radula; calcareous spicules or shell; open circulatory system (except cephalopods)
Nemertea *(ribbon worms)*	*c.* 1200	0.5 mm to 55 m (0.02 in. to 180 ft)	Marine; benthic (some pelagic forms) and inside other organisms (but possibly commensal)	Eversible proboscis not linked to the gut; proboscis contained within a long, dorsal cavity (rhynchocoel)
Brachiopoda *(lamp shells)*	*c.* 390	1 mm to 10 cm (0.04 to 4 in.)	Marine; benthic	Shell secreted by mantle; shell valves (typically asymmetrical) correspond to dorsal and ventral surface; pedicle extends from ventral shell valve and anchors animal to substrate; lophophore
Phoronida *(horseshoe worms)*	*c.* 10	1 to 50 cm (0.4 to 20 in.)	Marine; benthic	Actinotroch larva; chitinous tube secreted by epidermis; lophophore
Gastrotricha *(gastrotrichs)*	*c.* 790	0.05 to 4 mm (0.002 to 0.16 in.)	Marine, freshwater; benthic (some pelagic forms)	Epicuticle composed of numerous layers; cuticle covers the entire body, including the cilia; ventral bands of cilia provide propulsion; transient reproductive tract
Platyhelminthes *(flatworms, etc.)*	*c.* 29,300	<1 mm to 30 m (<0.04 in. to 100 ft)	Marine, freshwater, terrestrial (restricted to 'aquatic' habitats on land) and other organisms	Complex gut with a single opening; no body cavity; syncytial epidermis in many groups; ciliated epidermis in free-living forms
Gnathostomulida *(jaw worms)*	*c.* 110	0.5 to 4 mm (0.02 to 0.16 in.)	Marine; benthic	No cuticle; epidermal cells bear a single cilium; no body cavity; gut with a single opening (some species have an anal pore); complex pharyngeal jaw apparatus
Rotifera *(rotifers)*	*c.* 2800	0.04 mm to ~80 cm (0.0016 to ~32 in.)	Marine, freshwater, terrestrial (restricted to 'aquatic' habitats on land) and other organisms	Ciliated corona; mastax (jaw apparatus); intracellular cuticle; retrocerebral organ
Micrognathozoa *(jaw animals)*	1	~0.15 mm (~0.006 in.)	Freshwater; benthic (surfaces of aquatic vegetation)	No cuticle; most of the epidermal cells are non-ciliated, ventral pads of cilia (ciliophores); complex pharyngeal jaw apparatus; gut with a single opening (temporary anal pore)

Glossary

Anterior
The head (front) of an animal, or towards that end.

Benthic
Of or relating to the bottom of an aquatic habitat.

Biomass
The weight of total living organisms or of a species population per unit of area.

Boreal forest
Forests that grow in the cold climatic zone to the south of the Arctic. They are found in northern Eurasia and North America.

Buccal
Of or relating to the mouth.

Calcareous
Referring to material composed partly or entirely of calcium carbonate.

Cf.
An abbreviation for the Latin word confer. Used in taxonomy when an identification is not confirmed.

Cilium (sing.) / Cilia (pl.)
A hair-like organelle projecting from a eukaryotic cell. They can be motile (for locomotion or moving material) or immotile (for sensory functions).

Cloaca
The common chamber and outlet into which the digestive, genital and excretory tracts open.

Coelom
A body cavity between the body wall and intestine that is lined with mesodermal tissue.

Commensal
An animal that lives very close to another and benefits while the host is unaffected.

Dioecy / Dioecious
Having separate sexes.

Dorsal
The back or upper surface of an animal.

Ecosystem
A biological community of interacting organisms and their physical environment.

Epiphyte / Epiphytic
Any plant that grows on another plant.

Epithelium
Membranous tissue composed of one or more layers of cells forming the covering of most internal and external surfaces of an animal's body and its organs.

Eusociality
Referring to any colonial animal species that live in multigenerational family groups where the vast majority of individuals cooperate to aid relatively few (or even a single) reproductive individuals.

Flagellum (sing.) / Flagella (pl.)
Lash-like organelles that move in a whip-like action for locomotion (e.g., the movement of sperm) and to create water movement (e.g., in the choanocytes of sponges).

Fossorial
Characterized by digging or burrowing.

Gamete
A mature, sexual reproductive cell (i.e., a sperm or egg).

Gravid
Carrying developing young or eggs.

Haemolymph
The circulatory fluid of many groups of animals, such as arthropods and most molluscs, that have an open circulatory system. It is analogous to blood and lymph in vertebrate animals, and is not confined in a system of vessels.

Hermaphrodite
An organism that has reproductive organs normally associated with both male and female sexes. Hermaphrodites can be simultaneous (female and male reproductive organs present at the same time) or sequential (the organism changes from one sex to the other during its life).

Hydrostat
A structure, either the entire body of an animal or an appendage, with a musculature that enables complex movements without a supporting skeleton.

Hydrostatic skeleton
The musculoskeletal system of soft bodied animals where the muscle sheath of the body wall encompasses a fluid filled cavity (or series of cavities) or very flexible tissue.

Introvert
A retractile head or anterior end of the body that can be either everted or protruded.

Meiofauna
The assemblage of diverse, tiny animals that lives among aquatic sediment.

Morphology
The science of structure. Includes cytology (the study of cell structure), histology (the study of tissue structure) and anatomy (the study of gross structure).

Ossicle
A small piece of calcified material forming part of the skeleton of an animal such as an echinoderm.

Ovovivipary
Reproduction where eggs hatch within the body of the parent or immediately after laying.

Parthenogenesis
Reproduction that involves the production of young by females not fertilized by males. A parthenogenetic egg may contain a full or half complement of chromosomes.

Pelagic
Of, relating to, or living in open water, particularly the open ocean.

Pharynx
A modified, typically muscular, section of the fore-gut of an animal between the mouth and the oesophagus.

Pheromones
A chemical substance released by one organism that influences the behaviour or physiological processes of another.

Posterior
Situated at or towards the rear of the body.

Proboscis
A pre-oral, prehensile and often retractable appendage that is usually the anterior-most part of the body.

Sedentary
Stationary, sitting, inactive; staying in one place.

SEM
A scanning electron microscope or an image taken with this device (scanning electron microphotograph).

Sessile
Attached at the base; fixed to one spot; not able to move about.

Seta (sing.) / Setae (pl.).
A bristle-like, chitinous epidermal structure.

Sp.
Species (sing.).

Spp.
Species (pl.).

Stimulus / Stimuli
Something external that influences an activity.

Totipotency
The ability of a single cell to differentiate into all the cell types that make up an organism.

Ventral
The lower surface (underside) of an animal.

Zooid
One of the distinct individuals forming a modular, colonial animal.

Further Reading

Aarnio, K., Bonsdorff, E., & Norkko, A. 'Role of *Halicryptus spinulosus* (Priapulida) in structuring meiofauna and settling macrofauna'. *Mar Ecol Prog Ser* 1998;163: 145–53.

Ball, E.E., & Miller, D.J. 'Phylogeny: the continuing classificatory conundrum of chaetognaths'. *Current biology* 2006;16(15): R593–6.

Biron, D.G., et al. 'Behavioural manipulation in a grasshopper harbouring hairworm: a proteomics approach'. *Proc Biol Sci* 2005;272(1577): 2117–26.

Borgonie, G., et al. 'Nematoda from the terrestrial deep subsurface of South Africa'. *Nature* 2011;474(7349): 79–82.

Boto, L. 'Horizontal gene transfer in evolution: facts and challenge'. *Proc. R. Soc. B* 2010;277: 819–27.

Brown, J.R. 'Ancient horizontal gene transfer'. *Nat Rev Genet* 2003;4(2): 121–32.

Chen, J.Y., et al. 'The first tunicate from the Early Cambrian of South China'. *Proc Natl Acad Sci USA* 2003;100(14): 8314–8.

Conway Morris, S., & Peel, J.S. 'The earliest annelids: Lower Cambrian polychaetes from the Sirius Passet Lagerstätte, Peary Land, North Greenland'. *Acta Palaeontologica Polonica* 2008;53(1): 137–48.

Danovaro, R., et al. 'The first metazoa living in permanently anoxic conditions'. *BMC Biol* 2010;8: 30.

Daponte, M.C., et al. 'Composition, density, and biomass of Salpidae and Chaetognatha in the Southwestern Atlantic Ocean'. *Bulletin of Marine Science* 2011;87(3): 437–61.

Dunn, C.W., et al. 'Broad phylogenomic sampling improves resolution of the animal tree of life'. *Nature* 2008;452(7188): 745–9.

Edgecombe, G.D., et al. 'Higher-level metazoan relationships: recent progress and remaining questions'. *Organisms, Diversity, and Evolution* 2011;11(2): 151–72.

Emschermann, P. 'Lime-twig glands: a unique invention of an Antarctic entoproct'. *The Biological bulletin* 1993;185(1): 97–108.

Fisher, C.R., et al. 'Methane ice worms: *Hesiocaeca methanicola* colonizing fossil fuel reserves'. *Naturwissenschaften* 2000;87(4): 184–87.

Fumagalli, M., et al. 'Parasites represent a major selective force for interleukin genes and shape the genetic predisposition to autoimmune conditions'. *J Exp Med* 2009;206(6): 1395–408.

Garey, J.R. 'The lesser-known protostome taxa: an introduction and a tribute to Robert P. Higgins'. *Integrative and Comparative Biology* 2002;42(3): 611–18.

Garey, J.R., & Schmidt-Rhaesa, A. 'The essential role of "minor" phyla in molecular studies of animal evolution'. *American Zoologist* 1998;38(6): 907–17.

Goto, T., & Yoshida, M. 'The mating sequence of the benthic arrow-worm *Spadella schizoptera*'. *Biol. Bull.* 1985;169: 328–33.

Gubanov, N.M. 'Giant nematoda from the placenta of Cetacea; *Placentonema gigantissima* nov. gen., nov. sp.'. *Proc. USSR Acad. Sci.* 1951;77(6): 1123–25.

Haritos, V.S., et al. 'Harnessing disorder: onychophorans use highly unstructured proteins, not silks, for prey capture'. *Proc Biol Sci* 2010;277: 3255–63.

Hejnol, A., et al. 'Assessing the root of bilaterian animals with scalable phylogenomic methods'. *Proc Biol Sci* 2009;276(1677): 4261–70.

Hickman, C.P., et al. *Integrated Principles of Zoology 15th edition*. New York: McGraw Hill, 2010.

Higgins, R.P., & Thiel, H. *Introduction to the Study of Meiofauna*. Washington D.C: Smithsonian Institution Press, 1988.

Holland, N.D., Kuhnz, L.A., & Osborn, K.J. 'Morphology of a new deep-sea acorn worm (class Enteropneusta, phylum Hemichordata): A part-time demersal drifter with externalized ovaries'. *J Morphol* 2012;273(7): 661–71.

Hou, X.G., et al. 'An early Cambrian hemichordate zooid'. *Current biology* 2011;21(7): 612–6.

Hunt, D.J., & Moore, D. 'Rhigonematida from New Britain diplopods. 2. The genera *Rhigonema* Cobb, 1898 and *Zalophora* Hunt, 1994 (Rhigonematoidea: Rhigonematidae) with descriptions of three new species'. *Nematology* 1999;1: 225–42.

Kajihara, H., & Lindsay, D. '*Dinonemertes shinkaii* sp. nov., (Nemertea: Hoplonemertea: Polystilifera: Pelagica) a new species of bathypelagic nemertean'. *Zootaxa* 2010;2429:43–51.

Kristensen, R.M. 'An introduction to Loricifera, Cycliophora, and Micrognathozoa'. *Integrative and Comparative Biology* 2002;42: 641–51.

Kristensen, R.M., & Funch, P. 'Micrognathozoa: a new class with complicated jaws like those of Rotifera and Gnathostomulida. *Journal of Morphology* 2000;246: 1–49.

Kuris, A.M., et al. 'Ecosystem energetic implications of parasite and free-living biomass in three estuaries'. *Nature* 2008;454(7203): 515–8.

Lecointre, G., & Le Guyader, H., 2006. *The Tree of Life: A Phylogenetic Classification*. Cambridge, MA: Belknap Press of Harvard University Press, 2007.

Lee, J.H., Kim, T.W., & Choe, J.C. 'Commensalism or mutualism: conditional outcomes in a branchiobdellid-crayfish symbiosis'. *Oecologia* 2009;159: 217–24.

Lewis, D.B. 'Some aspects of the ecology of *Fabricia sabella* (Ehr.) (Annelida, Polychaeta)'. *Journal of the Linnean Society of London, Zoology* 1968;47: 515–26.

Lindberg, D.R. 'Monoplacophorans and the origin and relationships of mollusks'. *Evolution: Education and Outreach* 2009;2: 191–203.

Maxmen, A. 'Evolution: a can of worms'. *Nature* 2011;470(7333): 161–2.

Nielsen, C. 'After all: Xenoturbella is an acoelomorph!'. *Evolution and Development* 2010;12(3): 241–43.

Osada, Y., & Kanazawa, T. 'Parasitic helminths: new weapons against immunological disorders'. *J Biomed Biotechnol* 2010;2010: 743–58.

Pechenik, J. *Biology of the Invertebrates 6th edition*. McGraw-Hill Companies, 2009.

Philippe, H., et al. 'Acoelomorph flatworms are deuterostomes related to Xenoturbella'. *Nature* 2011;470(7333): 255–8.

Reinhard, J., & Rowell, D.M. 'Social behaviour in an Australian velvet worm, *Euperipatoides rowelli* (Onychophora: Peripatopsidae)'. *Journal of Zoology* 2005;267: 1–7.

Roberts, L.S., & Janovy Jr, J. *Foundations of Parasitology 8th Edition*. McGraw-Hill Science/Engineering/Math, 2008.

Rohde, K. *Marine Parasitology*. Victoria, Australia: CSIRO Publishing, 2005.

Rumpho, M.E., et al. 'Horizontal gene transfer of the algal nuclear gene psbO to the photosynthetic sea slug *Elysia chlorotica*'. *Proc Natl Acad Sci USA* 2008;105(46): 17867–71.

Rundell, R., & Leander, B. 'Masters of miniaturization: Convergent evolution among interstitial eukaryotes'. *BioEssays* 2010;32(5): 430–37.

Ruppert, E.E., Fox, R.S., & Barnes, R.D. *Invertebrate Zoology: A Functional Evolutionary Approach, 7th Edition*. Belmont, California: Brooks/Cole Thompson Learning, 2004.

Schierwater, B., et al. 'Concatenated analysis sheds light on early Metazoan evolution and fuels a modern "Urmeta-zoon" hypothesis'. *PLoS Biol* 2009;7(1): 0036–44.

Størmer, L. '*Gigantoscorpio willsi*, a new scorpion from the lower carboniferous of Scotland and its associated preying microorganisms'. *Norske Videnskaps-A kad. Oslo. I. Mat.-Naturfl. Kl* 1963;8: 171.

Stanley, G.D., & Stürmer, W. 'The first fossil ctenophore from the lower devonian of West Germany'. *Nature* 1983;303(518).

Stanley, G.D., & Stürmer, W. 'A new fossil ctenophore discovered by X-rays'. *Nature* 1987;328(6125): 61–63.

Suzuki, T.G., et al. 'Phylogenetic analysis of dicyemid mesozoans (phylum Dicyemida) from innexin amino acid sequences: dicyemids are not related to Platyhelminthes'. *J Parasitol* 2010;96(3): 614–25.

Syed, T., & Schierwater, B. 'The evolution of the Placozoa – A new morphological mode. *Senckenbergiana Lethaea* 2002;82(1): 315–24.

Tait, N.N., & Norman, J.M. 'Novel mating behaviour in *Florelliceps stutchburyae* gen. nov., sp. nov. (Onychophora: Peripatopsidae) from Australia'. *Journal of Zoology* 2001;253: 301–08.

Telford, M.J. 'Xenoturbellida: the fourth deuterostome phylum and the diet of worms'. *Genesis* 2008;46(11): 580–6.

Thomas, F., et al. 'Biochemical and histological changes in the brain of the cricket *Nemobius sylvestris* infected by the manipulative parasite *Paragordius tricuspidatus* (Nematomorpha). *Int J Parasitol* 2003;33(4): 435–43.

Thuesen, E.V., Goetz, F.E., & Haddock, S.H. 'Bioluminescent organs of two deep-sea arrow worms, *Eukrohnia fowleri* and *Caecosagitta macrocephala*, with further observations on Bioluminescence in chaetognaths'. *The Biological bulletin* 2010;219(2): 100–11.

Thuesen, E.V., & Kogure, K. 'Bacterial production of tetrodotoxin in four species of Chaetognatha'. *Biological Bulletin* 1989;176(2): 191–94.

Tunnicliffe, V., & Wilson, K. 'Brachiopod populations: distribution in fjords of British Columbia (Canada) and tolerance of low oxygen concentrations'. *Mar Ecol Prog Ser* 1988;47: 117–28.

Wallace, R.L. 'Rotifers: exquisite metazoans'. *Integ. and Comp. Biol.* 2002;42: 660–67.

Weinstein, P., & Austin, A.D. 'The host relationships of trigonalyid wasps (Hymenoptera: Trigonalyidae), with a review of their biology and catalogue to world species'. *Journal of Natural History* 1991;25: 399–433.

Woese, C.R., Kandler, O., & Wheelis, M.L. Towards a natural system of organisms: proposal for the domains Archaea, Bacteria, and Eucarya. *Proc Natl Acad Sci* 1990;87(12): 4576–9.

Wilson, N.G., et al. 'Field collection of *Laevipilina hyalina* McLean, 1979 from southern California, the most accessible living monoplacophoran'. *Journal of Molluscan Studies* 2009;75: 195–97.

Zhang, Z.-Q. 'Animal biodiversity: An introduction to higher-level classification and taxonomic richness'. *Zootaxa* 2011;3148: 7–12.

Zimmer, C. 'The Most Popular Lifestyle on Earth'. *Conservation Magazine* 2008;9(4).

Acknowledgments

In putting this book together, many people have lent their expertise, and Martin V. Sørensen, Jon Richfield, Arthur Anker, Kevin Lee, Alexander Semenov and Greg Rouse have suggested additions and changes to the manuscript and/or gone out of their way to provide images and advice.

Ian Denholm, Greg Edgecombe, Bernhard Egger, Jon Eisenback, Christian Emig, Daphne G. Fautin, Peter Funch, Hidetaka Furuya, Vladimir E. Gross, Alexander Gruhl, Rick Hochberg, Tohru Iseto, Reinhardt M. Kristensen, Gretchen Lambert, Georg Mayer, Klaus Rohde, Bernd Scheirwater, Andreas Schmidt-Rhaesa, Billie J. Swalla, Christiane Todt, Clint Turbeville, Jean Vacelet and Greg Wray all very kindly reviewed sections of the manuscript and made very helpful suggestions. Thanks also to Jane Miller for reading parts of the text and to Gloria Piper for helping with the glossary.

Sourcing good quality images was a major part of putting this book together and I am indebted to all those who provided photographs: Cynthia Abgarian, Ingo Arndt, Paul Bertner, Sarah Bourlat, Carlo Brena, Julian Cremona, Bernhard Egger, Jon Eisenback, José E. Fernández Alfaya, Ivan Fiala, Adrian Glover, Vladimir Gross, Alexander Gruhl, Rokus Groeneveld, Lynn M. Hansen, Marshall Hedin, Alan Henderson, Rick Hochberg, Jens Høeg, Ian Hope, David J. Hunt, Tohru Iseto, Sönke Johnsen, Hiroshi Kajihara, Reinhardt M. Kristensen, Bob Lester, Larry Madin, Georg Mayer, Alvaro E. Migotto, Oldrich Nedved, Arne Nilssen, Kazuo Ogawa, Josep Lluis Peralta, Michael Plewka, Ekaterina Raikova, Tomas Rak, Eric Roettinger, Bernd Schierwater, Thomas Shahan, David Spears, Robyn Stutchbury, Ria Tan, Max Telford, Christiane Todt, Brian Valentine, Bruno Vellutini, Dave E. Walter and Alex Wild. Special thanks also to Phil Miller for his numerous illustrations and to Laura Schwartz (www.laura-schwartz.com) for producing online animations.

Many of the photographers have more material available online:
Cynthia Abgarian (www.cgillsphotos.com)
Arthur Anker (artour_a @ Flickr),
Alan Henderson (www.photography.minibeastwildlife.com.au),
Kevin Lee (www.diverkevin.com),
Alvaro E. Migotto (cifonauta.cebimar.usp.br),
Michael Plewka (www.plingfactory.de/Science/Biohome.html),
Eric Roettinger (www.kahikaiimages.com),
Alexander Semenov (Alexander Semenov @ Flickr),
David Spears (www.cloudshillimaging.com),
Ria Tan (wildsingapore @ Flickr),
Dave E. Walter (www.macromite.wordpress.com), and
Alex Wild (www.alexanderwild.com).

I would also like to thank the following people, who have all helped in one way or another during the course of preparing this book for publication: Wendy Banks, Alan Cressler, Renata Cunha, Thomas Dahlgren, Paulo J. P. dos Santos, Casey Dunn, Diego Fontaneto, Per Flood, Hidetaka Furuya, Gonzalo Giribet, Antonio Guillén, Steve Haddock, Masayuki Hatta, Euichi Hirose, Jon Hood, Hiroshi Kajihara, Keiichi Kakui, Nick King, Nikolaos Lampadariou, Francesca Leasi, Dhugal Lindsay, Tristan Manco, Brian Miller, Jenny Miller, Claudia Mills, Birger Neuhaus, Jon Norenburg, Otto Müller P. Oliveira, Pete Olson, Karen Osborn, Hans-Jürgen Osigus, David Patterson, Pedro Pedro, Luitfried Salvini-Plawen, Stefano Schiaparelli, Daisuke Shimada, Yoshihisa Shirayama, Adam Simmons, Gary Smailes, Jon Spaull, Volker Storch, Vanessa Sullivan, Nina Svane-Mikkelsen, Antonio Todaro, Tom Trott, Nuria Vázquez, Dieter Waloßek, Ian Welby, Jonathan Wojcik, Hiroshi Yokoyama and Zhi-Qiang Zhang. In addition I would also like to thank the staff at Thames & Hudson for making *Animal Earth* possible, particularly Jamie Camplin, Avni Patel and Ben Plumridge.

Sources of Illustrations:
Title page: Vilaine Crevette, Shutterstock. **Preface**: Tomas Rak. **Introduction**: 1 Phil Miller; 5 Ross Piper; 6 Phil Miller; 7–9 Bruno Vellutini; 10 Srdjan Draskovic, Shutterstock; 11 Arthur Anker; 12 Carlo Brena; 13 Dirk Ercken, Shutterstock; 14 Alexander Semenov; 15 Lebendkulturen.de, Shutterstock; 16 Arthur Anker; 17 Kevin Lee; 18 Peter Scoones, Science Photo Library; 19 Sönke Johnsen; 20 Attern, Shutterstock; 21 Eye of Science, Science Photo Library. **Ctenophora**: 1 Alexander Semenov; 2 Eric Roettinger; 3 Alexander Semenov; 4 Alvaro Migotto; 5 Kevin Lee; 6–7 Alexander Semenov; 8. Kevin Lee; 9. Larry Madin; 10 Kevin Lee; 11 Wim van Egmond, Visuals Unlimited, Inc., Science Photo Library; 12 Eric Roettinger; 13–14 Alvaro Migotto; 15 Eric Roettinger; 16 Kevin Lee; 17 Alvaro Migotto; 18–20 Kevin Lee. **Porifera**: 1 Stephan Kerkhofs, Shutterstock; 2 Matthew Oldfield, Science Photo Library; 3 David Fleetham, Visuals Unlimited, Science Photo Library; 4 Dennis Sabo, Shutterstock; 5 Daryl H, Shutterstock; 6–7 Michael Patrick O'Neill, Science Photo Library; 8 Ken M. Highfill, Science Photo Library; 9 Alexander Semenov; 10 Steve Gschmeissner, Science Photo Library; 11 Eye of Science, Science Photo Library; 12 NOAA, MBARI; 13 Alexis Rosenfeld, Science Photo Library; 14 Arthur Anker; 15 Andrew J. Martinez, Science Photo Library; 16 Vilaine Crevette, Shutterstock. **Placozoa**: 1 Eric Roettinger; 2–3 Hans-Juergen Osigus, Schierwater Laboratory; 4 Eric Roettinger. **Cnidaria**: 1 Alexander Semenov; 2 Arthur Anker; 3 Ivan Fiala; 4 Ivan Fiala; 5 Alexander Gruhl; 6 Kazuo Ogawa; 7–8 Ekaterina Raikova; 9 Kevin Lee; 10 Georgette Douwma, Science Photo Library; 11 David Caron, Look at Sciences, Science Photo Library; 12–14 Michael Plewka; 15–16 Alexander Semenov; 17 Arthur Anker; 18 Alexander Semenov; 19 David Spears, Clouds Hill Imaging; 20–21 Alexander Semenov; 22 Kevin Lee; 23–25 Alexander Semenov; 26 Arthur Anker; 27 Undersea Discoveries, Shutterstock; 28 Kevin Lee; 29 Alexander Semenov; 30 Kevin Lee; 31 David Wrobel, Visuals Unlimited, Science Photo Library; 32–33 Kevin Lee; 34–35 Alexander Semenov; 36 Kevin Lee; 37–39 Alexander Semenov. **Cephalochordata**: 1 Arthur Anker; 2 Alvaro Migotto; 3 Arthur Anker. **Tunicata**: 1 Reinhard Dirscherl Visuals Unlimited, Science Photo Library; 2–3 Kevin Lee; 4 Wim van Egmond, Visuals Unlimited, Inc., Science Photo Library (inset: Phil Miller); 5–6 Kevin Lee; 7 Wim van Egmond, Visuals Unlimited, Inc., Science Photo Library; 8 Alvaro Migotto; 9–10 Kevin Lee; 11 Larry Madin; 12–13 Kevin Lee; 14 Eric Roettinger; 15–18 Kevin Lee; 19 Wim van Egmond, Visuals Unlimited, Inc., Science Photo Library; 20–21 Kevin Lee. **Craniata**: 1 Trevor Kelly, Shutterstock; 2 Steve Byland, Shutterstock; 3 Photosani, Shutterstock; 4 Argonaut, Shutterstock; 5 Craig K. Lorenz, Science Photo Library; 6 Reinhard Dirscherl, Visuals Unlimited, Science Photo Library; 7 Oleg Seleznev, Shutterstock; 8 Scientifica, Visuals Unlimited, Science Photo Library; 9–10 Kevin Lee; 11 Valeriy Evlakhov, Shutterstock; 12–13 Arthur Anker; 14–15 Kevin Lee; 16 Cbpix, Shutterstock; 17 Tatiana Belova, Shutterstock; 18–22 Kevin Lee; 23 Ross Piper; 24 Alexander Kupfer; 25 Laurie J. Vitt; 26 Arthur Anker; 27 Paul Bertner; 28 Ross Piper; 29 Alexander Semenov; 30–31 Paul Bertner. **Hemichordata**: 1 Eric Roettinger; 2–3 Arthur Anker; 4 Greg Rouse; 5–6 Phil Miller; 7 Greg Rouse; 8 Eric Roettinger. **Echinodermata**: 1–2 Arthur Anker; 3 Kevin Lee; 4 Sönke Johnsen; 5 Wim van Egmond, Visuals Unlimited, Inc., Science Photo Library; 6 Alexander Semenov; 7 David Caron, Look at Sciences, Science Photo Library; 8–12 Arthur Anker; 13–14 Kevin Lee; 15 Alexander Semenov; 16 Kevin Lee; 17–19 Alexander Semenov; 20–21 Arthur Anker; 22–23 Alexander Semenov; 24–25 Arthur Anker; 26 Ken Lucas, Visuals Unlimited, Science Photo Library; 27 Kjell B. Sandved, Science Photo Library; 28–29 Alexander Semenov; 30 British Antarctic Survey, Science Photo Library; 31 Arthur Anker. **Xenoturbellida**: 1–2 Max Telford; 3 Sarah Bourlat/Hiroaki Nakano/Max Telford. **Acoelomorpha**: 1 Eric Roettinger; 2 Bernhard Egger; 3–5 Christiane Todt; 6–8 Kevin Lee. **Chaetognatha**: 1 Steve Gschmeissner, Science Photo Library; 2 Phil Miller; 3 Ingo Arndt; 4 Alexander Semenov. **Nematoda**: 1 Jon Eisenback; 2 Michael Plewka; 3 Ian Hope; 4 Science Source, Science Photo Library; 5 Ross Piper; 6 Science Source, Science Photo Library; 7 Ian Hope; 8 Biomedical Imaging Unit, Southampton General Hospital, Science Photo Library; 9 Dr Richard Kessel & Dr Gene Shih, Visuals Unlimited, Science Photo Library; 10 David J. Hunt; 11 Arthur Anker; 12 Martin V. Sørensen; 13–14 Eye of Science, Science Photo Library. **Nematomorpha**: 1 Alan Henderson; 2–3 Phil Miller; 4 Malcolm Storey. **Tardigrada**: 1–2 Eye of Science, Science Photo Library; 3 Microfield Scientific Ltd., Science Photo Library; 4 Eye of Science, Science Photo Library; 5–6 Vladimir Gross; 7 Microfield Scientific Ltd., Science Photo Library; 8 Carolina Biological Supply Co., Visuals Unlimited, Inc., Science Photo Library; 9–10 Eye of Science, Science Photo Library. **Onychophora**: 1 Paul Bertner; 2 Georg Mayer; 3–5 Paul Bertner; 6 Robyn Stutchbury, Peripatus Productions Pty Limited, 7 Paul Bertner; 8 Morley Read, Shutterstock. **Arthropoda**: 1 David Spears, Clouds Hill Imaging; 2 Solvin Zankl, Visuals Unlimited, Science Photo Library; 3 Tomas Rak; 4 Arthur Anker; 5 Brian Valentine; 6 Terry Priest, Visuals Unlimited, Inc., Science Photo Library; 7 Milena_, Shutterstock; 8 Kevin Lee; 9–10 Dave E. Walter; 11 Alex Wild; 12 Tomas Rak; 13 Kevin Lee; 14 Dave E. Walter; 15 Tomas Rak; 16 Arthur Anker; 17 Tomas Rak; 18 Audrey Snider-Bell, Shutterstock; 19 Alexander Semenov; 20–21 Arthur Anker; 22 Ross Piper; 23–25 Arthur Anker; 26 Alexander Semenov; 27 Alex Wild; 28 Ross Piper; 29 Alex Wild; 30 David Spears, Clouds Hill Imaging; 31–34 Arthur Anker; 35 Tomas Rak; 36 Arthur Anker; 37 Tomas Rak; 38 David Spears, Clouds Hill Imaging; 39 Alex Wild; 40 Arthur Anker; 41 Wim van Egmond, Visuals Unlimited, Inc., Science Photo Library; 42–43 Ross Piper; 44 Paul Bertner; 45 Alex Wild; 46–47 Arthur Anker; 48 Thomas Shahan; 49–50 Arne Nilssen; 51 Bob Lester; 52 Jens Høeg; 53 D. Roberts, Science Photo Library; 54–55 Arthur Anker; 56 Adam Petrusek; 57 Alexander Semenov; 58 Arthur Anker; 59 Alex Wild; 60 Paul Bertner; 61 Ross Piper/David Spears Clouds Hill Imaging; 62 Paul Bertner; 63 Arthur Anker; 64 Wim van Egmond, Visuals Unlimited, Inc., Science Photo Library; 65 Alexander Semenov; 66 Alex Wild; 67 Brian Valentine; 68 Thomas Shahan. **Priapulida**: 1–2 Greg Rouse; 3–4 Phil Miller; 5 Martin V. Sørensen. **Loricifera**: 1 Phil Miller; 2–5 Martin V. Sørensen; 6 Phil Miller. **Kinorhyncha**: 1 Ross Piper; 2 Martin V. Sørensen; 3 Phil Miller; 4 Ross Piper; 5–7 Martin V. Sørensen. **Ectoprocta**: 1 Gerd Guenther, Science Photo Library; 2 Eric Roettinger; 3 Science Vu, Visuals Unlimited, Science Photo Library; 4 Phil Miller; 5–8 Alexander Semenov; 9 Eric Roettinger; 10 Aaron O'Dea; 11 Franz Neidl. **Entoprocta**: 1 Tohru Iseto; 2 Greg Rouse; 3–4 Alvaro Migotto; 5 Ross Piper; 6 Tohru Iseto; 7–8 Phil Miller. **Cycliophora**: 1 Phil Miller; 2–5 Matthias Obst. **Dicyemida**: 1 Phil Miller; 2 Hidetaka Furuya. **Orthonectida**: 1 Phil Miller; 2 Hidetaka Furuya. **Annelida**: 1 Alexander Semenov; 2 Arthur Anker; 3 Lynn M. Hansen; 4 Josep Lluis Peralta; 5–8 Arthur Anker; 9 Mircea Bezergheanu, Shutterstock; 10 Ross Piper; 11 Alexander Semenov; 12 Eye of Science, Science Photo Library; 13–15 NOAA-OE; 16 Adrian Glover; 17 Greg Rouse; 18 Melinda Fawver, Shutterstock; 19–20 Arthur Anker; 21–22 Alexander Semenov; 23–24 Arthur Anker; 25–26 Alexander Semenov; 27 Arthur Anker; 28 Alexander Semenov 29 Ingo Arndt; 30 Alexander Semenov; 31 Arthur Anker; 32 Rokus Groeneveld; 33 Alexander Semenov; 34 Michael Plewka; 35 Arthur Anker; 36 Alexander Semenov; 37 Arthur Anker; 38 Greg Rouse; 39 Kevin Lee; 40 Alexander Semenov; 41 Kevin Lee; 42 Arthur Anker; 43–44 Wim van Egmond, Visuals Unlimited, Inc., Science Photo Library; 45 Ingo Arndt; 46 Alexander Semenov; 47 Phil Miller. **Mollusca**: 1 Alexander Semenov; 2 Ross Piper; 3 Bluehand, Shutterstock; 4 Kevin Lee; 5 Arthur Anker; 6 Andrew J. Martinez, Science Photo Library; 7 David Fleetham, Visuals Unlimited, Science Photo Library; 8 José E. Fernández Alfaya; 9 Arthur Anker; 10 Wendell R Haag; 11 Arthur Anker; 12 Greg Rouse; 13 Arthur Anker; 14 Alexander Semenov; 15 Arthur Anker; 16 Alexander Semenov; 17 Cynthia Abgarian; 18 Greg Rouse; 19 Kevin Lee; 20–22 Arthur Anker; 23 Thomas Shahan; 24 Arthur Anker; 25 David Shale, Clouds Hill Imaging; 26 Louise Murray, Science Photo Library; 27 David Shale, Clouds Hill Imaging; 28 Ross Piper; 29 Alexander Semenov; 30 Arthur Anker; 31 Julian Cremona; 32 Christiane Todt; 33–34 Greg Rouse; 35–38 Alexander Semenov; 39–46 Arthur Anker; 47–48 Kevin Lee; 49–50 Arthur Anker; 51 Dr T.E. Thompson, Science Photo Library; 52 Arthur Anker; 53 Kevin Lee. **Nemertea**: 1–2 Arthur Anker; 3 José E. Fernández Alfaya; 4 Hiroshi Kajihara and Dhugal Lindsay; 5–8 Arthur Anker; 9 Franz Neidl, 10–11 José E. Fernández Alfaya; 12–13 Arthur Anker; 14 Kevin Lee; 15 José E. Fernández Alfaya. **Brachiopoda**: 1 Eric Roettinger; 2 Arthur Anker; 3–4 Alexander Semenov. **Phoronida**: 1 Kevin Lee; 2 Franz Neidl; 3 Ria Tan; 4 Kevin Lee; 5 Ria Tan; 6 Kevin Lee; 7 Phil Miller. **Gastrotricha**: 1–5 Michael Plewka; 6 David Scharf, Science Photo Library; 7 Michael Plewka; 8 David Scharf, Science Photo Library. **Platyhelminthes**: 1 Dan Exton, Shutterstock; 2 Michael Plewka; 3 Eye of Science, Science Photo Library; 4–5 Oldrich Nedved; 6 Arthur Anker; 7 Michael Plewka; 8 D. Kucharski & K. Kucharska, Shutterstock; 9 David Spears, Clouds Hill Imaging; 10 Arthur Anker; 11 Michael Plewka; 12 Dr Keith Wheeler, Science Photo Library; 13 Kevin Lee; 14 Paul Bertner; 15 Arthur Anker; 16–17 Paul Bertner; 18–19 Pan Xunbin, Shutterstock; 20 Klaus Rohde; 21 Juan Gaertner, Shutterstock; 22 M.I. Walker, Science Photo Library; 23 Jubal Harshaw, Shutterstock; 24–26 Kevin Lee; 27 Arthur Anker. **Gnathostomulida**: 1 Phil Miller; 2–3 Martin V. Sørensen; 4 Rick Hochberg; 5 Martin V. Sørensen; 6 Rick Hochberg; 7 Martin V. Sørensen. **Rotifera**: 1–2 Michael Plewka; 3 Dave E. Walter; 4–7 Michael Plewka; 8 Martin V. Sørensen; 9–13 Michael Plewka. **Micrognathozoa**: 1 Phil Miller; 2–3 Martin V. Sørensen; 4 Reinhardt M. Kristensen.

Index

Page numbers in *italic* refer to the illustrations